ENERGY AND THE ENVIRONMENT:
Forging the Link

American Council for an Energy-Efficient Economy
Series on Energy Conservation and Energy Policy

Series Editor, Carl Blumstein

Energy Efficiency in Buildings: Progress and Promise
Financing Energy Conservation
Energy Efficiency: Perspectives on Individual Behavior
Electric Utility Planning and Regulation
Residential Indoor Air Quality and Energy Efficiency
Energy-Efficient Motor Systems:
A Handbook on Technology, Program,
and Policy Opportunities
State of the Art of Energy Efficiency:
Future Directions
Efficient Electricity Use: A Development Strategy for Brazil
Energy Efficiency and the Environment: Forging the Link

ENERGY EFFICIENCY AND THE ENVIRONMENT:

Forging the Link

Edited by
EDWARD VINE
DRURY CRAWLEY
PAUL CENTOLELLA

American Council for an Energy-Efficient Economy
Washington, D.C. and Berkeley, California
in cooperation with:
Universitywide Energy Research Group
University of California
1991

Energy Efficiency and the Environment: Forging the Link

Copyright © 1991 by the American Council for an Energy-Efficient Economy.

Reproduction in whole or in part is strictly prohibited without prior written approval of the American Council for an Energy-Efficient Economy (ACEEE), except that reasonable portions may be reproduced or quoted as part of a review or other story about this publication, and except where specific approval is given in the text. All rights reserved.

Published by the American Council for an Energy-Efficient Economy,
1001 Connecticut Avenue, N.W., Suite 801, Washington, D.C. 20036
and 2140 Shattuck Avenue, Suite 202, Berkeley, California 94704.

Cover art copyright © 1990 M. C. Escher Heirs/Cordon Art - Baarn - Holland

Cover design by Wilsted & Taylor
Book design by Paula Morrison
Book typeset by Wilsted & Taylor
Printed in the United States of America by Edwards Brothers, Inc.

Library of Congress Cataloging-in-Publication Data

Energy efficiency and the environment : forging the link / edited by Edward Vine, Drury Crawley, Paul Centolella.
    434 p.   23 cm.
    "Based on papers prepared for the ACEEE 1990 Summer Study on Energy Efficiency in Buildings"—Introd.
    Includes bibliographical references and index.
    ISBN 0-918249-12-0 : $26.00
    1. Buildings—Energy conservation.   2. Energy conservation—
Environmental aspects—United States.   I. Vine, Edward L.   II. Crawley, Drury,
1957–   .   III. Centolella, Paul.   IV. ACEEE Summer Study on Energy
Efficiency in Buildings (1990)
TJ163.5.B84E5415   1991
333.79′16—dc20

                    91-34254
                    CIP

NOTICE

This publication was prepared by the American Council for an Energy-Efficient Economy. Neither ACEEE, nor any person acting on behalf of ACEEE: a) makes any warranty, expressed or implied, with respect to the use of any information, apparatus, method, or process disclosed in the publication or guarantees that such use may not infringe privately owned rights; b) assumes any liabilities with respect to the use of, or for damages resulting from the use of, any information, apparatus, method, or process disclosed in this publication.

♻ Printed on recycled paper.

Contents

Preface		xiii

Contributing Authors		xv

Introduction	Energy Efficiency and the Environment: Forging the Link *by Ed Vine, Dru Crawley, and Paul Centolella*	1
	Issues Raised	1
	Summary	6
	Acknowledgments	6

Chapter 1	Energy Efficiency and Greenhouse-Gas Emission Reductions: Some International Results *by Paul Schwengels and William J. Pepper*	9
	Introduction	9
	Reference Scenario	11
	Analysis of Response Options	14
	Conclusions	26
	References	26

Chapter 2	Lay Perspectives on Global Climate Change *by Willett Kempton*	29
	Introduction	29
	Interviews	30
	Global Warming Incorporated into Existing Concepts	33
	Perceptions of Weather	39
	Support for Environmental Protection	45
	Environmental Values	47
	Reactions to Policy Options	56
	Conclusions	62
	Acknowledgments	66
	References	66

Chapter 3   **Environmental Improvement and Energy Efficiency in Buildings: Opportunities to Reduce CO$_2$ Emissions** *by Erich Unterwurzacher and Genevieve McInnes* **71**

Introduction 71
Pollutant Emissions and Energy Demand Trends 72
Technology Options 79
Barriers to the Efficient Use of Energy 83
Review of Policy and Program Options 84
Conclusions 87
References 87

Chapter 4   **Carbon Dioxide Emissions and Energy Efficiency in U.K. Buildings** *by George Henderson and Les Shorrock* **89**

Introduction 89
Energy Use in Buildings 90
Carbon Dioxide Emissions from Fuel
    Consumption 91
Carbon Dioxide Emissions from Building
    Energy Use 93
Improvements to Energy Efficiency in Existing
    Buildings 94
Reductions in CO$_2$ Emissions Through Energy-
    Efficiency Improvements 96
Current Trends in CO$_2$ Emissions 99
Other Greenhouse Gases 101
Conclusions 101
Acknowledgment 102
References 102

Chapter 5   **Environmental Benefits of Energy Efficiency: Impact of Washington State Residential Energy Codes on Greenhouse-Gas Emissions** *by Richard Byers* **105**

Introduction 105
The Washington State Residential Building
    Energy Code 106

Contents — vii

Cost and Energy Savings for Energy Code
    Insulation Levels   107
Estimation of Greenhouse-Gas Reductions   110
Economics of Energy Code and Cost of
    Environmental Benefits Based on Individual
    Homes   112
Characteristics of the Washington State
    Housing Stock   114
Statewide Energy, Peak Electrical Load, and
    Equivalent Carbon Dioxide Savings   115
Conclusions   117
Appendix: Accuracy of the SUNDAY Model   117
References   118

Chapter 6 **The CO_2 Diet for a Greenhouse Planet:**
Assessing Individual Actions
for Slowing Global Warming
by John M. DeCicco, James H. Cook,
Dorene Bolze, and Jan Beyea   **121**
Introduction   121
Our Greenhouse Planet   122
What Do You Emit?   125
Reducing Greenhouse Emissions   131
Conclusion   140
Appendix   141
References   142

Chapter 7 **Emissions Impacts of Demand-Side**
Programs: What Have We Achieved So Far
and How Will Recent Policy Decisions
Change Program Choices?
by Richard S. Tempchin, A. Joseph Van den
Berg, Vera B. Geba, Curtis S. Felix,
and Marc W. Goldsmith   **145**
Introduction   145
Steps Toward Estimating the Impacts of Electric
    Utility Demand-Side Management Programs
    on Power Plant Emissions   146
Carbon Dioxide Reduction Through
    Electrification   151
Conclusions   158
References   158

viii — Contents

Chapter 8 **The Global Climate Change Issue—
What It Is, Where It Is Going, and
How It Will Impact Utility DSM**
by Bonnie B. Jacobson and David W. Kathan **161**
Introduction 161
Summary of Global Climate Change 161
Policy Options 162
Legislative Proposals 168
Research Activities Now in Progress 169
Impact on Electric Utilities 172
Impacts on Utility Demand-Side Management 174
Summary 175
References 176

Chapter 9 **Integrated Resource Planning
and the Clean Air Act**
by Daniel M. Violette and Carolyn M. Lang **177**
Introduction 177
The New Rules 178
Compliance with the New Rules 180
Integrated Resource Planning and Compliance 183
Winning Approval 185
Conclusion 187
References 187

Chapter 10 **Consideration of Environmental Externality
Costs in Electric Utility Resource Selections
and Regulation**
by Richard L. Ottinger **189**
Background 190
Why Utilities and Commissions Should
   Consider Environmental Externality Costs 191
Costs to Be Included 192
Major Issues in Damage Risk Valuation 196
State Incorporation of Environmental
   Externalities 201
Proposed Incorporation Methods 207
Recommendations for Incorporation 210
Next Steps 210
Acknowledgments 211
References 211

Contents — ix

Chapter 11   **Valuation of Environmental Externalities in Energy Conservation Planning**
by Paul L. Chernick and Emily J. Caverhill   **215**
Introduction   215
Definition of Environmental Externalities   216
Scope of an Externalities Analysis   217
Valuation Methods   218
Conclusion   226
References   227

Chapter 12   **Incorporating Environmental Externalities in Integrated Resource Planning: One Utility's Experience**
by Dean S. White, Timothy M. Stout, and Mary Sharpe Hayes   **229**
Introduction   229
Background   230
The New England Electric Methodology   232
Application of Methodology to Integrated Resource Plan   239
Limitations and Planned Revisions to the Methodology   243
Conclusion   244
A Final Note   244
Acknowledgments   245
References   245

Chapter 13   **The Inclusion of Environmental Goals in Electric Resource Evaluation: A Case Study in Vermount**
by Stephen Bernow and Donald Marron   **249**
Introduction   249
Description of the Overall Study   250
Selection of Electric Resources   250
Selection of Environmental Loadings to Be Evaluated   252
Development of Environmental Loadings Coefficients   253
Valuation of Environmental Loadings: Evaluation Methodologies   258

x — Contents

Derivation of Emission Avoidance Costs Using
the Environmental Standards Approach 261
Results 263
Conclusion 266
Acknowledgments 267
References 267

Chapter 14   **Air Pollution Projection Methodologies:**
Integrating Emission Projections with
Energy Forecasts
by Michael R. Jaske **271**
Introduction 271
Current Practice in Emissions Projections 272
Integrated Energy/Emission Projections 274
Air Quality Scenarios Project 280
Further Development 285
Acknowledgments 286
References 287

Chapter 15   **Conserving Energy to Reduce SO$_2$**
Emissions in Ohio: An Evaluation Using
a Multiobjective Electric Power
Production Costing Model
by Benjamin F. Hobbs and James S. Heslin **289**
Introduction 289
Methodology for Estimating the Output and
Emissions of Generating Units 291
Assumptions 293
Results 296
Conclusions 299
Acknowledgments 300
References 300

Chapter 16   **Building Energy Consumption and the**
Environment: What Past, Present, and
Future Commercial Buildings Energy
Consumption Surveys Can Tell Us About
Chlorofluorocarbons
by Julia D. Oliver and Eugene M. Burns **305**
Introduction 305

Contents — xi

	The CBECS	308
	What the 1986 CBECS Can Tell Us About CFC	
	Usage in Buildings	309
	What the 1989 CBECS Can Tell Us	316
	What Future CBECS May Tell Us	317
	Summary and Conclusions	318
	Acknowledgments	318
	References	318

Chapter 17　**Measured Cooling Savings from Vegetative Landscaping**
by Alan K. Meier **321**

Introduction	321
Heat Gain Paths Influenced by Vegetative Landscaping	322
Simulations to Estimate Cooling Savings	322
Case Studies	324
Conclusions	329
Acknowledgments	332
References	332

Chapter 18　**Simulating Effects of Turf Landscaping on Building Energy Use**
by James R. Simpson **335**

Introduction	335
Results and Discussion	336
Comparison of Measurements and Simulations	340
Conclusions	340
Appendix: Methods	343
References	347

Chapter 19　**Economic Modeling for Large-Scale Urban Tree Plantings**
by E. Gregory McPherson **349**

Introduction	349
Components of and Inputs to the Model	350
Results	358
Conclusions	365
References	367

xii — Contents

Chapter 20   **A Synopsis of *Cooling Our Communities:***
The Guidebook on Tree Planting and
Light-Colored Surfaces
by Joe Huang, Susan Davis, and
Hashem Akbari   **371**
Introduction   371
Background of Project   372
Heat Island Mitigation Strategies   373
Implementation Issues   375
Conclusion   378
References   379

Preface

Ralph Cavanagh, *Natural Resources Defense Council*

Every two years, one of the world's most important environmental protection inquiries is convened in California under the auspices of the American Council for an Energy-Efficient Economy (ACEEE). I draw this conclusion even though the participants are not formally invited to discuss "the environment" at all; the focus is energy-efficiency technologies and strategies for realizing their extraordinary promise.

The latest ACEEE convocation produced this book, which makes explicit a connection that many in the environmental community have long celebrated. Energy efficiency is quite simply the most powerful engine of environmental protection ever devised. It is our best hope for stabilizing the global climate, abating acid rain, retaining wild rivers, and fending off a host of residual environmental insults. It may even allow President Bush and his lieutenants to fashion their long-promised White House assault on the greenhouse effect, although at this writing progress remains paralyzed by inexplicable inabilities to distinguish efficiency from pain and economic stagnation.

The best antidote for that kind of confusion is books like this, and the public education campaigns that books like this inspire. Moreover, it bears recalling that of late the locus of energy policy and energy solutions in the United States has shifted far from Washington, D.C.; the linkage between energy efficiency and the environment is already being forged in statehouses, utility boardrooms, code adoption hearings, and scientific laboratories far removed from any office of the federal government.

Some of those responsible for this shift are also contributors to this volume, and I close with a word about them collectively. Over the decades ahead, I believe that heroic international exertions will forestall the global environmental horrors that figure prominently in this book. When a grateful posterity assembles the roster of those responsible, the creators of the energy efficiency revolution will head the list. Their message will be as powerful as that of a John Muir or a David Brower, and their legacy will be just as enduring.

Contributing Authors

Hashem Akbari
Lawrence Berkeley Laboratory
Berkeley, California

Stephen Bernow
Tellus Institute
Boston, Massachusetts

Jan Beyea
National Audubon Society
New York, New York

Dorene Bolze
National Audubon Society
New York, New York

Eugene M. Burns
U.S. Department of Energy
Energy Information Administration
Washington, D.C.

Richard Byers
Washington State Energy Office
Olympia, Washington

Emily J. Caverhill
Resource Insight, Inc.
Boston, Massachusetts

Paul L. Chernick
Resource Insight, Inc.
Boston, Massachusetts

James H. Cook
National Audubon Society
Islip, New York

Susan Davis
Lawrence Berkeley Laboratory
Berkeley, California

John M. DeCicco
American Council for an Energy-
Efficient Economy
Washington, D.C.

Curtis S. Felix
Energy Research Group, Inc.
Waltham, Massachusetts

Vera B. Geba
Energy Research Group, Inc.
Waltham, Massachusetts

Marc W. Goldsmith
Energy Research Group, Inc.
Waltham, Massachusetts

Mary Sharpe Hayes
Tennessee Valley Authority
Knoxville, Tennessee

George Henderson
The Building Research Establishment
United Kingdom

James S. Heslin
Energy Management Associates, Inc.
Atlanta, Georgia

Benjamin F. Hobbs
Case Western Reserve University
Cleveland, Ohio

Joe Huang
Lawrence Berkeley Laboratory
Berkeley, California

Michael R. Jaske
California Energy Commission
Sacramento, California

xvi — Contributing Authors

Bonnie B. Jacobson
ICF Resources, Inc.
Fairfax, Virginia

David W. Kathan
ICF Resources, Inc.
Fairfax, Virginia

Willett Kempton
Princeton University
Princeton, New Jersey

Carolyn M. Lang
RCG/Hagler, Bailly, Inc.
Boulder, Colorado

Genevieve McInnes
International Energy Agency
Paris, France

E. Gregory McPherson
U.S. Department of Agriculture,
Forest Service
Northeastern Forest Experiment
Station
Chicago, Illinois

Donald Marron
Massachusetts Institute of Technology
Boston, Massachusetts

Alan K. Meier
Lawrence Berkeley Laboratory
Berkeley, California

Julia D. Oliver
U.S. Department of Energy
Energy Information Administration
Washington, D.C.

Richard L. Ottinger
Pace University Law School
White Plains, New York

William J. Pepper
ICF Information Technology, Inc.
Fairfax, Virginia

Paul Schwengels
U.S. Environmental Protection
Agency
Climate Change Division
Washington, D.C.

Les Sharrock
The Building Research Establishment
United Kingdom

James R. Simpson
University of Arizona
Tucson, Arizona

Timothy M. Stout
New England Power Service Company
Westboro, Massachusetts

Richard S. Tempchin
Edison Electric Institute
Washington, D.C.

Erich Unterwurzacher
International Energy Agency
Paris, France

A. Joseph Van den Berg
Edison Electric Institute
Washington, D.C.

Daniel M. Violette
RCG/Hagler, Bailly, Inc.
Boulder, Colorado

Dean S. White
New England Power Service Company
Westboro, Massachusetts

Introduction

Energy Efficiency and the Environment: Forging the Link

During the 1990s, increased resources will be devoted to solving environmental problems at the local, state, national, and global levels. One of the most important solutions to these problems will be increasing energy efficiency in homes, the workplace, and transportation. *Energy Efficiency and the Environment: Forging the Link* addresses the specific linkages between energy conservation and environmental issues such as global warming, air pollution, acid rain, and ozone depletion.

The key issues in this book are how energy-efficiency measures and programs can alleviate environmental problems and how planners can factor environmental externalities into their selection and use of resources. The chapters in this book—written by leading researchers, program analysts, and policymakers—are based on papers prepared for the ACEEE 1990 Summer Study on Energy Efficiency in Buildings. The Summer Study attracts international and interdisciplinary participants to exchange information and share ideas on the efficient use of energy in buildings. For the first time in the history of the conference, papers were requested on the potential contribution of energy efficiency in buildings to environmental protection.

Issues Raised

Global environmental risks, growing interest in market-based environmental regulation, and integration of environmental and energy planning have focused attention on energy efficiency as a low-cost pollution-prevention strategy. These factors are channeling the public's

2 — Introduction

general concern about the environment into specific concerns about improvements in energy efficiency.

Efficiency can be a cost-effective strategy for reducing both greenhouse-gas emissions, particularly carbon dioxide (CO_2), and the acid rain precursors sulfur dioxide (SO_2) and nitrogen oxides (NO_x), which are released primarily by energy production and use. Efficiency also addresses the range of other air pollution, land use, waste disposal, and water quality concerns associated with energy supply alternatives.

Global Climate Change

Modern industrial development has greatly increased emissions and atmospheric concentrations of greenhouse gases. As a result of these gases' persistence in the atmosphere, even stabilizing emissions of greenhouse gases would not halt the increase in atmospheric concentrations.

Greenhouse gases absorb infrared energy radiated from the Earth's surface, warming the lower atmosphere. While uncertainty remains regarding the importance and timing of the impact of this phenomenon, a growing body of scientific opinion has concluded that a significant increase in average global temperature and associated climate changes are likely within the next few decades.

Energy production and use accounts for nearly 60% of current contributions to greenhouse gases. To the extent that conservation programs are cost-effective without reference to their environmental benefits, such programs represent a zero-cost or negative-cost means to reduce CO_2 emissions.

In the United States, much of the debate has been framed by those who emphasize scientific research and those who argue for a "win-win" policy, pursuing measures that are cost-effective on other grounds. While such a policy could significantly reduce the growth in CO_2 emissions, it may be insufficient to reverse increases in atmospheric concentrations. This limitation raises difficult questions. What is our responsibility to our children with respect to making long-lasting changes in the Earth's atmosphere that have, at best, uncertain implications?

The first two chapters devoted to global climate change introduce the topic from two different perspectives. In chapter 1, Schwengels and Pepper review national studies of potential policy responses that have been submitted to the United Nations Intergovernmental Panel on Climate Change. These studies quantify both significant reductions in the rate of growth in CO_2 emissions—and, in some cases, stabilization or net reductions in atmospheric concentration—that could be achieved through policy-induced efficiency improvements. In chapter

Introduction — 3

2, Kempton explores public concepts of global warming, using ethnographic interview techniques that provide information unavailable through conventional opinion surveys. The degree to which ethical questions regarding energy use and global warming have yet to surface in the American consciousness may be the result of public misconceptions.

The next two chapters examine in greater detail the role of energy efficiency as a core component of expected policy responses to global climate change. In chapter 3, Unterwurzacher and McInnes quantify CO_2 emissions for various end-use sectors in Organization for Economic Cooperation and Development (OECD) countries and describe the realistic potential for efficiency improvements in each sector. In chapter 4, Henderson and Shorrock examine the technical potential of energy efficiency in buildings for limiting CO_2 emissions in the United Kingdom. They conclude that emissions could be reduced by 25% through cost-effective efficiency measures, if present levels of demand for services were maintained. To achieve potential CO_2 reductions requires implementing specific regulatory or incentive mechanisms such as building codes or appliance rebates. In chapter 5, Byers provides a detailed analysis of the impact of the Washington State residential energy code on CO_2 emissions. In chapter 6, DeCicco and his co-authors provide a "CO_2 diet" that evaluates individual contributions to global climate change.

Market-Based Environmental Regulation

The United States has relied on a "command-and-control" system of environmental regulation. This approach requires detailed implementation plans for specific sources of emissions and uniform standards based on available control technology. Regulation has emphasized control rather than prevention of pollution. Regional and global environmental problems are shifting the focus of environmental regulation to market-based approaches that emphasize pollution prevention. Acid rain control provisions in the 1990 Clean Air Act include a market-based system for reducing utility SO_2 emissions. These provisions create tradable SO_2 tonnage allowances. Since it is the total quantity of emissions that matters under the allowance system, a source can achieve its emission limitation through emission controls or by using conservation programs to avoid emissions (as well as by trading allowances or using previously banked allowances). This permits energy efficiency to compete with other emission control strategies.

The next group of chapters raises questions on regulatory policy options and how we look at utility demand-side programs and environmental regulation. In chapter 7, Tempchin and his co-authors describe

4 — Introduction

efforts to quantify emission reductions achieved through electric utility demand-side programs. While electric utilities focus on documenting peak load impacts, many environmental benefits are primarily associated with reduced kilowatt-hour consumption. In chapter 8, Kathan and Jacobson describe the range of market-based, regulatory, and research approaches to climate policy under discussion in the United States. And in chapter 9, Violette and Lang examine the implications of market-based environmental regulation and the opportunities for promoting energy efficiency using the 1990 Clean Air Act Amendments as a starting point.

Integration of Energy, Environmental, and Economic Planning

Energy, environmental, and economic policy is often fragmented, with different organizations responsible for energy, for environmental protection, and for development activities. For example, at least 16 departments and major agencies of the United States government are involved in activities related to greenhouse gas emissions. While the Domestic Policy Council has active working groups on climate issues, there is no Executive Office council that brings together all of the major agencies involved in climate change policy. At the state level, environmental protection agencies often set standards with insufficient information regarding how to cost-effectively achieve environmental objectives. Public utility commissions simply accept these standards as constraints. There is frequently little or no coordination among a state's environmental regulator, utilities commission, and development agency. As the costs of fragmented regulation grow, government and utility planners are beginning to take steps that address fragmentation.

Utility planners and regulators are placing greater emphasis on incorporating environmental costs into resource planning decisions and proceedings. In chapter 10, Ottinger reviews resource planning activities that consider environmental externalities at 26 state commissions, the Northwest Power Planning Council, and the Bonneville Power Administration. Ottinger notes that most regulatory commissions accept nonquantitative methods for determining environmental externality costs in utility resource planning, although a growing number have either calculated specific externality cost estimates or attached a proxy adder to the polluting resource costs (or bonus for nonpolluting resources). These commissions have used control or pollution mitigation costs, rather than societal damage costs, in their regulatory computations. Ottinger supports a methodology based primarily on

Introduction — 5

assessing damage costs where adequate studies exist to permit quantification.

Two other approaches, which have been vigorously debated in New England, are contrasted to the use of damage costs. In chapter 11, Chernick and Caverhill explore an implied valuation or "shadow-pricing" approach, based primarily on control costs for meeting existing emission control requirements. And in chapter 12, White and his co-authors have used an environmental screening matrix based on both qualitative judgments and quantitative factors.

Finally, in chapter 13, Bernow and Marron present a fourth approach that considers the control costs of achieving specified environmental targets.

The next two chapters describe analytical efforts to bridge institutional fragmentation. In chapter 14, Jaske describes work at the California Energy Commission to integrate energy and emissions forecasts. In chapter 15, Hobbs and Heslin explain a method for evaluating acid rain control options by modeling dispatching and control options based on trade-offs among emissions and cost.

The U.S. Department of Energy (DOE) is also exploring how to incorporate environmental considerations in its energy efficiency programs. The 1987 Montreal Protocol requires reductions in the production and importation of chlorofluorocarbons (CFCs). The Clean Air Act of 1990 requires a phase-out of most uses of CFCs, including all working fluids in cooling systems and foaming agents in insulation production, by the end of the decade. Other ozone-depleting chemicals will be phased-out over a longer time frame. In chapter 16, Oliver and Burns identify information on CFC use that is now available from DOE's Commercial Buildings Energy Consumption Survey and further information that can be secured in future surveys.

Landscaping and Urban Heat Islands

Increased concern about global climate change has helped draw attention to a small but persistent group of researchers who have been advocating urban tree plantings as a technique to reduce the magnitude of the heat island effect and the related costs of urban air conditioning.

In chapter 17, Meier reviews the published literature examining the temperature and energy impacts of changes in vegetation around structures. Only a small number of studies, primarily in hot climates, has been published. Meier addresses questions raised by the studies and points out the difference in physical processes involved in the cooling reported. In chapter 18, Simpson reports results of a case study examining vegetation strategies near identical scale-model structures.

6 — Introduction

The results are informative, since they show savings resulting from processes other than shading.

In chapter 19, McPherson presents a careful cost-benefit analysis of a large-scale urban tree planting project in the urban southwest. The paper projects substantial energy and environmental benefits with a benefit-cost ratio of 2.6 over the 40-year lifetime of the project. And in chapter 20, Huang and his co-authors discuss the issues that must be considered in developing programs to mitigate the effects of urban heat islands.

Summary

A new field of research is emerging at the nexus between energy and environmental issues. The papers presented here reflect exciting and significant work. They also underscore how much more we need to learn to meet the environmental challenges facing the energy sector and the global community.

Given the breadth and quality of the information it contains, this book, we believe, will become an important resource for professionals—in utilities, government, research institutions, and universities—who are attempting to understand the significant contribution of energy efficiency in buildings to environmental protection. We hope that readers will be stimulated by the book to develop their own ideas for contributing to this area.

Acknowledgments

As noted previously, earlier versions of these papers were presented at the ACEEE 1990 Summer Study on Energy Efficiency in Buildings. As co-chairs of this conference, we wanted the content of these papers to be available to a wider audience, and, with the encouragement of Carl Blumstein, president of ACEEE, we decided to publish this book. We are especially grateful for the reviewers of the conference papers and for one of the Environment Panel leaders who participated in the review process: Dave Grimsrud.

We are indebted to copy editor Stephen Frantz, whose careful reviews and revisions reconciled the writing styles of the authors, and to managing editor Glee Murray and production manager Michelle Stevens, who guided the process of turning the edited manuscripts into a book.

Finally, we are grateful to the many organizations that supported the 1990 Summer Study and made the preparation of this book possible. Sponsors were Bonneville Power Administration, California

Energy Commission, Electric Power Research Institute, Gas Research Institute, Lawrence Berkeley Laboratory, New England Power Service Company, New York State Energy Research and Development Authority, Ontario Hydro, Pacific Gas and Electric Company, Pacific Northwest Laboratory, San Diego Gas and Electric Company, Southern California Edison Company, U.S. Department of Energy, and Western Area Power Administration. Contributors were American Public Power Association, Brookhaven National Laboratory, California Institute for Energy Efficiency, Central Maine Power Company, Libbey Owens Ford Company, National Energy Program Evaluation Conference, Niagara Mohawk Power Company, North Carolina Alternative Energy Corporation, Northeast Utilities, Oak Ridge National Laboratory, RCG/Hagler, Bailly, Inc., Rochester Gas and Electric Corporation, Seattle City Light, Southern California Gas Company, Texas Governor's Energy Office, The Fleming Group, and Wisconsin Power and Light Company.

Ed Vine
Dru Crawley
Paul Centolella

Chapter 1

Energy Efficiency and Greenhouse-Gas Emission Reductions: Some International Results

Paul Schwengels, *Climate Change Division, U.S. Environmental Protection Agency*

William J. Pepper, *ICF Information Technology, Inc.*

Introduction

Global climate change has recently become one of the key environmental issues driving debates over energy policy domestically and internationally. This chapter summarizes recent preliminary analyses of the potential for reducing energy-related greenhouse-gas emissions in 21 countries and focuses primarily on the role of energy efficiency.

Atmospheric concentrations of several greenhouse gases—notably carbon dioxide (CO_2), methane, chlorofluorocarbons (CFCs) and related chemicals, and nitrous oxide—are increasing rapidly (IPCC 1990). These gases all share a property of absorbing infrared energy radiated from the Earth's surface, warming the lower atmosphere. Figure 1-1 illustrates the relative contributions of the principal gases to increases in the absorption of infrared energy from 1980 to 1990. While greenhouse-gas concentrations will most likely continue to increase, considerable uncertainty exists about the rate and magnitude of global warming associated with these increases.

The process of stabilizing atmospheric concentrations of greenhouse gases is difficult and would require large reductions in emissions. The activities that generate emissions—primarily energy production and consumption—are extremely important to national

10 — Chapter One

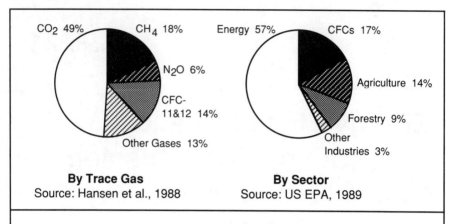

Figure 1-1. Sources of Greenhouse Warming

economies and populations accustomed to certain levels of personal comfort. Energy production and consumption account for nearly 60% of the increases in absorption of infrared energy from 1980 to 1990 (Figure 1-1), and emissions from these activities are expected to increase over time. Due to the long atmospheric lifetimes of most greenhouse gases, large reductions—60–80%—in current global emissions levels will be necessary to stabilize atmospheric concentrations of greenhouse gases or even reduce the rate of increase of these concentrations.

A balanced response to global warming is appropriate. A major research effort is under way but may take decades to resolve scientific uncertainties. The United States and other industrialized nations have indicated interest in near-term actions to reduce the risk of climate change in parallel with efforts to improve scientific understanding. Initial actions in the United States include a commitment to phase out CFCs, the president's proposed reforestation program, and energy-efficiency initiatives recently announced by the Department of Energy.

Recent international discussions on climate change have been coordinated primarily through the Intergovernmental Panel on Climate Change (IPCC), established by the World Meteorological Organization (WMO) and the United Nations Environment Program (UNEP). Information is being developed and exchanged in a collaborative process involving most of the major interested governments and international organizations. The Response Strategies Working Group (RSWG) has focused on policies to reduce emissions of greenhouse

Energy Efficiency and Greenhouse Gas Emission Reductions — 11

gases. Within the RSWG, the Energy and Industry Subgroup (EIS) has coordinated and compiled studies of future energy use, energy-related CO_2 emissions, and options for reducing these emissions. In developing its initial report, this group was able to assemble results of 21 individual country case studies of energy and carbon emissions. These studies were developed by the national governments of the countries or by independent analysts within the countries. The countries include Australia, Brazil, Canada, China, Federal Republic of Germany, Finland, France, Hungary, India, Indonesia, Japan, Mexico, Netherlands, Norway, Poland, Republic of Korea, Switzerland, Union of Soviet Socialist Republics, United Kingdom, United States, and Venezuela.

Reference Scenario

Each of the studies submitted to the EIS included a scenario of future energy use and emissions that assumed no major policy initiatives to respond to global climate change concerns. This scenario is called the "reference" scenario; the scenarios based on policy options for increasing energy efficiency are called "policy" or "option" scenarios. An expert group compiled the results of the country reference scenarios, supplemented with regional data from the International Energy Agency, and integrated them into a single, global reference scenario (US/Japan Expert Group 1990). This global reference scenario is illustrative and falls within the range of plausible futures. Future energy consumption and emissions are difficult to predict reliably, especially as distant as 2025, due to the inherent uncertainty surrounding key variables such as population growth, level and structure of economic activity, carbon intensity of energy systems, and development and dissemination of new technologies.

Common assumptions for the studies were suggested by the EIS, but a great deal of latitude was provided to account for country-specific situations and for modeling capabilities within the countries. Oil prices were assumed to reach $31.10 per barrel (1987 U.S. dollars) in the year 2000 and climb to $44.30 per barrel by 2030. The EIS suggested that moderate or mid-range estimates of economic growth be used but allowed each country to specify this rate of growth. Table 1-1 summarizes the economic growth rates used in the studies and aggregated on a regional basis. None of the country reference scenarios assumed major technological breakthroughs in energy efficiency or noncarbon energy production technologies.

Modeling approaches varied considerably by country. Those countries participating in the EIS used energy models developed by

12 — Chapter One

Table 1-1. Reference Scenario Results

	CO$_2$ per capita (tons C/capita)			GDP growth (1985 = 100)			Primary energy use (exajoules)			CO$_2$ emissions (billion tons C)		
	1985	2000	2025	1985	2000	2025	1985	2000	2025	1985	2000	2025
Global totals	1.06	1.22	1.56	100	160	330	328.2	462.1	776.9	5.15	7.30	12.42
Developed	3.12	3.65	4.64	100	155	280	234.7	308.0	434.6	3.83	4.95	6.94
North America	5.08	5.73	7.12	100	148	241	85.4	108.2	142.1	1.34	1.71	2.37
Western Europe	2.11	2.29	2.68	100	147	249	54.7	64.8	81.3	0.85	0.98	1.19
OECD Pacific	2.14	3.01	3.68	100	177	341	19.2	29.6	42.2	0.31	0.48	0.62
USSR & Eastern Europe	3.19	3.78	5.02	100	163	346	75.5	105.4	169.0	1.33	1.78	2.77
Developing	0.36	0.51	0.84	100	185	563	93.4	154.0	342.3	1.33	2.35	5.48
Africa	0.29	0.32	0.54	100	152	483	13.5	21.0	52.9	0.17	0.28	0.80
Centrally Planned Asia	0.47	0.68	1.15	100	214	799	31.2	47.0	91.9	0.54	0.88	1.80
Latin America	0.55	0.61	0.91	100	152	360	19.1	27.5	55.0	0.22	0.31	0.65
Middle East	1.20	1.79	2.41	100	219	667	8.0	19.2	43.2	0.13	0.31	0.67
South and East Asia	0.19	0.32	0.64	100	194	603	21.6	39.3	99.2	0.27	0.56	1.55

Annual rates of change

	GDP growth		Energy intensity		Carbon intensity	
	1985 2000	2000 2025	1985 2000	2000 2025	1985 2000	2000 2025
Global totals	3.2%	2.9%	−0.9%	−0.8%	0.0%	0.0%

Energy Efficiency and Greenhouse Gas Emission Reductions — 13

their governments. In some cases, the studies provided to the EIS were based on recently completed or ongoing studies, developed for other reasons, of the countries' future energy use. The models used by the Netherlands included a set of optimization, simulation, and penetration models and was supplemented with the Markal Model for the last policy options. The developing countries used an end-use model that was developed by the Lawrence Berkeley Laboratory and that is more an accounting framework used to organize expert judgments by the analysts from the different countries.

Table 1-1 and Figures 1-2 and 1-3 summarize primary energy consumption and carbon emissions from this reference scenario through the year 2025. The results show a rapid increase in primary energy consumption globally and a corresponding rapid growth in carbon emissions—well over a doubling of both global energy consumption and carbon emissions by 2025. This growth in emissions presents a major challenge to all who are concerned with designing and evaluating climate change response options and illustrates the importance of any policy options that can be implemented to reduce the growth of energy consumption.

However, many of the participating countries already are implementing or planning policy initiatives to reduce the energy intensity of their economies and have incorporated these expected results into their reference scenarios. China, for example, already places a high priority on programs to reduce energy intensity (energy use per dollar of GNP), primarily for economic reasons, and plans to continue to reduce energy intensity at a rate of 2.7%/yr for the entire projection period. These plans combine rapid economic growth with an aggressive effort to improve energy efficiency, such as nearly doubling the fuel efficiency of automobiles and motorcycles and of steel and cement production. Similarly, the Federal Republic of Germany (FRG) and France are planning or have in place programs that are expected to markedly reduce the energy intensities of their economies regardless of any international agreements on climate change. In the FRG, these policies include fossil-fuel taxes combined with environmental restrictions and introduction of improved technology and technological procedures. In France, these policies include continued development of nuclear energy and a strong emphasis on improving energy efficiency.

In the reference scenario, significant absolute growth in primary energy use and carbon emissions occurs in all regions, but rates of growth differ, reflecting differences in regional economic growth rates, population growth, energy policies, and resource endowments. Fossil fuels continue to dominate energy supply in all regions. Energy use and carbon emissions grow fastest in the developing countries where

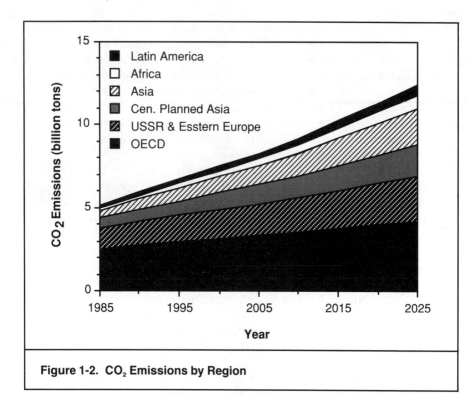

Figure 1-2. CO_2 **Emissions by Region**

both population growth and economic growth are greater than in industrialized countries. The more rapid economic growth allows developing countries to keep pace with more rapid population growth while improving living standards. In spite of these increases, carbon emissions per capita in the developing countries stay well below those in the industrialized countries. As shown in Table 1-1, annual per capita carbon emissions for the developing countries as a whole increase only from about 0.4 tons per capita in 1985 to about 0.8 tons per capita in 2025. During the same period, annual emissions in North America increase from about 5 tons to over 7 tons per capita.

Analysis of Response Options

The following summary and analysis of the response options reflect the different types of response options evaluated in the country studies, the wide range of approaches used, and the different ways of presenting the results. These differences precluded the development of an inte-

Energy Efficiency and Greenhouse Gas Emission Reductions — 15

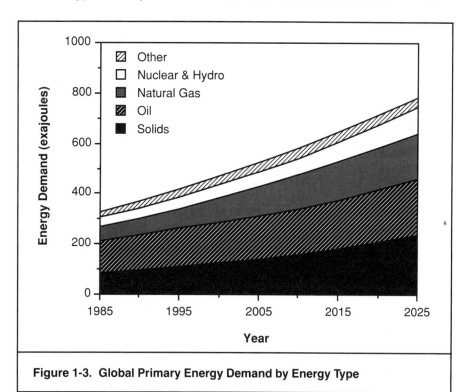

Figure 1-3. Global Primary Energy Demand by Energy Type

grated, global options scenario. Instead, the results of the options scenarios are reviewed for broad regions where similarities or patterns are found. In every country for which options analysis was available, energy efficiency was identified as a key component. Table 1-2 summarizes the results of the individual country analyses.

The approaches used by the analysts in the countries depended primarily on the tools available to those analysts. Most of the country studies provided by national governments used tools already available and in some cases built their analysis from studies already in progress and being performed for the national government. For example, the Canadian response options scenarios utilized a recently completed and detailed evaluation of Canadian energy demand and supply options. These options were incorporated within a regionally and sectorally detailed energy model, which was developed and maintained for the Canadian Department of Energy, Mines, and Resources.

The study for the FRG was commissioned by the Federal Ministry of Economics as part of their ongoing efforts to understand future

16 — Chapter One

Table 1-2. Results of Country-Specific Policy Analysis

Country	Average GDP growth rate	Primary energy use (exajoules)			Carbon emissions (billion tons C)		
		Initial[a] Year	Forecast[a] Ref.[b]	Pol.[c]	Initial Year	Forecast Ref.	Pol.
Western Europe							
FRG	2.3%	11.4	11.0	9.9	197	184	164
France	2.4%	8.5	11.4	10.5	94	117	95
Netherlands	2.2%	2.6	4.1	n/a[d]	40	71	28
Norway	n/a	1.2	1.4	1.3	10	12	10
Switzerland	2.0%	1.2	1.1	n/a	14	12	n/a
United Kingdom	2.4%	9.0	15.4	n/a	158	242	n/a
Other OECD							
Australia	3.1%	3.4	5.3	n/a	67	109	n/a
Canada	2.6%	10.2	16.1	n/a	114	175	107
Japan	3.5%	18.7	28.1	26.2	294	422	363
USSR & Eastern Europe							
USSR	3.2%	54.4	117.2	59.9	899	1,752	773
Poland	2.8%	5.3	11.4	5.7	119	255	100
Hungary	2.0%	1.3	1.9	1.2	21	30	18
Developing							
Brazil	3.2%	6.0	15.2	9.6	41	129	63
China	5.6%	29.0	86.0	73.2	503	1,719	1,360
India	4.9%	8.1	36.6	29.0	98	620	480
Indonesia	3.0%	1.8	9.3	7.7	22	141	111
Rep. of Korea	5.9%	2.4	10.2	7.5	44	166	103
Mexico	3.5%	4.5	12.0	9.1	68	199	133
Venezuela	4.3%	1.7	7.1	6.5	23	67	52

Notes:
[a] Data for each country study are provided for initial year and forecast year, which are as follows: FRG (1987, 2010), France (1988, 2010), Netherlands (1985, 2030), Norway (1985, 2000), Switzerland (1988, 2010), United Kingdom (1985, 2020), Australia (1985, 2005), Canada (1985, 2005), Japan (1988, 2010), and all others (1985, 2025).
[b] Ref.—reference scenario
[c] Pol.—policy scenario
[d] n/a—not available

Energy Efficiency and Greenhouse Gas Emission Reductions — 17

developments in energy supply and demand. The approach used top-down analysis to define and validate assumptions on economic and population growth and combined this analysis with detailed bottom-up methodologies to estimate energy consumption.

The studies for the developing countries were sponsored by the U.S. Environmental Protection Agency (EPA) and coordinated by the International Energy Studies Group, Energy Analysis Program, at the Lawrence Berkeley Laboratory using a detailed energy end-use model developed by the group. In these studies, the country representatives used data available from the countries to characterize economic growth, structural change in the economy, future energy demand, and improvements in energy efficiency. The studies for the USSR and Eastern European countries were also sponsored by the U.S. EPA and coordinated by Bill Chandler at the Pacific Northwest Labs. For the USSR, the study utilized a modeling system developed in the USSR and available to the country analyst performing the study.

Western Europe

Within the Organization for Economic Cooperation and Development (OECD) countries, several Western European countries produced options analyses with somewhat similar results, suggesting a broad pattern for the region. Assumptions of future economic growth for these countries are similar, and the policy scenarios suggest significant potential for reducing the energy intensity (energy use per dollar of GNP) of the economies over time and stabilizing or even reducing emissions. These scenarios differ for some of the countries, notably France, the FRG, and Switzerland, which assume major programs to promote energy efficiency, resulting in significant declines in energy intensity in the reference scenarios.

In France, efficiency improvements have been a major goal of national energy policy since the early 1970s. From 1973 to 1988, CO_2 emissions in France actually declined 26% while GDP increased 40%. About one-third of the decline is due to conservation policies, with the other two-thirds resulting from rapid expansion of nuclear power. In the reference scenario, aggregate energy efficiency improves at a rate of 1%/yr due to structural changes in the economy and to continuation of the current government policies and programs to reduce energy consumption.

In the policy scenario, the French government analysts estimated that the rate of energy efficiency improvement could be increased to 1.5%/yr through more aggressive government policies—primarily regulations in the buildings sector; taxes in the transportation sector; and expanded information, research, and incentive programs across all

18 — Chapter One

sectors. These added policies reduce growth in primary energy consumption during 1988–2010 from 35% in the reference scenario to 24% in the policy scenario. With additional fuel substitution measures, the policy scenario holds CO_2 emissions roughly constant through 2010.

The reference scenario for the Federal Republic of Germany assumes a continuation and expansion of current government policy, which encourages energy conservation and efficiency improvements. In addition, the study assumes a slight decline in population by 2010. The impact of unification with the 'German Democratic Republic has not been included in this analysis. Largely as a result of government tax policies that increase energy prices and programs in order to promote upgrading of energy-consuming equipment, primary energy use in 2010 will be 3% less than 1987 levels. Combined with greater use of natural gas to replace coal and oil, CO_2 emissions decline by 7% by 2010.

The policy scenario provided by the FRG focuses almost exclusively on improving efficiency. Through a combination of additional energy conservation promotion measures and doubling of taxes on fossil fuel producers and consumers, energy demand declines a further 10% below the levels in the reference scenario for 2010, or 13% below 1987 levels. In addition, the energy pricing policies produce increases in renewable energy production. Renewable energy represents 1.8% of primary energy use in the reference scenario in 2010 and 2.4% in the policy scenario. The combined effect is to reduce CO_2 emissions to 20% below 1987 levels by 2010.

In the reference scenario provided by the Netherlands, energy consumption increases 47%, and CO_2 emissions increase 64%, by 2030. Even though average energy intensity in the economy declines 27% during the period, economic growth causes increases in energy use, and greater reliance on coal causes more than proportional increases in CO_2 emissions.

The first set of response options evaluated is a series of efficiency and conservation measures. In the residential and commercial sectors, these measures include efficiency standards for new dwellings and gas appliances and compulsory standards for electric appliances. Investment subsidies will be increased threefold, from 150 million to 450 million guilders, and will focus on retrofitting existing buildings, on combined heat and power production, and on wind energy. Development and demonstration budgets for conservation techniques and renewables will be increased from 120 million to 200 million guilders.

In the transportation sector, policies promote modal shifts: a 75%

Energy Efficiency and Greenhouse Gas Emission Reductions — 19

increase in the use of cars and other road vehicles by 2010 in the reference case is reduced to an increase of less than 35% by promoting the use of bicycles for short distances and of public transport for longer distances. Energy use is further reduced by a 25% improvement in the energy efficiency of vehicles. These goals are achieved by reviewing income tax deductions of transport costs and strengthening investments in public transport.

Policies also increase recycling of basic materials. All measures, combined, offset growth in CO_2 emissions through 2000 and reduce energy consumption 22% below 2030 levels in the reference scenario but still 16% higher than primary energy use in 1990. A range of additional advanced efficiency technologies, combined with greater use of natural gas, nuclear power, and renewable energy technologies, reduces CO_2 emissions 9–26% below 1990 levels in 2030.

Other OECD Countries

A few case studies from industrialized market economies outside of Western Europe present a slightly different picture. In Australia, Canada, and Japan, overall rates of growth in energy consumption and carbon emissions tend to be higher than those projected for Western Europe. Analysis of response options is very preliminary in all three countries but suggests considerable difficulty in achieving stabilization or absolute reductions in carbon emissions. Nonetheless, all of these countries do identify some cost-effective, near-term opportunities for improving efficiency of energy use as a key component of response options.

For Australia, options include efficiency improvements in residential appliances, retrofitting and improved design of residential buildings, and improved energy intensity in industry. Although a response option scenario was not developed, the study estimated that implementation of these energy-efficiency measures along with increased use of natural gas and renewable energy would reduce carbon emissions a maximum of about 11% from reference-scenario emissions for 2005.

The study for Canada examined about 140 individual energy policy measures aimed at reducing carbon emissions and focused primarily on end-use energy efficiency improvements and electricity generation. In aggregate, economically attractive end-use measures reduce reference-case CO_2 emissions by 10% (72 million metric tons of CO_2) in 2005. Additional technically feasible end-use measures reduce emissions in 2005 an additional 10%. The transportation sector accounts for 50% of the efficiency-related emission reductions, with industrial measures contributing 22% and measures in the residential

20 — Chapter One

and commercial sectors each accounting for about 14%. All economic measures, including those for electricity generation, reduce reference-case CO_2 emissions by 27% in 2005. All technically feasible measures reduce emissions 12% from 1989 levels by 2005.

The impact of policies in the study for Japan reflects assumptions about rapid economic growth combined with considerable improvements in energy intensity in the reference scenario. Preliminary evaluation of policies identified some cost-effective additional efficiency improvements, which could reduce energy consumption 7% from the reference-case levels in 2010. In combination with other fuel-switching measures, the overall impact of policies is to reduce emissions in 2010 by 14% of reference-case emissions, a level still 25% greater than 1988 emissions.

Even though the United States did not submit a policy scenario to the EIS, a case study was developed using the same type of approach used with the studies for the USSR and Eastern Europe (Chandler 1990). The reference scenario in the study assumes an average annual increase in GNP of 2.5% through 2010 with population growing from 240 million in 1985 to 288 million in 2010. Petroleum prices grow an average of 4.8% annually. The scenario assumes that no changes are made in industry standards or energy regulations beyond those already agreed upon. Energy demand grows from 84.5 exajoules (EJ) in 1988 to 101.3 EJ in 2010.

An energy-efficiency scenario incorporates cost-effective measures that reduce energy use in 2010 to 92.7 EJ. In the building sector, households would pay an additional $5 billion for efficiency measures resulting in a decrease in energy use of 14% in 2010 and realizing energy savings of $21 billion. A cost-effective 22% efficiency gain is projected to be available in automobiles by 2010 with no or little reduction in performance. Replacing existing steelmaking plants with efficient existing technology could reduce energy use by 42% at a cost well below the cost of the energy saved. Opportunities in other sectors include completely replacing wet- with dry-process cement manufacturing, cogeneration, increased application of computer control systems, and automated process control.

USSR and Eastern Europe

The USSR and Eastern Europe currently account for about one-quarter of global carbon emissions. These nations rank among the most energy intensive in the world, typically about 80% more energy intensive than Western European countries. Energy use and CO_2 emissions grow significantly in the reference scenarios, which assume slow implementa-

Energy Efficiency and Greenhouse Gas Emission Reductions — 21

tion of policies currently being considered or enacted in these countries to promote restructured, market economies. These policies could cause significant reductions in energy consumption and carbon emissions and vary by country. They include elimination of energy price subsidies, reduced emphasis of energy-intensive industry in future development, and use of higher-quality fuels.

In the reference scenarios, economic growth in the region increases demand for consumer amenities toward Western European levels. Structural shifts toward production of consumer goods and away from the current emphasis on basic materials production reduces energy intensity. As energy supply currently acts as a constraint on economic growth in these countries, more aggressive policies to promote structural change would allow more rapid economic growth. All three of the country studies available for this region emphasize two common themes: (1) structural shifts are necessary for economic growth and will reduce growth in energy consumption, and (2) additional energy conservation measures can also significantly reduce energy consumption at less than the cost of supplying additional energy.

In the USSR reference scenario, continuation of current trends in energy use results in a doubling in energy consumption by 2025. Rapid progress in restructuring the economy would produce significantly higher economic growth but lower carbon emissions 25% from the reference levels for 2025. This result is due primarily to the observation that economic programs under the current structure promote continued further development of energy supply at costs much greater than the costs of reducing energy demand through conservation. A range of identified additional measures to improve end-use energy efficiency also reduces energy consumption and carbon emissions at less than the marginal cost of additional energy supply. Successful structural change combined with aggressive implementation of cost-effective efficiency measures could allow significant economic growth while holding carbon emissions stable at roughly current levels. Efficiency measures, in order of annual energy savings, include use of regulated electric drives, efficient lighting, use of gas turbines and combined-cycle plants, low-capacity multifuel boilers, centralized ovens with greater efficiency, insulation of steam supply networks, control and measurement of energy use, switching small boilers to high-grade fuels, switching low-efficiency ovens to large boilers, improved gas compressors in pipelines, shift from harvesters to site threshing, advanced technologies for industrial heating, scrap recycling in the steel industry, insulation of cattle breeding buildings, improved electricity trans-

22 — Chapter One

mission, automation of heating stations, replacing wet cement clinker with dry methods, and improvments in brick production (Chandler 1990).

In Poland, the situation is in many ways similar to that in the USSR. The high energy intensity of the economy combined with the heavy reliance on coal results in high emissions of CO_2 as well as severe local air and water pollution problems. The country is already in the process of shifting from central planning to a market system and attempting to conserve and protect environmental resources as well.

In the reference scenario, both energy consumption and carbon emissions double by the year 2025. A successful transition to a market economy could reduce growth in emissions by 75% by means of less energy-intensive economic activity and some shifting from coal to natural gas due to removal of heavy current subsidies on coal prices. A range of identified cost-effective energy-efficiency measures could reduce energy consumption at no net economic cost. In the steel industry, these measures include adoption of continuous casting, continuous hot-rolling and finishing lines, replacement of open-hearth furnaces by the basic oxygen furnace, and proper use of electric arc furnaces. In the building industry, they include replacing wet-process cement kilns with semi-dry and dry processes, broader use of fly ash from power stations as a cement additive, full recycling of glass scrap, and better management of direct-fired processes in ceramics manufacture. In food processing, potential efficiency improvements include the replacement of coal-fired boilers with oil and gas boilers and major switching from coal to natural gas and liquid fuels. In other sectors, options include improved thermal performance of building envelopes, more efficient appliances, upgrading signaling systems on railways, utilizing motor control devices on locomotives, and more efficient intercity buses. These measures in combination with structural change could stabilize carbon emissions at current levels through 2025.

In Hungary, the economy is also very energy-intensive, despite the fact that the country imports about half of its energy, with energy prices close to market levels. In the reference scenario, continuation of current trends results in a growth in energy consumption and CO_2 emissions by 50% by 2025. Structural change reduces this growth to 33%. Successful implementation of additional economic energy-efficiency measures, combined with structural change, reduces carbon emissions to 20% below current levels by 2025. Efficiency measures include replacement of pig iron production facilities with the most efficient available models that nearly halve energy use per ton of pig iron produced. Similar measures can reduce energy use in steel making by one third, in clinker production by 40%, and in brick production by

Energy Efficiency and Greenhouse Gas Emission Reductions — 23

36%. Very few state-owned residential flats contain thermostats, so indoor temperatures must be regulated by opening windows. Correcting this problem and insulating distribution pipes could lead to a 30% reduction in energy use in the buildings sector.

Developing Countries

Country-level scenarios for seven key developing countries account for a substantial majority of the current and projected emissions from developing countries as a whole. These countries include Brazil, China, India, Indonesia, South Korea, Mexico, and Venezuela. While considerable variation exists between countries, several trends are apparent. In all of these countries, relatively rapid growth in population and economic activity leads to substantial increases in CO_2 emissions despite the fact that all but one of the countries analyzed assumed significant reductions in energy intensity in the reference scenario.

The analysts in developing countries identified increased energy efficiency as a precursor to economic development, irrespective of global environmental problems. Thus, the reference scenarios incorporate measures to promote efficiency improvements in most countries. These measures are most dramatic in China, where energy intensity declines at an average rate of 2.7%/yr through 2025. This decline results from a continuing commitment of the government to implement aggressive policy measures to increase efficiency, primarily for economic reasons. These measures include improved management, energy pricing, restructuring of the economy, technical improvement of out-of-date devices, and utilization of more efficient equipment for new production facilities (80% of China's production capacity would be replaced by 2025 under the projected economic growth assumptions).

Long-term economic growth expected in developing countries will also, over time, change the structure of economic activity and alter the share of value added provided by the manufacturing and service industries and, within manufacturing sector, the share of value added by energy-intensive and other industries. In the reference scenarios for developing countries, energy intensity declines as production of less energy-intensive nonprimary goods becomes a larger share of total production. This decline occurs despite an increase in both overall and per capita production of steel, aluminum, paper, cement, and other highly energy-intensive primary goods.

The current stock of capital goods in developing countries is relatively small and often very energy-inefficient. As an example, steel manufacturing consumes 39 gigajoules (GJ) per ton of output in China and India compared to 18 to 20 in Japan and the United States. Cement

24 — Chapter One

manufacturing consumes 4.8 and 5.6 GJ/ton respectively in China and India compared to around 3 GJ/ton in developed market economies. This capital stock should grow rapidly as economies grow, providing a significant opportunity for penetration of more efficient technologies, suggesting that large improvements in energy efficiency may be possible in much of the developing world.

Table 1-3 shows the improvements in unit energy consumption that are assumed in the scenarios for some key end uses for individual countries. The table shows improvements assumed in the reference scenarios as well as total efficiency improvements after implementation of response options. Average energy consumption for cars today, for example, tends to be lower for Asian countries than for Latin America, due in part to the average size of cars. The potential for reductions in unit energy consumption is, therefore, larger in Latin America than in Asia because reductions in size can be achieved.

On the other hand, where current unit energy consumption is low, faster economic development, particularly in Asia, could increase demand for amenities, leading to acquisition of larger cars, more appliances, and greater saturation of both. This trend can offset some of the gains from more efficient end-use technologies.

As stated above, rapid economic growth exceeds the significant improvements in energy intensity assumed in reference scenarios for all developing countries. Energy consumption in all of these countries grows dramatically by the year 2025, with increases ranging from about 150% for Brazil to over 350% for Indonesia. In general, CO_2 emissions grow proportionally with energy use. However, all of these countries have identified opportunities for additional efficiency improvements that could reduce the rates of growth in energy consumption. The options scenario for Venezuela is the least optimistic, with growth in energy consumption reduced by only 10% from the reference scenario. Other countries' estimates are generally in the range of 20–40% reductions in the growth of energy consumption. Brazil provides the most significant scenario of policy-induced efficiency improvements, resulting in a decrease of 60% in growth in energy consumption by 2025.

Costs

Estimates of the costs of reducing CO_2 emissions through energy efficiency varied considerably between studies, but most of the studies concluded that significant reductions in energy use could be achieved at costs less than or close to the costs of the energy supply. Differences existed in how these reductions could be achieved and did not look at

Energy Efficiency and Greenhouse Gas Emission Reductions — 25

Table 1-3. Assumptions for Unit Energy Consumption in Developing Countries

	Biomass cooking (GJ/yr)			Refrigerators (kWh/yr)			Cars (liters/100 km)		
	1985	Ref.[a]	Pol.[b]	1985	Ref.[a]	Pol.[b]	1985	Ref.[a]	Pol.[b]
China	42.0	25.0	22.0	400	348	348	10.2	6.2	5.2
India	30.6	27.6	24.5	300	200	150	9.4	6.2	5.2
Indonesia	16.5	13.2	9.9				9.4	6.2	5.2
Rep. of Korea	12.6	10.3	10.3	389	622	467	10.2	9.4	7.3
Brazil	53.0	53.0	42.4	666	1000	400	13.1	7.3	5.3
Mexico							19.6	6.7	5.5
Venezuela	33.5	26.8	20.1	900	540	540	23.5	7.1	5.9

	Trucks (liters/100 km)			Steel (GJ/ton)			Cement (GJ/ton)		
	1985	Ref.[a]	Pol.[b]	1985	Ref.[a]	Pol.[b]	1985	Ref.[a]	Pol.[b]
China	3.10[c]	1.71[c]	1.55[c]	39	21	19	4.8	2.6	2.4
India	24	19	16	39	33	28	5.6	5.0	4.2
Indonesia	25	18	13						
Rep. of Korea	22	16	14	23	17	14	3.8	2.9	2.5
Brazil	23	18	9	26	24	13	4.1	3.7	2.1
Mexico									
Venezuela	50	25	15	30	26	24	5.7	4.6	4.1

Notes:
[a] Ref.—reference scenario for 2025
[b] Pol.—policy scenario for 2025
[c] Units for trucks in China are megajoules per ton-kilometer

the costs of implementing policies to reduce energy use nor at the welfare loss due to other factors.

Potential reductions and the cost of these reductions depended heavily on current patterns of energy use. Structural change in the USSR, Poland, and Hungary are expected to result in economies with less emphasis on energy-intensive industry and with large improvements in unit energy efficiency. Structural change is expected to result in reduction of net costs and to accelerate economic growth. In developing countries, improved energy efficiency is necessary for rapid economic growth. Rapid expansion of the industrial base in these countries provides tremendous opportunities to improve the efficiency

26 — Chapter One

of these industries at no net cost, and full access to advanced, energy-efficient technologies is a central theme to many of the developing country policy scenarios.

Emission reductions in developed market economies and the costs of these reductions vary considerably. The study for Japan suggests that the most efficient technologies will be used anyway and that policies result in only minor improvements in energy efficiency. These results contrast with results from studies for Canada and some Western European countries, studies that suggest that the most energy-efficient technologies are not always used and that large improvements can be achieved with no or at least very low net cost.

Conclusions

The analyses summarized in this paper represent first steps in developing a better understanding of the possible response options available to address concerns about increasing greenhouse-gas concentrations and potential climate change. A great deal of additional analysis and documentation is needed. Many of the studies have evaluated economic costs and specific policy implementation measures in a cursory fashion, if at all. More careful comparison and standardization of assumptions and methodologies between countries is needed and would improve our ability to generalize and compare strategies across regions and countries.

Nevertheless, even the preliminary results available to date suggest strongly that a significant potential exists for energy-efficiency improvements over the next few decades beyond what would be induced by market forces. The available case studies suggest that much of this incremental efficiency improvement could be achieved at no net costs to national economies and could enhance economic growth. In addition, the studies suggest that very significant reductions in the rate of growth in CO_2 emissions—and in some cases stabilization or net reductions—could be achieved through policy-induced efficiency improvements. However, given the potential for rapid growth in emissions, major policy-induced measures would be necessary to reduce global emissions of CO_2 enough to stabilize atmospheric concentrations.

References

Chandler, W. 1990. *Carbon Emission Control Strategies, Case Studies in International Cooperation.* Washington, D.C.: World Wildlife Fund & The Conservation Foundation.

EPA. See U.S. Environmental Protection Agency.

Hansen, J., A. Lacis, I. Fung, D. Rind, S. Lebedeff, R. Ruedy, and G. Russell 1988. "Global Climate Changes as Forecast by the Goddard Institute for Space Studies' Three-Dimensional Model." *Journal of Geophysical Research* 93: 9341–64.

IPCC 1990. See IPCC Working Group 1.

IPCC Working Group 1 1990. "Policymakers' Summary of the Scientific Assessment of Climate Change. Report to IPCC from Working Group 1." Draft. Bracknell, U.K.: IPCC Group at the Meteorological Office.

U.S. Environmental Protection Agency 1989. "Policy Options for Stabilizing Global Climate. Draft Report to Congress." Washington, D.C.: Office of Policy, Planning, and Evaluation.

US/Japan Expert Group 1990. "Integrated Analysis of Country Cases Studies. Report to the Energy Industry Subgroup, Response Strategies Working Group, Intergovernmental Panel on Climate Change." Draft. Geneva, Switzerland: Intergovernmental Panel on Climate Change.

Paul Schwengels manages the international energy program in the Climate Change Division, U.S. Environmental Protection Agency. He has conducted and managed analysis of energy-related environmental policy issues, including acid rain and global climate change. He holds a B.A. from Parsons College and an M.A. from the American University.

William J. Pepper is vice president of ICF Information Technology, Inc. He has specialized in evaluation of energy and climate change policy issues. He received a B.S. from the University of Maryland and an M.A. from Temple University.

Chapter 2

Lay Perspectives on Global Climate Change[1]

Willett Kempton, *Center for Energy and Environmental Studies*

Introduction

Despite extensive media attention since the summer of 1988, global climate change remains a challenge to lay comprehension. The scientific issues are staggeringly complex, with major predictions still debated by climatologists. Nevertheless, a social consensus on climate change policy will be required because the costs of both prevention and adaptation are very large, and preventive responses are inherently collective. Citizens' comprehension and value judgments are therefore significant in the political decision-making process.

This chapter attempts to document current lay thinking on global climate change, based on extensive, open-ended interviews with a small group of U.S. residents from diverse walks of life. Topics include personal weather observations, the environment, and global warming, with emphasis on informants' ideas of what global warming is, the possible policy responses, and the values that they would bring to bear in making decisions.

One goal of this work is to make lay perspectives accessible to a broad range of professionals involved with climate change, including atmospheric scientists, policymakers, educators, and science journalists. Lay thinking has been made inaccessible and foreign to those most active in the policy debate, due to the very jargon and established scientific models that experts must internalize to understand the climate and policy issues. Another goal is to provide a revealing case

---

[1] This chapter is appearing simultaneously in the journal *Global Environmental Change*. Copyright is held jointly by Butterworth-Heinemann and the American Council for an Energy-Efficient Economy.

30 — Chapter Two

study for anthropologists and cognitive scientists concerned with how people assimilate new concepts by adjusting them to fit preexisting ones.

It would be foolish to expect lay people to think like atmospheric scientists or policy analysts about global climate change. Even if specialists have more accurate knowledge, it is useful to document lay concepts, since this information can be used to design more effective public communication. Also, if public support or rejection of a proposed policy is founded on misconceptions, this should be of concern to interests on all sides of the policy debate. Finally, it is revealing to draw out the values underlying the upswing in environmental concern observed in recent public opinion polls, and to show how these values affect views of an environmental issue that is abstract, global, and has its primary effects in the future.

Interviews

The interviews use ethnographic methods, which are recommended when the interviewer initially has little understanding of native concepts (see Agar 1980; Spradley 1979). These methods contrast with survey research, in which the questions and range of possible answers are already known and precise answer frequencies are measured. Ethnographic interviewing methods use open-ended questions, follow-up probes for topics raised by informants, and paraphrases for verification. Interviewers must be highly experienced for a successful outcome. Probe questions are essential to understand what answers really mean and to discover unfamiliar concepts, although using them necessarily means that some answers will not be comparable across the sample of informants. Ethnographic methods were developed by anthropologists to study foreign cultures, but the author has found in prior work that an equally important application is in bridging the gap between the layperson and the scientist in understanding science policy issues.[2]

The interviews consisted of three parts. First, the interviewer asked questions about weather observations and the environment in

---

[2] Differences between survey and ethnographic methods are clearly seen by comparing the present report with a survey of 1,200 respondents (RSM 1989). The survey was conducted at approximately the same time (October–November 1989) by Vince Breglio, former chief pollster for the Bush-Quayle campaign. The survey never asked about three-quarters of the points found to be important in the ethnographic interviews. On the other hand, when both studies asked a similar question, the survey provides far more reliable national percentages. In this paper it is one of several national surveys that are cited periodically to check the representativeness of the small ethnographic sample.

general, then assessed the informant's recall of information on global warming. Second, the interviewer gave a short presentation on global warming to provide background information. The presentation was designed to be similar in length and detail to an in-depth article in a weekly news magazine. Third, informants were asked for reactions to the presentation and to a set of policy proposals. Since it followed the interviewer's presentation, the third part of the interview provided a rough gauge of how people assimilate new knowledge about global warming and of how they may react in the future if the causes and consequences become better known. (See Kempton 1990 for the interview protocol.)

Fourteen interviews were analyzed for this paper. They were typically conducted jointly by the author and a research assistant; most took place in Hamilton Township, a demographically diverse community bordering on Trenton, New Jersey.[3] The first two interviews, used in part to test the interview protocol, were conducted in Princeton with a married couple who were acquaintances of a research assistant. Of the remaining twelve informants, four were approached in public places (a park or shopping center), and eight were approached in their homes. Houses were picked by external appearance to span a diversity in neighborhoods and housing costs. We did not prearrange interviews but just rang doorbells. After warning prospects that the interview might take 45 minutes, we were refused interviews by approximately two-thirds of those approached.

Table 2-1 summarizes the demographics of the subject population. The sample includes equal numbers of males and females. Annual household income ranges from $13,000 to $62,000, and educational levels range from ninth grade to college graduate (the informant in the second of the two test interviews holds a Ph.D.). In Table 2-1, "religion" is the answer to "What is your religious background?" and "Are you active now?" "Political" is "How would you describe yourself politically?" (not necessarily a party) and their 1988 presidential vote. Notice that income is reported for the household, whereas all other data are for the individual informant. Questions coded "na" were not asked. All informant names are pseudonyms.

All interviews except the first test interview were taped. This paper draws its data from our verbatim transcripts, occasionally supplemented by field notes, totaling 102 single-spaced pages (60,000

---

[3] After this set of 14 interviews, the protocol was refined and expanded. A second set of interviews is underway, with an additional 26 recently completed in other U.S. states (Maine, Pennsylvania, West Virginia), for a total of 40. Those interviews, now transcribed but not yet fully analyzed, are consistent with the results reported here.

32 — Chapter Two

Table 2-1. Demographic and Social Data on the 14 Informants Interviewed for this Study

#	Name	F/M	Age	Ed.	Household income	Occupation; race/ethnicity; religion; political
1	Ellen	F	33	BA	$60,000	freelance writer; white; inactive Catholic; "liberal," Dukakis
2	Eddie	M	36	PhD	$60,000	university admin.; white; inactive Jewish; "liberal," Dukakis
3	Susan	F	57	2 col	$26,000	network administrator; white; active Catholic; na
4	Paige	M	39	HS	$53,000	industry "production"; black; "nonaffiliate"; na (probable Democrat)
5	Walt	M	73	HS	$13,000	retired line worker in factory; white; active Catholic; na
6	Doug	M	32	BS	$47,000	pharmaceutical research; white; active Catholic; independent, Bush "democrat for everything else"
7	John	M	73	9th gr	$16,000	retired supervisor for wire mfgr.; Hungarian-born, naturalized; active Catholic; independent, Bush
8	Joe	M	31	HS	$17,000	factory welder; black, Liberian national; active Pentecostal; alien (nonvoter)
9	Jenny	F	39	HS	$62,000	high school teacher; white; active Catholic; "liberal," Dukakis
10	Wilbur	M	62	9th gr	(refused)	retired fireman; white; inactive Methodist; independent, Bush
11	Amanda	F	33	BA	$55,000	housewife, was special ed teacher; white; active Protestant; Dukakis
12	Cindy	F	32	HS	$48,000	housewife; white; active Protestant; Democrat, Bush
13	Jane	F	81	HS	$14,000	retired, clerical; white; active Protestant; independent, Bush
14	Tara	F	36	3 col	$40,000	sales rep, sports equipment; white; inactive Baptist; "leaning Republican," Bush

words). Quotations presented here use the following conventions. Quotes are word-for-word transcriptions from tapes, except that redundant "uh"s, repetitions, and false starts have been removed. Nonstandard grammar and word choice are preserved, marked with [*sic*] when they might otherwise cause confusion. (Readers unfamiliar with verbatim transcripts should be warned that sentences from spoken conversation are rarely as complete and syntactically well formed as written text.) Underlining indicates emphasis by the informant, and ellipsis points (". . .") indicate material I have deleted. Brackets [] denote my post-interview clarification or paraphrase, as deduced from the context, from prior statements by the informant, from intonation, and so forth. Interviewer questions and statements are given in italics, to distinguish them from informant responses.

Global Warming Incorporated into Existing Concepts

After several questions about perceptions of weather, we asked, "Have you heard of the greenhouse effect?" We chose this term over "global warming" or "global climate change" as more commonly used by the media and more widely recognized by the public. Ten of our 12 informants had heard of it, as had both test informants.[4] Our sample did not over-represent those familiar with the phenomenon, since our 10 of 12 is virtually identical to the proportion found in a national probability sample: 79% had heard of the greenhouse effect (RSM 1989).

We next asked those who had heard of the greenhouse effect, "What have you heard about that?" The responses were seriously at variance with the scientific models of global warming. When looking at one or two interviews, the responses seem idiosyncratic. Inspection of the entire set of interviews, however, reveals repeated and systematic transformations from scientific models. As described below, I conclude that new information on global warming is being fitted into four prior concepts: stratospheric ozone depletion, tropospheric (near surface) air pollution, plant photosynthesis, and seasonal and geographic temperature variation. This process is of interest both theoretically, as an example of how cultures assimilate new information, and practically, since it has a significant effect on public support for various policy options.

---

[4] Nine recognized the specific term "greenhouse effect." A tenth did not recognize that term but had heard that, as he put it, "in the 21st century it's going to be unbearably warm."

34 — Chapter Two

Stratospheric Ozone Depletion

The most widely shared transformation from scientific to lay knowledge is caused by categorizing global warming as a subset or effect of ozone depletion. For example:

> *Have you heard about the greenhouse effect?* Is that what they're talking about the ozone layer? . . . They had last year for a, create the hot spell? . . . But I couldn't understand that, last year we had it, what made it change this year? We don't have it quite as severe . . . *What other things can you remember . . . ?* Well that was about the only thing . . . through the gases and that in the cans, you know, pressurized cans.—Wilbur

While the man quoted above had only a ninth-grade education, the same confusion is seen in the following quote from a college graduate, who expresses an interest in environmental and health news and who warned us that she was atypical because she talks with scientists about the greenhouse effect:

> Most people say burning fossil fuels is changing the climate because we are making the ozone layer disappear, that's the layer that protects us from the sun's harmful rays. This will greatly affect the climate over the next 100 years.—Ellen

There are some interdependencies among climate change, tropospheric ozone pollution and stratospheric ozone depletion. For example, CFCs cause stratospheric ozone depletion as well as global warming, and ozone itself is a greenhouse gas. However, these are secondary, tertiary, or lesser effects—the science has clearly been altered when "the ozone layer" or "pressurized cans" are given as the primary, or the only, cause of global warming.

We were able to observe the merging of the concepts of ozone and greenhouse effect right before our eyes during one interview, with John. Since John initially said he had heard nothing of the greenhouse effect or of a warming trend, his subsequent discussion of these topics was presumably derived from what we said. Shortly after our presentation, we asked about automobile technology that did not cause the "greenhouse effect." He replied that they had already changed the formulation of gasoline to "cut down the ozone effect."

Why did he respond to our question about "greenhouse effect" with "ozone effect"? We had mentioned the greenhouse effect about eight times by this point in the interview, whereas we mentioned ozone depletion only once, and then only to note that we were not going to discuss it.[5] I conclude that even in this semi-controlled environment, any talk about pollution affecting the stratosphere makes most people think "ozone" no matter how much one mentions CO_2 and the greenhouse effect. Moreover, many beliefs about ozone depletion are incorrect: many informants assigned the blame to aerosol cans (a notion that is a decade out of date in the U.S.), and some confused stratospheric ozone depletion with urban tropospheric ozone pollution. Nevertheless, it appears that the ozone hole has arrived as a concept in the U.S. public's consciousness, but the greenhouse effect is entering primarily as a subset of the ozone hole phenomenon, the closest model available.

Tropospheric Air Pollution

The second confounding prior concept is tropospheric pollution. "Air pollution" is a well-established category, and greenhouse gas emissions are conceived as just another instance of that category. This results in misconceptions about, for example, health effects:

> Well, I like warm weather, personally, but I think it's wrong for what humans are doing to the atmosphere. *In what way?* With all the aerosols and the ozone and so forth . . . that's being projected up into the atmosphere. . . . *[If you like warm weather, why do you say the greenhouse effect would be wrong?]* Well, I think it's wrong because at the same time, we are ingesting and breathing in all these different chemicals that are being put into the atmosphere.—Susan

One consequence of regarding greenhouse gases as pollution is that traditional pollution controls seem a solution:

> We're just going to have to probably . . . find out where most of the pollution is coming from. [For the sources that are] industrial, have an incredibly fine filter . . . where you prevent most of this excess CO_2 from going into the atmosphere.—Doug

> *Do you think the United States should do anything about [the greenhouse effect]?* Yes, . . . trying to find nonpolluting fuels. There should be more money poured into mass transit. There should be strong controls on industrial pollution. *Like today's pollution controls but stricter?* Oh, much stricter.—Ellen

The above suggestions are traditional solutions to pollution. Some, like mass transit, could ameliorate both problems simultaneously. But

---

[5] This explicit mention of ozone depletion, saying it was not the focus of the interview, was added to the mid-interview presentation after the first few interviews revealed confusion of ozone depletion and the greenhouse effect.

36 — Chapter Two

measures such as filters and strengthening today's "pollution controls" would not reduce global warming.

Another problem with the air pollution model is that it focuses on industrial smokestacks and vehicle sources. While burning of fuels is 57% of the problem (Lashof and Tirpak 1989), the lay air pollution model obscures the roles played by seemingly nonpolluting sources such as farming, ranching, leaky refrigeration equipment, and indirectly, energy end-use inefficiencies.

A terrifying, if idiosyncratic (only reported by one person), interpretation is that descending stratospheric ozone would force out fresh air. This interpretation may derive from a confusion with urban (tropospheric) ozone pollution:

> What have you heard about the greenhouse effect? I really don't understand that much, but I know that the ozone is getting, coming down more on account of, you know, the exhaust and all that kind of stuff. [later] Do you remember hearing about what kind of effects the greenhouse effect might have? Well, I know it wouldn't be a good effect, I don't know how. [pause] It makes me think we're going to be, not have any air to breathe if it keeps gettin' closer into our air. That's the way I understand it.—Jane

Of the nine informants who provided enough detail to discern, six carried inappropriate pollution concepts into their inferences about global warming. (Two of the three who did not—Eddie and Doug—were the two who read scientific journals on the subject.) In many cases the distinction does not matter, since many measures simultaneously reduce CO_2 and pollution. In some cases, however, pollution-reduction technologies actually increase greenhouse gas emissions. The point is that the pollution model is a source of many misconceptions about both global warming and possible policy responses.

Photosynthesis

New information about the greenhouse effect is also incorporated into prior knowledge of plant photosynthesis, taught early in school and thus widely shared. Several informants understood and could recite the idea that trees absorb CO_2 and produce oxygen. From this foundation, some had reinterpreted the media descriptions of forest destruction causing increased CO_2 levels into the idea that we would exhaust all atmospheric oxygen. This reaction was not as widespread as the ozone depletion and pollution models, but it was a very frightening prospect:

> Well it has to do, I think, with the climate changing as a result of the atmosphere, atmospheric changes that are presently going, and cutting down all of our woods takes away a certain chem-, the oxygen, or some-

thing, that is required for us to have good air quality, and uh, it's kind of scary.—Tara

That's what scares me. *What?* When they cut all the forests down, they say, pretty soon we're not going to have any oxygen to breathe. Why do they let them do that?—Cindy

Science educators deserve credit for successfully imparting the concept of photosynthesis to generations of students. However, in this case people are misled by the concept (or by what they remember of it), for two reasons. First, CO_2 is currently only 0.034% of the volume of the atmosphere, compared with oxygen at 21% (Harte and Socolow 1971: 281). The contemplated doubling in 50–100 years would raise CO_2 to 0.068%, still less than 1% of oxygen. Second, the contribution of growing plants to atmospheric oxygen is almost entirely cancelled out by decay of plants after their death. Atmospheric oxygen is increased only when those plants are buried before decomposing, estimated at 0.04% of the oxygen produced each year and equalling 1 part in 15,000,000 of atmospheric oxygen (Broecker 1970: 1531). A fully mature standing forest adds little oxygen to the atmosphere, although clearing and not replacing that forest releases all its stored carbon into the atmosphere. In short, large increases in CO_2 or large losses of forests will not significantly decrease atmospheric oxygen. There may be other excellent reasons to stop deforestation—such as species habitat preservation or carbon storage—but running out of oxygen is not one of them.

Experienced Temperature Variation

The fourth source of lay perspectives is preexistent knowledge of seasonal and geographical temperature variation. North Americans are familiar with winter-summer temperature swings of 100°F, frequent daily temperature swings of 20°F, and major geographical differences in temperature. From this experience, an average temperature rise of less than 10°F does not seem very harmful. This was seen by reactions to our presentation like:

That doesn't sound so bad, does it? Three to nine degrees [warming in 100 years].—Cindy

One person argued that future generations would acclimate to the new temperatures:

I wouldn't want to live in that, from what I am accustomed to. But it relates, I'm sure, to what you're accustomed to. If you live in Alaska, or you live at the North Pole, and you like it, [then] you like it.—Tara

38 — Chapter Two

Similarly, there is anecdotal evidence from a climatologist that people underestimate the temperature change needed for major climate-induced effects: Schneider (1988) notes, "Many people are surprised to find out that the ice age was only about 9°F colder than the present average Earth temperature." Schneider based this observation on reactions and questions of audiences at his public lectures and testimony (Schneider, personal communication)—a subpopulation presumably more informed than the average citizen.

Unlike the former three sources of misconceptions, knowledge of seasonal and geographic temperature variation distorts only in scale, not by applying concepts inappropriately. Nevertheless, it is clearly more difficult to understand what all the fuss is about when one's direct experience—from both travel and local temperature cycles—teaches that 3°F to 9°F is "not much" temperature change.

A related difference between scientific and lay understanding is the significance of higher average temperatures. When a scientist, especially a climatologist or biologist, refers to concern about "global warming," he or she implicitly is referring to multiple geophysical and ecosystem effects. To the lay ear, "global warming" without further elaboration simply means "hotter weather."[6] Prior to our presentation, warmer weather was the first mentioned and, for several, the only remembered consequence of the greenhouse effect:

> *What have you heard about [the greenhouse effect]?* This portion of the country basically is going to become warmer over a certain length of time. That's really all I know about it. . . . *Have you heard about any other effects . . . ?* Not that I can recall.—Susan

While middle-class white informants like Susan did not see warmer weather as a great problem, it was more troublesome to an informant familiar with life in the inner city:

> *If the weather gets a lot warmer, do you think it would be good, bad, or neutral?* I think it would be bad; I think it would be terrible. *Why?* Well, I think people react differently in warm weather than when it's cooler. I think it has an effect on attitudes—behavior. . . . I mean in the prison system especially, where the people are just, you know, stuck in there, and they've got to let off steam. So, sure. *So you think in prison it makes people more violent?* Sure, but outside the prisons, too, 'cause I even see it at work; you know, when the weather is extremely warm, people tend to be, you know, a little hot-tempered. I think, you know, their blood boils. And when the blood boils in the body, it goes to the head,

---

[6] This is an argument for referring to the anticipated changes as "global climate change," which is really more the point than "global warming," per se.

and next thing you know, there's, you know, an explosion. . . . I've seen them react that way.—Paige

Without endorsing all the explanatory mechanisms advanced here, I am pointing out that some social groups may legitimately perceive hotter weather as having personal and social costs. Nevertheless, Paige was like the other informants in conceiving of global warming in terms of the personally familiar effects of hot summer weather rather than the more profound, but unfamiliar, environmental consequences.

Durability of Existing Concepts

In summary, the four models lay people apply to global warming illustrate the difficulties in putting a new concept on the layperson's cognitive map. Global warming is not caused by ozone depletion, and it differs greatly from traditional pollution in effects and solutions. CO_2 increases do not threaten to suffocate terrestrial mammals, and major environmental and biological change would result from average temperature changes far smaller than seasonal swings. In time, given continued media coverage, school lessons on this topic, and public discussion, American citizens' understanding will surely improve. We can take heart from my evidence that the public has assimilated the aerosol-ozone link. In the meantime, however, the scientific content of media discussions, no matter how accurate the words and graphs employed, will continue to be distorted as citizens try to fit this strange new phenomenon into their known world.

Perceptions of Weather

People directly perceive weather, not climate. Lay weather observations have been mentioned by others in raising the political question of whether climate change will cause citizens to demand action on global warming. Before addressing this political question, we must first ask three prior cognitive questions: (1) how do people integrate casual weather observations and draw conclusions from them; (2) if the climate begins to change, would the average citizen notice; and (3) would ordinary people find plausible the claim that climatic warming was caused by anthropogenic gases?

To address these cognitive questions, the interview protocol briefly covered informants' personal observations about weather patterns. These questions were the very first on the protocol, so that the questions concerning environmental pollution and the greenhouse effect would not contaminate them. The data suggest that people overemphasize both the human effects on climate and the extent to which climate is already changing.

40 — Chapter Two

Human Effects on Weather

We began our interview asking, "What factors would you say affect the weather?" This question was intended as a quick check on the informant's knowledge of meteorology. We got some expected answers, such as:

Sunspots, volcanic activity, earth movements.—Ellen

The jet stream is the biggest factor concerning the weather.—Jenny

A more common type of answer was totally unexpected. People saw diverse nonnatural causes of weather change. Several informants mentioned "pollution" as the primary factor:

Pollution affects the weather. That's all I can think of.—Paige

Well, burning, like these rain forests, and these western forest fires. Spraying for insecticides and stuff like that. *Mmm-hum.* And, herbicides, like on the farms, to prevent weeds from growin'. . . . But the main thing is burning and auto pollution, stuff like that.—Walt

One man thought the weather had become more violent or erratic, a change he attributed to atomic bomb testing:

I have an answer, but it's, [pause] I don't know, I've always felt that when they had that bomb I think it had an awful bearing on the change of our weather. The A-bomb. They had those tests . . . just seemed like here things have changed ever since. It's become more torrent, the weather here in the past few years. . . . *When you say more torrent, you mean like more changing?* Violent, violent, yeah. The weather is very changeable. They say that didn't have nothing to do with it, but I still feel that it did somewhere along the line.—Wilbur

Lay attribution of weather change to atomic bomb testing was in fact widespread in the early 1960s (Kimble 1962). A *U.S. News & World Report* article of the time began: "Many remain convinced that man—especially American man—has knocked the weather out of kilter with space and atomic experiments. Officials say no." The same article recalled that during World War I, unusually wet weather in Europe and the U.S. was widely attributed to extensive artillery fire (1963: 46,48).

In our interviews, three people (20% of our sample) mentioned space shots as affecting the weather:

I have my own private theory. [pause] *What's that?* That every time they shoot something up in space it disturbs things up there! *There could be something to that.*[7] I've been told I have no foundation for that, but it

---

[7] This follow-up comment violates my interviewing ethics because I stated something I do not believe to be true. I did so because I was dumbfounded, yet I needed to respond quickly and positively to encourage the informant to expand on her "private theory."

just seems every time something happens we get this strange type of weather. . . . *Like what?* . . . Well, for instance, tornadoes were very rare in this section of the country . . . tornadoes, and violent-type storms . . . It used to be rather calm here.—Susan

Well, I don't know what the hell they're doin' up on the moon and shooting those things up there. I think they're disturbing the atmosphere. So much rain we've had, so much rain. [These comments were made in response to a request for ideas as to what could be done about the greenhouse effect.]—John

These quotes suggest a propensity for people to believe that the weather is affected by human activities, especially human activities that occur in the atmosphere and are regarded as unnatural or immoral (space shots, atomic bombs, pollution).[8] Although no one specifically mentioned greenhouse gases ("burning" and "pollution" were closest), the concept of human activities changing the weather would seem to be readily assimilable, especially if those activities are seen as unnatural or immoral. In this sense, U.S. culture almost seems to have been designed in advance to readily believe in anthropogenic global climate change.

There is a corollary to my finding that unnatural or immoral human activities are seen as changing the weather: the cultural belief system would be consistent if the same people also considered it improper for humans to attempt to change the weather. I find evidence that some Americans do believe this, based on an earlier set of surveys.

In nonurban areas of four states, surveys asked whether respondents believed that "cloud seeding probably violated God's plans for man and the weather." Agreement ranged from 30% to 48% (Farhar 1977: 289). An organized religion framework was not essential to this perspective, since the answer covaried with two secular statements: "Man should take the weather as it comes and not try to alter it to suit his needs or wishes" and ". . . cloud seeding programs are very likely to upset the balance of nature" (Farhar 1976). Would similar concerns

---

[8] Overall, when asked "What factors affect the weather?" five mentioned exclusively human activities, three mentioned both human and natural phenomena, and only four listed exclusively natural phenomena. (Two didn't answer the question.) In part, the pollution answer may have been enhanced because we initially described the interview topic as "weather and the environment." There is also an unintended ambiguity in the question word "affect," as indicated by one informant's paraphrase of our question as "things that could change the natural weather patterns." It may be worth asking about this topic again with a reworded question. Nevertheless, these potential objections do not alter my judgment from the existing data that people have a tendency to believe that human activities affect the weather.

42 — Chapter Two

be cited in reaction to polluting a river or paving a meadow for a shopping center? Surely fewer would express their objections to such non-weather modifications in terms of "God's plans" or the "balance of nature." Although I do not have the ideal direct and explicit data from either my own interviews or the cited surveys, I am arguing that the weather (and hence climate) is a natural system with special cultural meaning and that there are tendencies, even in our technology-based society, to feel that it is wrong for humans to tamper with it.

Belief That the Weather Has Changed

Many people believe that the weather has *already* changed. Some reported their own personal observations of warming, most typically noting milder winters:

> I know out in Wisconsin it's much more warm than it was years ago. . . . Yes, 'cause a couple of years ago they [her son's family] took up cross-county [*sic*] skiing, and the man said they had to close down [for lack of snow]. . . . They don't have snow at all like they used to. It is much warmer.—Jane

> We used to have snowdrifts all the way up the telephone poles, all the way up the side of the house. Now, you get a couple of inches and they close the schools. *When do you think that changed . . . ?* I guess in the last ten years, there was a big change. *[Are you thinking about this now for the first time?]* No, I've thought about it, we've discussed that. [Mother indicates agreement.] You know when we look at old pictures and stuff we say, "Look at all that snow."—Cindy

In the following case, I suspected that a warming trend was reported in part because the informant had heard that it was expected:

> . . . what I've noticed in the weather patterns around here is that we seem to be having hotter summers, actually, it seems to be much more hotter [mentions a television show on the greenhouse effect]. *And you're saying that you noticed yourself that it's gotten hotter, besides what you heard on the TV.* Yeah. *You personally, is that?* I think that the number of hot days we've had has been significantly increasing. . . . I've noticed the amount of days. And it seems that also as far as the winters are concerned, we don't seem to be having cold winters as we used to, used to get when I was younger, it seems. So it actually seems, we seem to be having mild winters, and hotter summers.—Doug

In contrast, two informants explicitly mentioned that they had not yet observed any greenhouse-induced warming. It is highly unlikely that laypeople could accurately discern a climate trend from casual weather observation of the modest global warming effects to date. Thus Doug's quote, and another like it from Tara, suggest that lay climate generalizations are like a Rorschach inkblot test—one sees the patterns that one is predisposed to see.

Another common observation was that weather patterns were becoming less predictable:

> Well your springs, springs are unpredictable any more, you can't really tell. As far as, you know, especially if you tried to plant a garden and that. . . . Now you can't really go by it. What was it? Somewhere in the middle, fifteenth of April when you could pretty near predict that frost was done. And you go ahead and plant. But now . . . you could have a cold spot, you know, it can seem like anytime you turn right around, it's snowing.—Wilbur

> I think our weather is more unpredictable now, uh, we've got more, modern, satellites and all to predict the weather and I think they're worse at predictin' it now than when they didn't have them. So, your seasons don't seem to run the same. I mean, they're less predictable for patterns. . . . Like now in the spring you get real hot weather like the summer, and then when the summer comes you got weather that's like the spring, and then in the middle of winter sometimes you get weather that's almost like summer. Uh, it isn't an even, even pattern. *When would you say you started noticing that . . . ?* Uh, I don't know, it's come on gradual.—Walt

In addition to warmer winters and more unpredictability, people who believed that unnatural or immoral human activity was influencing the weather often perceived an increase in storms or violent weather, as mentioned in the previous section.

My finding that a majority reported noticing a change in the weather is consistent with what little historical and survey research I have found on perceived weather changes. For example, during the American colonial period there was a widespread, publicly articulated belief that the American climate had grown warmer. Ludlum (1987) cites people's accounts that winters did not seem as harsh as those described by their grandparents. One early published explanation (Williamson 1771) attributed the warming to human alteration of the New World: felling trees allowed temperate marine winds to penetrate from the east more deeply into the country and bared soil to receive and store some solar heat. Another popular explanation was that "the rise of urban communities with heated buildings and smokepots was leading to a milder climate, as they claimed had occurred in Europe" (Ludlum 1987: 257). The attribution of a warming trend to deforestation, heated buildings, and smokepots rings curiously familiar in the present. My point is not with these specific effects, but that we have a historical propensity to perceive weather change, whether or not it is occurring, and to attribute it to human perturbations.

44 — Chapter Two

For a more contemporary example, a 1977 survey in Illinois asked, "Have you noticed any particular changes in the weather during the time that you've lived here?" Sixty-two percent answered yes (Farhar et al. 1979: Tables B17, B18).[9] Asked how they observed this change, respondents overwhelming cited direct personal observation of the weather (55% direct observation, 16% talking to others, 16% TV/radio, 9% newspaper/magazine, and 1% or fewer for each other response).

Climate is long-term and manifested in many instances of weather. It is not surprising that lay people do not compute accurate statistical trend lines based on personal observation. What is surprising is that the majority claim to have personally observed weather changes during their lifetimes. In the case of my present data (but not the Illinois study), the frequently reported observation of warming may in fact be accurate, since the majority of my informants, and the majority of U.S. residents, live in areas that have experienced ambient temperature increases due to urban growth. This effect, referred to as urban heat islands, is carefully factored out by climatologists studying the weather record, but it is the strongest signal present in casual personal observations.

In sum, these data do not provide any basis for the concern expressed by some climatologists (for example, Schneider 1990) that people would not notice global warming or would not attribute it to human activities. On the contrary, the evidence presented here suggests that at any given historical moment, the majority of the population believe that they have observed a change in the weather. Warmer winters and more variable and more violent weather seem to be common observations. The current publicity about global warming simply gives people an appropriate framework within which they can interpret their own observations. Further, many people are predisposed to attribute changes in weather to unnatural human activities. These factors together clarify why a potentially implausible hypothesis—that human activities will warm the entire planet—has been so readily picked up by the general public.

---

[9] The study concerned a local weather change that actually had occurred in the St. Louis area over the prior 25 years: increases in growing-season precipitation, thunderstorms, and hail. However, actual weather change does not explain the finding, since only 11% reported the changes that actually did occur. The most commonly reported change was less rain, perhaps due to the immediately prior summer being unusually dry. Also, answers were not significantly different in the control area where the weather change had not occurred (Farhar et al. 1979: D3).

Support for Environmental Protection

Recent national surveys show strong support for environmental protection. Consider, for example, a survey question that has been asked consistently since 1981: "Do you agree or disagree with the following statement: Protecting the environment is so important that requirements and standards cannot be too high, and continuing environmental improvements must be made regardless of cost." This statement was phrased in such strong terms "to identify people who felt strongly about the environment" (*N.Y. Times*, 22 April 1990: 24), presumably expected to be a minority. American adults agreeing with this statement have climbed steadily from about 45% in 1981 to 79% in 1989. Those disagreeing dropped from 42% to just 12% over the same period. This shift was sufficiently dramatic to make the top of the front page of *The New York Times* (2 July 1989, A1), a distinction earned by few nonelectoral poll results.

Recent surveys also find that people report a willingness to make personal payments for environmental protection: Cambridge Reports found 65% willing to spend $100 yearly additional taxes for "toxic waste cleanup," and (in a different survey) willing to pay a mean of $9 extra monthly to purchase products "that did not harm the environment" (summarized in Americans for the Environment *et al.* 1989).

Consistently with the surveys, our ethnographic interviews also found near-universal support for strong environmental protection. It is interesting to first examine the one informant, John, who was the most negative about environmental protection: "Well, environment, there's nothing you can do about it. I'm not going to upset myself over it. That's all. I'm not in no position, I have no authority." In response to a question about Americans changing the way we live, he said:

> We can't change the way we live. Nobody's going to make anybody. *We could drive cars less or something like that.* . . . No way, not for me.
> —John

John was cynical and pessimistic about solving environmental problems, and certainly was not going to inconvenience himself. Nevertheless, he acknowledged pollution as a serious problem and thought "people with authority" should be doing something.

Our interviews provide a sense of the experiences and ideas behind the survey percentages. The following statements were made after we raised the topic of environmental protection by asking, "Would you say that protecting the environment is extremely important?" All those asked agreed with the statement (the first four infor-

46 — Chapter Two

mants were not asked). We then followed up by asking "Why?" Most answers cited not abstract principles, but concrete examples they could relate to personally:

> Because there are a lot of chemicals they're making today and when we breathe we take in the chemical and when we take them, we destroy the body, you know. So it's very important.—Joe

> Well, in as far as environment, you're talking about our lands, our greens, our trees and a lot of the animals are nice, . . . the air we breathe, you know. I think it's important, it's something that we've got to look forward to, you know.—Wilbur

> I mean, New Jersey is the worst place to even breathe in, for people [working] in industry, with the industrial odors and industrial chemicals and all that! And then people add their cigarettes to that! I mean, where's the air? Where is it? I don't have any!—Paige

> The thing that really got to me was when, I think it was two years ago, we stayed down at the shore for a week and the last day we were there, they closed the beaches, and I thought, "Boy, if you can't even swim in this water, it's got to be really bad."[10]—Cindy

Citizens have immediate experience with environmental problems, and many have seen degradation of the local environment within their lifetimes. As noted by Hayes (1982: 529), an important part of citizen support for environmental protection is local: the desire to preserve local recreational and scenic places and to protect one's own health. When asked about environmental protection in the abstract, citizens raise these specific, concrete examples.

A few informants went beyond their local knowledge and made more abstract statements about reasons for protecting the environment:

> When you destroy your environment, you destroy your, it's like burning down your home, destroying where you live. And it ain't gonna be replaced as fast as they're destroyin' it.—Walt

> Because, how many places do we have to go once we destroy this one, you know? I think it's kind of primary. Don't you think?—Jenny

> We've got to get more people politically involved to really save the environment before it just goes down the tubes, because I don't believe this is a disposable world, even though a lot of people feel that way.

---

[10] This statement was made to explain why she believed news reports about the greenhouse effect. The line of argument was that she has personally seen one form of broad-scale environmental degradation, ocean pollution. Therefore, she can more readily believe reports from others about less observable problems such as the greenhouse effect.

[Who?] . . . people that are very narrow-minded and don't really, see the beauty in the world that nature has to offer.—Doug

Why [is protecting the environment important]? Well, because future generations depend on how we care for our environment.—Amanda

These statements may help us understand part of the extensive support for environmental protection in this country. As has been said of politics, we might say environmentalism is local. Many citizens have seen environmental degradation close to home. But this is surely not the whole story. In the next section I address the values that underlie support for environmental protection.

Environmental Values

Do Americans feel that the environment should be protected for moral reasons? Moral foundations for environmental protection can be found in the animistic religions of small traditional cultures living close to the land. By contrast, it has been argued that the Judeo-Christian view of creation is intrinsically anthropocentric, considering humanity to have "transcendence of, and rightful mastery over, nature" (White 1967). Within such a moral framework, protection of nature would be justified only as "resource management."

Several current religious and moral writings take a contrary position, arguing that man should consider nature to have intrinsic value (Berry 1988; Rolston 1988). Some scholars also make the descriptive claim that the intrinsic rights of nature have ample roots in Western culture (Nash 1989).

The increasing U.S. environmental ethic of the 1960s was explained in social-psychological terms in a seminal article by Heberlein (1972). Heberlein saw in the 1960s a developing "land ethic," a concept derived from Leopold: "A thing is right when it tends to preserve the integrity, community, and beauty of the natural environment. It is wrong when it tends otherwise" (Leopold 1949). Heberlein attributes this development in American ethics to increases in two factors: public awareness of consequences of pollution and attribution of responsibility for environmental destruction to specific individuals.

Dunlap and Van Liere (1977) accept much of Heberlein's argument, but point out that his research includes only situations in which environmental destruction will harm other humans. Thus, they argue, Heberlein has found not a land ethic or environmental norms but simply another application of the interpersonal golden rule ("do unto others as you would have them do unto you").

48 — Chapter Two

Eliciting Environmental Values

What do ordinary citizens say when asked to explicitly address these issues? Environmental values proved to be difficult to elicit. In discussing interview plans with colleagues, I successfully conveyed my intent with the term "environmental morals." This terminology did not work in interviews. For example, in response to asking whether the bad consequences of the greenhouse effect were a moral issue, Ellen said, "It's not a moral thing." Yet later, in describing why she pays attention to environmental news, she said the following (note that we pick up her terminology and incorporate it into subsequent questions, a key method for ethnographic interviews):

> I have a personal philosophy—environmental and conservation values are an important part of it, therefore I would be concerned about [news regarding the environment]. *Describe your environmental values.* I think everyone is on this planet for a very short period of time and we have a duty when we leave to leave something for our children, to leave the Earth in as good or <u>better</u> shape than we found it. If we keep raping the land . . . by the time my child [now an infant] is 50 there will be one environmental disaster after another and I don't want him to have to face that. *Is this 'duty' in a moral or religious sense? In other words, why should we care?* That is what Ronald Reagan would say: "Why should we care?" I'm not a religious person but I have some moral things that guide my life, and I <u>do</u> think it's a moral duty.—Ellen

In some interviews (Doug, Jane), our questions using the term "morality" produced an answer that went directly off to another topic—an indication that the question, as phrased, did not make sense. Based on unsuccessful experiences asking about "morals," we tried asking about environmental "values" instead and were careful to incorporate words used by the informant in subsequent questions.

The following interview segment illustrates how people tend to answer initially in terms of the public debate and logical self-interest rather than personal values. I quote from this interview at length to illustrate the layering of justifications for environmental concern.

> *Would you say that you have environmental values?* . . . Well, I don't know that I do. I'm not one, I mean not at this point have I joined any of these environmental groups, although I have thought about it. And I think I have a couple of applications sitting on the table right there, to send in my $25 or whatever . . . I'm not taking an active part in it, but I'm generally concerned about it. *Well it isn't just whether you're active or not, but when you say you're "concerned" about it, what makes you concerned?* . . . the next 30 or 40 years that I'm going to be on this Earth, I wouldn't want to think that I could die from air pollution . . .

everybody should be concerned . . . *Why* . . . *?* Well, for their health.—Tara

Had I not pursued this topic further, I would have to conclude that Tara's concern for the environment was based solely on protecting her own health. Later in the interview, she again referred to having 30 or 40 years of life remaining, speculating that people with children would be more concerned than she is.

Let me follow up on that, let's just say, hypothetically: You don't have children; you know you're not going to have children. Would it still be of concern if we could say: "For 30 or 40 years there's going to be very little effect, it's really something that's going to be happening" [she completes sentence] "happening further down the line." No, 'cause that's being very selfish. . . . I mean all these years, the country . . . you take pride and like to talk history, where we started and how we began. And to think a hundred years from now, there's nobody here? . . . *Let me just ask why you say that. Why not be selfish?* It's just not the [pause]. Well, I don't know, how can I answer that? Uh, that's just my opinion. . . . *I mean some people might say that goes against what my parents taught me, or . . . that's not what the Lord wants us to do. Do you have any sort of system you refer to when you think about what's right or wrong?* Oh, I see what you mean. Well, Christian. Uh, my faith, I guess if you want to say. . . . I think we should be concerned about the next guy, and I guess this was the way I was brought up in values and, "try not to be selfish," and that to me would be a very selfish way to be. . . . We're all really a product of what went on a hundred years ago, at least. And it's just sort of the nature of the way it's been and should continue.—Tara

This quote shows that when I pursued the topic—more doggedly with Tara than with the other informants—her initial value-free statements, "I don't know that I do [have environmental values]" and "I wouldn't want to . . . die from air pollution" gave way to a mix of value systems: American history, national pride, family teaching not to be selfish, and Christian values.[11]

---

[11] Most of these reported values seem reliable because the questioning was nondirective, and the informant readily elaborated once she acknowledged her underlying values. I am less confident of her "Christian" response, since it followed a more explicit interviewer prompt and she provided no further elaboration.

Another methodological issue is that informants sometimes give the normatively correct answer in an interview, but act differently in the privacy of the voting booth or marketplace. Since Tara has not yet mailed in her environmental group applications waiting on the table—her own definition of being "active"—her values have not yet motivated her to act.

50 — Chapter Two

Preserving the Environment for Our Descendants

As several earlier quotes suggested, one's descendants loom large in thinking about environmental issues. This was, in fact, the strongest value basis to emerge in the interviews. Although none of our questions asked about children, all but two informants mentioned children as a justification for environmental protection.[12]

> Our children have to live with it. . . . I'm very concerned what their future is going to be like and what their life is going to be like.—Jenny

> I guess I do have quite a interest, and actually I guess an anxiety about it [global warming] in sort of a way that, uh, here's my daughter [he gestures to his daughter] and I think to myself, "Well, gee, what kind of world is she going to be living in?" I mean, I really think about that, and a lot. In other words, what can I do to make my kid's life better besides givin' her a good home? I mean, I can do all those things and still the environment could be totally, a terrible place to live. . . . —Doug

The second quote above is interesting in that it defines parental responsibility as extending beyond the traditional realms to also include environmental protection.

Two people explicitly volunteered that their concern for descendants was multigenerational (again, this was not one of our questions, and we assume more would have agreed had we explicitly asked about it):

> I think when you have kids and you think about long term, you know, like we might be comfortable now, but what are they going to have when they get older? I think basically, [that is what] makes you decide what's important and what's not important that much. . . . It might not happen for another hundred years, but maybe not for them, but their kids. . . . —Cindy

> [Informant is discussing people who he feels are not sufficiently concerned about the environment.] They're complacent, I guess. They figure that it ain't gonna be in my lifetime, is I think what, I think what a lot of 'em [pause] They don't figure that, that it might be in their kids' or grandkids' or [pause] great-grandkids'. It's gonna be in somebody's lifetime, that's for sure.—Walt

---

[12] Children were not mentioned by Eddie, who intellectualized the problem, nor by Paige, who had plenty of his own reasons for environmental protection. Paige did say we should solve environmental problems "for the future of this country." In our current interviews, we ask informants whether they have children, and we ask more explicitly about values relating to descendants.

Lay Perspectives on Global Climate Change — 51

Even John, the cynic, in the midst of a diatribe that he is old and doesn't care about the environment, pauses long enough to say:

. . . So, in other words, I'm not concerned. Although, I'm concerned about the children. Poor little kids. I lived my life, that's enough.—John

Based on these interviews, I would identify a value given to future generations in general, and to one's descendants in particular, as Americans' most widely and strongly held point of reference for environmental values.[13]

Given the importance of descendants' well-being, it is curious that this value has not played a more central role in public discussion and scholarly investigation regarding global environmental change. The most thorough treatment of environmental values and descendants comes from philosophers (MacLean and Brown 1983). To resolve tricky problems with arguments about benefits accruing to future generations, MacLean argues that meaning, as well as happiness, determines the quality of life and that "a commitment to securing resources and opportunities for future generations . . . is an appropriate way of expressing our belief that the society and culture that matter to us are important enough to survive into the future" (MacLean 1983: 194). In other words, knowing that the society will endure makes our lives more meaningful and thus of higher quality. MacLean offers one plausible explanation for my finding of a strong value for preserving the environment for one's descendants, though I have no direct interview data to identify this as the cause over other mechanisms.

A Species Preservation Ethic. Since the biological effects of climate change are poorly understood even by many in the technical community, I quote briefly from an authoritative review:

Of all the impacts climate disruption may have, its potential effects on biotic resources are the least understood and perhaps most profound . . . practically every environmental variable to which plants and animals are sensitive is affected directly or indirectly by climate. Paleoecological studies reveal that climatic changes in the past have been accompanied by profound changes in community structure and composition, by large shifts in the distribution of species, by extinctions, and by evolutionary adaptation to new climates. Organisms living today will likely respond to anthropogenic global warming in the same ways, save

---

[13] Other analysts have identified immediate desires for health, scenic outdoors, and quality of life as key bases for support of environmental protection (for example, see Hays 1987). That descendants emerged more frequently in these interviews may be due to the interview's focus on the long-term greenhouse effect. Nevertheless, the strength of feeling expressed for descendants far exceeded that expressed for other factors.

52 — Chapter Two

that evolutionary adaptation is unlikely to keep pace with the rate of warming . . . (Lester and Myers 1989/90: 196; also see Peters and Lovejoy 1988).

I suspect that the above reference to biotic effects being "least understood" refers to the intelligentsia rather than the lay public, but I similarly found that only one of our informants mentioned species loss when initially asked about effects of global warming. (They did mention it after our presentation, as I discuss below.)

Apart from the direct economic costs, I hypothesized that—if informed—citizens might be concerned about species loss per se. To explore this possibility, the interview's factual presentation included description of potential extinctions and changed range of species. Our summary list of possible effects mentioned species loss, and we asked informants to rank and discuss their concern about it versus other effects.

In a formal sense, the preservation of species is already a national value, since it is recognized in U.S. law (the Endangered Species Act). Despite public law, my interviews suggest that the value of species preservation is neither fully conceptualized nor valued in the abstract by the majority of the public. When the question was raised, informants most typically argued for species preservation by appeals to human uses or human enjoyment. We asked only six of the twelve informants directly about the value of species (the others were not asked because this was a follow-up question mentioned only if they picked species loss as an important effect).

Our presentation mentioned migrations and species extinctions, and we subsequently asked which of all the effects informants were most concerned about. Species loss and range shift were usually mentioned second or third, often second after crop damage/higher food costs. While this response indicates a strong level of concern, only one informant had a clear, abstract sense of the uniqueness and intrinsic value of species. This was Walt, a retired factory worker who reported a strong interest in nature from childhood onward and who subscribed to several wildlife and environmental magazines. When asked which of the effects of global warming in our presentation he would be most concerned about, he was the only one to mention species loss as his first concern:

Well . . . [I'm concerned if] your plant and animal life becomes extinct, because that's something you can't replace. And that's what they're doin' by destroyin' our rain forest. That's where all your new medicines would come from . . . you know, life has a chain, and when they destroy insects they destroy bird life, and destroying bird life destroys other life, that's a chain reaction. . . . —Walt

Lay Perspectives on Global Climate Change — 53

His concern seems to derive primarily from his respect for nature. He considers nature to be owned commonly ("our rain forest"), and as having a precious endowment of species ("can't replace"), with complex interdependencies ("chain reaction"). Some of his other points may come from the wildlife and environmental magazines he reads (for example, the argument of potential medicines from the rain forest has been made recently by many environmental groups).

Walt was the only one of the twelve who ranked species loss as the effect he was most concerned about. Even Doug, who like Walt was environmentally well informed and sensitive, ranked rising food costs ahead of extinctions and justified species preservation in more anthropocentric terms:

> . . . probably also plants and animals extinctions would probably be very important too. *How's that? It wouldn't affect you directly like food costs* . . . let's say we're dealing with 6,000 known species. We're startin' to really lose a lot of major promises, especially plants. You talk about deforestation, for one thing. There might be certain plants that might be cures for cancer, because, hey, we gotta make farmland, or our Third World country's growin', we don't care about this particular plant. And, because that species of plant no longer exists, you just probably destroyed a cure for some known disease. And also, with animals [I'm concerned about animal extinctions], being an animal-lover and all that stuff.—Doug

Five of the twelve described themselves as having a personal interest or empathy with animals or as being an "animal lover." Informants usually related species extinctions most directly to animals they had personally seen in their neighborhoods or in zoos:[14]

> I'm an animal lover . . . you always feel sorry . . . when you hear about them becoming extinct, you know like in Africa where they're killing the elephants for the ivory and all like that. You know, it's terrible. . . . [later] You need birds around. You can hear 'em singing, it makes you feel better, you know. [laugh] I used to have a bird feeder. . . . Before they built all that [points to townhouses], I used to hear the tree toads, you know, in the spring. . . . —Jane

Another informant with strong environmental feelings, Paige, also cited human benefit to justify preservation of green areas (not explic-

---

[14] In response to reading this section, an entomologist reported a similar bias among biologists: "There is even a problem among scientists. Most of the scientists have focused their concern on the large, beautiful animals and plants and little concern is directed toward the microscopic organisms and other small organisms, like insects. Yet the structure and functioning of the ecosystem is dominated by the little things" (David Pimental, Cornell University, letter of 12 Sept. 1990).

54 — Chapter Two

itly preservation of species), in this case presenting a distinct inner-city perspective:

> . . . some things in the environment you just need, and if it's not there for your need, you're going to suffer . . . you need to feel grass, to see the greenery out here, somehow you need it to keep this thing [points to head] going. You know, it's like a cement jungle. I mean, inner-city Trenton, they're crazy. I mean, if they've got one blade of grass in front of their house, they're doing good. . . . I think it makes them, you know, violent. . . . It's like a concrete jungle out there. And, I mean, it's not that it's a ghetto . . . they don't see anything pleasant, and therefore they're not going to be too pleasant themselves.—Paige

In one case, biological ecology was linked to religious beliefs. Cindy first joked that losing some animals, like snakes, "wouldn't bother me." But she became more serious when I more explicitly asked:

> *How about other animals that we don't particularly depend on, like squirrels or maybe some animal that's out in the Midwest that people just don't see that often?* They're all here for a reason [pause] they gotta be food for something. [laugh] . . . *could you elaborate?* Well, if they're food for a certain animal and they become extinct then they'll try to get something else. Maybe us! [laugh] . . . It's just, it'll unbalance everything. My daughter asked me the other day, "Why did God make mosquitos?" and I said, "Well, the birds need food." She said, "But Mommy, they suck our blood!" [laugh] And I said, "Well, you'll have to ask Him that when you get there."—Cindy

For Cindy, a teleological belief in the divine purpose of creation is intertwined with scientific biological models of species interdependence in a way that I could not clearly separate despite my attempts in subsequent interview questions.

Another informant said she would be concerned about species loss (which she picked second, after higher food prices) because she would regard it as a warning:

> . . . the plants and animals becoming extinct is to me just a very [pause] a very strong indication that something is extremely wrong. *Could you elaborate?* That the Earth is not providing the needs for this animal group or plant group to live, and why not? *Why would you be concerned with that?* . . . because I would think that it would eventually effect the entire food chain.—Amanda

Five informants, including those quoted above, addressed abstract questions of ecosystems and species extinctions, using terms like the "chain of life" or "cycle," and said that loss of one species could cause unpredictable further changes. I interpret these quotes as displaying not only knowledge of species interdependence but also a respect for it. Science education appears to have succeeded in imparting this concept to a sufficiently large segment of the population that almost half (five of twelve) of our informants mentioned species interdependence without explicit prompting. From this limited comparison, informant knowledge of photosynthesis and species interdependence versus the greenhouse effect, it appears that the lay public has a more sophisticated understanding of biology than of atmospheric science.

Several informants expressed reservations about how much species extinctions should count against other human concerns. For example, Wilbur, when asked why he said he was concerned about species loss, said:

> Well, I have a personal interest in animals. Like I said, I was a hunter and I've always been around animals. *Why this special concern for animals?* Well, again, I really want to keep this world, the Earth, alive for the kids. I hate to think our kids would never be able to see the things [the species] we've seen in our lifetimes. But I can see both sides, you know. If it isn't something we use, I'm not sure I'd feel that strongly. . . . It's really a pain to worry about something that may never happen or something that we may not be able to do anything about. . . . If you look very carefully at the insect population, every little thing affects their lifestyle [*sic*]. You've got to look very closely at the cycle, it's so easy for us to just screw it up. . . . —Wilbur

In sum, most of my informants lacked an established, clearly conceptualized value for species preservation, and they justified species preservation in anthropocentric terms. Only one person mentioned these biological effects when initially asked about global warming. When we covered this information in our presentation, the connection was one of the most difficult to understand.

We might conclude that a general species preservation value is restricted to scientists, who are different from the remainder of their own society in valuing information for its own sake. But that conclusion probably is unfair—surely the ordinary person should be expected to pick food supply and higher food prices before the abstract concept of extinction and to justify species preservation with reference to the personal value of familiar species. It would seem that the foundations for a species preservation ethic are there—in the future generations value, the common "animal lover" statements, and the general pro-environmental attitudes. The public does not yet understand that small changes in global temperature can cause massive biological changes, but this idea is only a small step from the existing appreciation for complex interdependencies among species and from the recognized sensitivity of the weather to human perturbations. It would take a concerted publicity effort to bring out these connections and to cultivate an

56 — Chapter Two

abstract species preservation ethic in the broader culture. To date, politicians and environmental groups who themselves hold this ethic have instead chosen to publicly argue more in pragmatic, economic, anthropocentric terms.

Reactions to Policy Options

Before reading our presentation, we asked those who had heard about the greenhouse effect, "Do you think the United States should do anything about this?" and followed up with "What?" The same question was asked after the presentation. At the end of the interview we solicited reaction to four specific policy proposals; the three proposals yielding data of general interest are discussed below.

Prior to our presentation, informants' policy suggestions seemed to be based primarily on the tropospheric air pollution model discussed earlier. For example, the following were given as policies to combat the greenhouse effect:

> They're trying . . . to cut down evaporation of gasoline in your car; tryin' to make people pool cars . . . laws that you can't [burn leaves in your backyard]. Just trying to cut down on your hydrocarbons and stuff that's being put in the air.—Walt

> I suppose, make companies, you know, not have that smoke come out, of course I don't know how they'd do that.—Jane

> Stricter rules regarding environmental pollution by individuals as well as corporations. Saving the forests at all costs. The Amazonian rain forest is essential to our existence on this planet.—Jenny

> I don't know . . . other than . . . cut down on aerosols. . . . —Susan

"Saving forests" is the only suggestion specific to global warming. The "cut aerosols" suggestion probably derives from stratospheric ozone depletion, but is incorrect in either case because U.S. manufacturers have already eliminated CFCs from aerosol spray cans. In short, prior to our presentation and suggested proposals, informants had virtually no knowledge of the potential global climate change policies actually being debated.

Energy Efficiency

In-depth media accounts have usually mentioned energy efficiency as the best way, or a primary way, to combat global warming (e.g. *Newsweek* 1988; Koppel 1988). Perhaps the most surprising policy-related finding from the interviews is the lack of connection between the greenhouse effect and energy efficiency. Despite ten of our twelve

Lay Perspectives on Global Climate Change — 57

informants having heard of the greenhouse effect, only one, Doug, cited specific energy solutions. (Additionally, both of the test informants had heard of the greenhouse effect, and both mentioned energy issues as a partial solution.)

To most informants, we were able to convey the connection of energy and global warming in our brief presentation without conceptual hurdles. But we were usually making new links that did not exist previously. Several commented on our chart showing that the largest component of the greenhouse effect was "burning coal, oil, and natural gas for energy":

> People are burning that much coal? Why? *Well, a lot of the electricity we get is from burning coal.* . . . Oh, oh, . . . I was still thinking, water power . . . I wasn't aware of it. All I knew was, you turn the light on, and that's it, you know, I don't know where it <u>came</u> from. . . .
> —Paige

> *What was new to you?* . . . Energy use, I guess I should have known. Maybe not that big of a percentage. Others I guess I was aware of.
> —Tara

> [interrupting discussion of fuels as greenhouse gas sources] *Is it?* That's what causing the greenhouse? *It's more than half the problem, yeah.* Burning gas? So that's like our heating and all that? *Right. Also gasoline, coal in power plants, oil in powerplants.* Well, there's nothing that the government will do about that. Nobody's going to stop working, right?—Cindy

This is my only finding that directly contradicts national survey data. In one survey, respondents were asked: "Some people say that global warming and the greenhouse effect are caused by the destruction of the great rain forests of the earth. Others believe it is caused primarily by fossil fuel emissions from power plants and exhaust gases from cars and trucks. Which is these two views comes closest to your own opinion?" In response, 60% said fossil fuels, and 27% said loss of forests. From these percentages, the report's overview summarizes that "American voters continue to be sensitive to the role of fossil fuels in the generation of atmospheric pollutants and their contribution to global warming." (RSM 1989). Given my small sample, I must concede that I may have just happened to draw a group ignorant on this one issue. However, until more definitive data are available, I am personally more inclined to believe that, given the two-way choice in the RSM survey question, respondents invoked their pollution model and chose "emissions . . . and exhaust gases." I am arguing that Americans connect global warming to pollution but not to energy consumption.

58 — Chapter Two

The quote above from Cindy illustrates a second problem in proposing energy efficiency as a solution to global warming. People do not understand what energy efficiency is, because they have had very little direct contact with it. When analysts talk about reducing energy use, lay people tend to interpret it as decreasing energy services. This problem was identified a decade ago (by myself and others)—conservation makes people think of curtailment and sacrifice rather than improved equipment or better management (Stern & Gardner 1981; Kempton *et al.* 1982). The conceptual problem is clearly still with us.

It's hard to believe that our way of life is affecting the environment as drastically as it is . . . I'm not sure how to change my lifestyle. . . . without going all the way back to colonial times. Or, I don't know, a different way to produce energy for us, which I'm sure they're developing.—Amanda

Doug also mentioned we could "go back to colonial times" and a "simpler society" as a way to deal with the greenhouse effect, a proposal he described as "somewhat serious." Several mentioned alternative energy sources; it seems to be much easier to conceptualize alternative energy sources replacing carbon sources than to conceptualize energy efficiency.

On the other hand, the one energy-efficiency proposal we specifically described was broadly supported. We suggested auto efficiency, saying that with 55 mpg cars, "This would mean that the next car you bought would be a little smaller, and it would not be a high-performance car." John's response was again the exception: "Well, I got a Cadillac. Fleetwood. I wouldn't feel safe in a little coffin. Matter of fact you look at these people that are killed. They're all little cars. Every one." John fits the media stereotype of Americans' "love affair with the automobile." But a more typical reaction from my informants was:

It [smaller, lower-performance cars] wouldn't bother me a bit; I'm a small-car person. For me a car is four wheels on a frame to get me from point A to point B. If you look at the general population, you'll see more smaller cars than large cars. If you want to impress someone, you rent a limo.—Susan

The support I found is consistent with polling results. For example, 83% of a national sample favored a 45 mpg fuel standard, when told that cars would cost $500 more but that the added cost would be recovered in gasoline savings over four years (RSM 1989).

In sum, analysts may favor energy efficiency as the fastest and most cost-effective path to American CO_2 reductions. However, anyone publicly advocating this solution will need to clearly spell out the connection with global warming. There are two independent conceptual hurdles: the public now neither understands energy efficiency nor believes that energy use is a major contributor to the greenhouse effect.

Adaptation without Prevention

We presented a proposal for adaptation without prevention, deriving from analysis of the economically best response to uncertainties in prediction and the high costs of preventing global warming. For example, Manne and Richels urge more research before attempting prevention, arguing that "it is unclear whether it would be justified to incur these costs [of prevention]" (1989). Similarly, Schelling has argued that we could adapt to global warming by expanding irrigation, building dikes, and developing new seed strains and that future generations may be better off if we invest in other productive technology and capital, rather than spending the same money on prevention (Schelling 1990a; Passell 1989). Schneider (1989: 777) claims that the adaptation strategy "tends to be favored by many economists." This claim would make sense, as the standard economic method of discounting future costs, given the long time frame, would greatly reduce the present value of the cost of damages, thus reducing the amount of money that could justifiably be spent today on prevention. A related justification for adaptation is a belief by some of its advocates that our grandchildren's society will be more wealthy than today's, as we are more wealthy than our grandparents (Schelling 1990b).[15]

In our 12 interviews, only one informant (John, the cynic) thought adaptation was a reasonable strategy. Many of the remaining 11 informants who expressed negative reactions did so with vehemence. Despite our framing of the adaptation strategy as a decision to deal with uncertainty about effects, some of which could be beneficial, many informants perceived adaptation as postponing or avoiding decision making:

> Well, that's more or less an avoidance proposal, and apparently, we have been avoiding—me included—have just avoided this up to now,

---

[15] Attempting to express these ideas briefly in lay language, our interview question was: "We cannot be 100% certain that the climate will change. Why spend money if we may not even need to? Anyway, some effects, like warmer winters, may be beneficial. We could wait and see what actually happens, then react to that. For example, if the Midwest becomes hotter and dryer, farmers could switch to crops that require less moisture, move north, or go into other businesses. Or, if sea level rises, populated areas could be diked, as has been done in Holland. What would you think about that?"

60 — Chapter Two

and I would rather not take a chance. If there is [another] proposal that could maybe, prevent this from happening I'd rather go with that. —Amanda

No. [vigorous shaking of head] Because it's just supporting "Let's wait and see." Let's not do anything and wait 'till the last minute and hope that we can solve it all in an hour or two. . . . *Suppose I could convince you that we really could do something at the last minute . . . throw up a bunch of dikes to take care of the sea level and shift those farmers to other crops. . . . Would you have any other objection?* I think we're clever and we can come up with quick solutions in times of desperate need, but I'm just not a risk taker. . . . We don't know what might happen so why not be ultra-conservative up front, rather than less prepared at the end. It's too critical.—Jenny

Taking a different tack, Paige made the analogy with military preparedness to refute our adaptation argument "Why spend money if we may not even need to?"

This is not a good proposal. You shouldn't wait and see what happens. No. I mean, they're putting money into arms for wars that may not happen, and cutting school lunches to put money into bombs. This [adaptation proposal] is not good.—Paige

One could argue that lay people do not understand the economic concepts of opportunity cost, the time value of money, and alternative investment opportunities. More sympathetically, I would argue that economic discounting applies to corporate planning but does not reflect economic decisions that involve passing on an intact environment to descendants. In the children/descendants reference frame, informants did not acknowledge discounting of future costs. My claim here is somewhat impressionistic and merits more systematic study, but no informants said anything like "the effects matter less because they're so far in the future" or "my kids will benefit more if money is invested in building up industry now rather than protection of the environment for 100 years from now." On the contrary, in the sensitivities of my informants, reflected in the passages quoted in this paper, I see a fundamental American value that is distorted by discounting the future cost of environmental damage.

In any event, whether due to failure to understand economic theory or due to a "descendants value" not modeled by discounting future costs, the negative reaction to the adaptation proposal was broad (11 of 12) and strong. Further, adaptation was perceived as avoidance of decision making rather than as a coherent, planned strategy. These factors suggest that advocating adaptation over prevention would be politically challenging.

Energy Tax

One policy proposal called for a 100% energy tax, on electricity, natural gas, and gasoline.[16] The 100% level was deliberately chosen to be extreme, to force informants to balance environmental benefit against significant personal expense. (I call a 100% energy tax extreme only in the U.S. political context, as European gasoline taxes are even higher, and those societies are functioning satisfactorily.) We translated the tax rise into relevant units by asking for the amounts of the informant's current bills and saying, "This would mean your bills would be ___." As we expected, many reactions were negative:

> It wouldn't stop people. It wouldn't stop me. . . . I commute to work, so I have to have gas. It's not a good idea; it wouldn't accomplish its object. . . . *Suppose it were spent on research to prevent the greenhouse effect?* I don't trust the government . . . I think it should be private industry, not the government.—Susan

> No, I wouldn't be interested in that proposal, and I wouldn't use less gas. . . . That would really anger me if I were taxed on my gas . . . just so I can take my children to school and visit my mother. [later] Well, it doesn't have to be 100%? *No, it doesn't. I'm taking an extreme figure.* OK. Higher [tax] on gasoline or higher tax on utilities, that would be fine with me, if that money were to be going towards [solving the problem]—Amanda

> No, no, no, no [laugh] . . . It doesn't matter how high gas would go, I would still buy it, just like cigarettes. . . . you have to go to work. *Suppose the 100% tax went straight into developing technologies that don't cause the greenhouse effect. Would that change your reaction at all?* [moans] That's hard, I don't think I'd like a 100% tax on anything. If they raised the rates a little bit and said, this money is going to prevent the greenhouse, yeah, I would go for that, but I still don't want 100%. —Cindy

In the above reactions, the negative response is immediately followed by a statement about inability to reduce personal gasoline consumption. Not one person mentioned getting a more fuel-efficient car to reduce gasoline costs. This is especially surprising considering that the immediately prior question mentioned a 55-mpg car and considering that fuel efficiency was the primary public reaction to fuel price increases in the 1970s. No one seems to remember that Americans

---

[16] Taxes were proposed on the retail cost of three fuels for simplicity in explanation, even though this is not a current policy proposal. A tax on carbon content may be a more sound policy for energy efficiency (Chandler and Nicholls 1990), but I judged that explaining that refinement would cost interview time with little gain in value of the data.

62 — Chapter Two

responded to the last price jump by buying much more efficient cars rather than reducing miles driven (Greene 1987).

Nevertheless, the latter two people quoted above would accept an energy tax if it were not as high as 100%. Others accepted the proposal with few objections or preconditions:

> I wouldn't be happy. [laugh] It's a lot more money, but you know, if the end result was that it was going to be beneficial, you live with it, you know.—Jenny

> My fuel bills [gasoline] each month are close to a couple hundred dollars. So that would be a major cut in my belt. I wouldn't _like_ that, but again if it could help. I don't know that they'd have to do a 100% increase, . . . I couldn't handle a 100% [gasoline] tax increase on the salary that I make now. . . . _[would it affect how much you use?]_ . . . my job [is] to call on accounts, and most of the time you can't do it by phone, I mean I do some work on the phone, but most of the time I'm out on the road so, whether it costs me $200 [or] $400 a month for gas, I'm gonna have to do it, I guess.—Tara

Since this question was designed to highlight the negative impact on the individual, by explicitly calculating their increase in bills, I found the reactions surprisingly mild. Overall, three informants accepted this policy immediately (five, including our two test informants), and three more would accept it if the tax revenues went to help the global warming problem (two of them insisted that the tax rate should be less than 100%). These informants were probably more receptive after our previous discussion of global climate change problems in the interview. Nevertheless, these reactions suggest that, properly framed, energy or carbon taxes may not be the political impossibility they are often assumed to be.

Another discrepancy between analyst and lay perspectives was in understanding the purpose of an energy tax. We began the tax proposal with " . . . people and companies use less when prices are higher." As noted above, several informants refuted this assertion by saying higher cost would not affect their consumption, while others seemed to ignore the assertion. Rather than a pricing mechanism, the tax was seen as a punishment or as a way of raising money. I conclude that perceived inelasticity of energy consumption contributes to the political opposition to energy taxes.

Conclusions

While fully 79% of the U.S. public has heard of the greenhouse effect, understanding of the causes and consequences by the lay public is quite

Lay Perspectives on Global Climate Change — 63

Table 2-2. Scientist and Lay Constructions of Causes and Major Concerns

Causes—scientist

Release of sequestered carbon
    Primarily, fossil fuels
    Secondarily, reduction of forested land area
Other gases: methane, CFCs, nitrous oxide, etc.

Causes—lay

Aerosol spray cans
The ozone hole
Cutting trees
Air pollution

Major concerns—scientist

Climate change has profound and unpredictable effects
Climate change has never occurred so quickly
Biological disruption and extinctions
Agricultural shifts
Possible sea level rise

Major concerns—lay

Our children will have to live with it
Sea level rise
Uncomfortably warm in summer
Depletion of atmospheric oxygen
Breathing greenhouse gases

---

different from that by scientists and policy analysts. Tables 2-2 and 2-3 offer a brief summary of many of the differences. Table 2-2 outlines scientist and lay constructions of the causes and major concerns. Table 2-3 compares their perspectives on evidence to date that climate change is occurring and on potential policies to deal with it. Table 2-3 is in a tabular comparison format because the entries are more directly comparable than those in Table 2-2.

In Tables 2-2 and 2-3, the "scientist" entries derive from my own reading of the literature and from discussions with those working on the scientific and policy problems. The lay entries should not be regarded as definitive, both because of the small sample and because evidence for some entries draws only on the minority of informants who happened to mention that topic. Also, the current lack of appro-

64 — Chapter Two

Table 2-3. Comparison of Scientist and Lay Views on Current Evidence and Policy Effectiveness

Evidence to date	Scientist Perspective	Lay Perspective
CO_2 increases	Clearest evidence	Not known
CFC increases	Yes	Yes
More extreme weather	No	Already apparent
Temperature increases	No or maybe	Warmer winters already apparent
Policy effectiveness		
Energy efficiency	Yes, very important	No, irrelevant
Stop using aerosol sprays	Irrelevant in U.S.	Yes!
Halt deforestation	Helpful	Extremely important
Reforestation	Limited potential	Yes!
Stricter pollution controls	Irrelevant	Yes
Carbon tax or fuel tax	Yes	Ineffective and unfair
Adaptation without prevention	May be cost-effective (minority view)	No! Avoidance of decisions by politicians

---

priate lay models should not be construed as lack of ability. Many of our informants did have a good working understanding of other scientific concepts such as species interdependence and photosynthesis.

The problem of public understanding of global climate change would probably work itself out over many years through science education in schools and media coverage. This process could be accelerated greatly if those who communicate with the public (science journalists, environmental groups, and others) specifically target some of the gaps and misleading prior models I have identified here. Particularly high-priority needs are connecting the greenhouse effect to energy consumption, developing a concrete concept of what energy efficiency is, realizing that small changes in mean global temperature could have large geophysical effects, and recognizing the sensitivity of the living world to climatic conditions. At the same time, people could be reassured that they need not worry about breathing greenhouse gases or running out of oxygen.

The present work is limited in that a small sample was interviewed, all from a single geographical area, and only members of the lay public were included, not those whose work affects the climate change issue. Each of these limits is being addressed in continuing or proposed work. We have recently completed interviewing people from other areas of the country and members of a few groups affecting climate change policy, including environmentalists, coal industry people, designers of energy-using equipment, and congressional staff. This paper has provided only a preliminary glimpse of neighboring domains outside our current research focus, including lay values for species preservation, beliefs about the weather, and perceived ineffectiveness of an energy tax. Many significant questions raised here have not been satisfactorily answered.

What are the political and policy consequences of these gaps in public comprehension? Some would argue that the public need only express concern, not understand all the details of global warming—the scientific, political, and industrial leaders can put together appropriate responses by themselves. I would challenge this argument for two reasons. First, just as citizen concern is needed to motivate leaders, so also is some citizen knowledge needed to keep leaders on the right track. My second apprehension about today's state of public knowledge is this: If world leaders decide to reduce greenhouse-gas emissions by two-thirds,[17] such a large reduction will require consumer and worker cooperation as well as citizen consent that major societal changes are worth the effort.

My findings help explain the political success of recent policy directions that may seem ineffective, even incomprehensible, to most scientists. I refer to the White House–directed policy combination of creating an ambitious (but underfunded) tree-planting program while staunchly resisting action on energy policy. Tables 2-2 and 2-3 suggest that this combination would be very well received by the public (compare scientist and lay perspectives on causes in Table 2-2 and on policy effectiveness in Table 2-3). I have not attempted a political analysis, and I can only speculate from the outside as to whether this policy course was set by cynical political consultants, by misconceptions of

---

[17] The Intergovernmental Panel on Climate Change states "We calculate with confidence that . . . the long-lived gases would require immediate reductions in emissions from human activities of over 60% to stabilize their concentrations at today's levels" (IPCC Working Group 1 1990: 1). Similar figures are given by a comprehensive U.S. report (Lashof and Tirpak 1989).

66 — Chapter Two

the elected officials and their political staff, or by some combination of these and other factors. The present analysis does suggest that our nation's current policy combination would be attractive to a person with the American lay understanding, whether that person is the one watching the television news or the one running the country.

It is easy to overemphasize the shortcomings of the lay public, such as the inappropriate models and the abundant factual omissions documented by this study. A more complete picture is that average Americans are aware that measurable environmental disturbance is now global in scale, they believe that something should be done about it (even if they are mostly wrong about what actions would be effective), and they want decisions to be based not only on costs and benefits but also on our responsibility to leave a healthy planet for our descendants. In these lay values I see hope, and the possibility of support for politicians ready to take a leadership position in the environmental area.

Acknowledgments

I am grateful to Apoorva Muralidhara and Dan Levi for assistance in conducting and transcribing the interviews. Prior drafts have been improved thanks to extensive comments by James S. Boster, William C. Clark, Kerry Cook, John DeCicco, Dan Deudney, Barbara C. Farhar, Jennifer Hartley, James Risbey, Robert Socolow, Paul Stern, and reviewers and editors for ACEEE and *Global Environmental Change*. I am grateful to Barbara Farhar and David Hart for bringing key data to my attention. This work was supported by grant BNS-8921860 from the National Science Foundation's Cultural Anthropology Program, and by a grant from the Hewlett Foundation to Princeton University's Center for Energy and Environmental Studies.

References

Agar, M. 1980. *The Professional Stranger: An Informal Introduction to Ethnography*. New York: Academic Press.

Americans for the Environment, L. Harris, V. Tarrance, and C. Lake 1989. "The Rising Tide: Public Opinion, Policy & Politics." Report released 20 April. Washington, D.C.: Americans for the Environment.

Berry, T. 1988. *The Dream of the Earth*. San Francisco: Sierra Club Books.

Broecker, W. 1970. "Man's Oxygen Reserves." *Science* 168 (26 June): 1537–38.

Chandler, W., and A. Nicholls 1990. *Assessing Carbon Emissions Control Strategies: A Carbon Tax or A Gasoline Tax?* ACEEE Policy Paper No.3. Washington: American Council for an Energy-Efficient Economy.

Dunlap, R., and K. Van Liere 1977. "Land Ethic or Golden Rule: Comment on 'Land Ethic Realized' by Thomas A. Heberlein." *Journal of Social Issues* 33(3): 200–207.

Farhar, B. 1976. "The Impact of the Rapid City Flood on Public Opinion About Weather Modification." *Pacific Sociological Review* 19(1): 117–44.

——— 1977. "The Public Decides About Weather Modification." *Environment and Behavior* 9(3): 279–310.

Farhar, B., J. Clark, L. Sherretz, J. Horton, and S. Krane 1979. *Social Impacts of the St. Louis Urban Weather Anomaly.* Final Report, Vol. 2. Boulder: Institute of Behavioral Science, University of Colorado.

Greene, D. 1987. *Research Priorities in Transportation and Energy.* Transportation Research Circular No. 323, September. Washington, D.C.: Transportation Research Board, National Research Council.

Harte, J., and R. Socolow 1971. *Patient Earth.* New York: Holt, Rinehart, and Winston.

Hays, S. 1987. *Beauty, Health, and Permanence: Environmental Politics in the United States, 1955–1985.* In collaboration with B. Hays. Cambridge, England, and New York: Cambridge University Press.

Heberlein, T. 1972. "The Land Ethic Realized: Some Social Psychological Explanations for Changing Environmental Attitudes." *Journal of Social Issues* 28(4): 79–87.

Intergovernmental Panel on Climate Change, Working Group 1 1990. Policymakers Summary of the Scientific Assessment of Climate Change. WMO, UNEP.

IPCC. See Intergovernmental Panel on Climate Change.

Kempton, W. 1990. *Lay Perspectives on Global Climate Change.* PU/ CEES Report No. 251, August. Princeton, N.J.: Center for Energy and Environmental Studies, Princeton University.

Kempton, W., C. Harris, J. Keith, and J. Weihl 1984. "Do Consumers Know 'What Works' in Energy Conservation?" *What Works: Documenting Energy Conservation in Buildings.* Ed. by J. Harris & C. Blumstein. (Selected papers from 1982 ACEEE Summer Study).

68 — Chapter Two

Washington, D.C.: American Council for an Energy-Efficient Economy.

Kimble, G. 1962. "But Somebody Does Something About It." *New York Times Magazine*, 8 July, 11 ff.

Koppel, T. 1988. *ABC News Nightline*. "The Greenhouse Effect." Broadcast on 7 September. Transcript available from Journal Graphics, New York.

Lashof, D., and D. Tirpak 1989. *Policy Options for Stabilizing Global Climate*. Draft report to Congress. Vols. 1, 2. Washington: United States Environmental Protection Agency, Office of Policy, Planning, and Evaluation.

Leopold, A. 1949. *A Sand County Almanac, with Essays on Conservation from Round River*. Oxford, England: Oxford University Press. (Reprinted 1970 by Ballantine Books, New York).

Lester, R., and J. Myers 1989/90. "Global Warming, Climate Disruption, and Biological Diversity," in *Audubon Wildlife Report*. New York: Academic Press.

Ludlum, D. 1987. "The Climythology of America." *Weatherwise* 40 (October): 255–59.

MacLean, D. 1983. "A Moral Requirement for Energy Policies." In *Energy and the Future*. Ed. by D. MacLean and P. Brown. Towatta, N.J.: Rowman and Littlefield.

Manne, A., and R. Richels 1989. "CO_2 Emission Limits: An Economic Analysis for the USA." Paper presented at MIT Workshop on Energy and Environmental Modeling and Policy Analysis, July, in Cambridge, Mass.

Nash, R. 1989. *The Rights of Nature: A History of Environmental Ethics*. Madison: University of Wisconsin Press.

Newsweek 1988. "Inside the Greenhouse." 11 July, 17 ff.

Passel, P. 1989. "Cure for Greenhouse Effect: The Costs Will Be Staggering." *New York Times*, 19 Nov.

Peters, R., and T. Lovejoy (eds.) 1989. *Proceedings of the Consequences of the Greenhouse Effect for Biological Diversity*. Washington, D.C.: World Wildlife Fund (Conference held 4–6 October 1988; brief summary in *Science* 242 [18 Nov.]: 1010–12).

Rolston, H. 1988. *Environmental Ethics: Duties to and Values in the Natural World*. Philadelphia: Temple University Press.

RSM, Inc. (Research/Strategy/Management, Inc.) 1989. "Global Warming and Energy Priorities: A National Perspective." Survey commissioned by the Union of Concerned Scientists. Unpublished manuscript available from UCS, Cambridge, Mass.

Schelling, T. 1990a. "Global Environmental Forces." In *Energy: Pro-*

duction, Consumption and Consequences. Ed. by John L. Helm. Washington, D.C.: National Academy Press.

———— 1990b. Public remarks at conference, "Energy and the Environment in the 21st Century." Massachusetts Institute of Technology, 26–28 March.

Schneider, S. 1988. "Doing Something about the Weather." *World Monitor, The Christian Science Monitor Monthly*. 1 (December): 28–37.

———— 1989. "The Greenhouse Effect: Science and Policy." *Science* 243 (10 Feb.): 771–81.

———— 1990. Lecture at Princeton University, 1 March.

Spradley, J. 1979. *The Ethnographic Interview*. New York: Holt, Rinehart, and Winston.

Stern, P., and G. Gardner 1981. "Psychological Research and Energy Policy." *American Psychologist* 36(4): 329–42.

U.S. News & World Report (author unspecified) 1963. "Is Man Upsetting the Weather?" 11 November, 46–49.

White, L. 1967. "The Historical Roots of Our Ecological Crisis." *Science* 155 (10 March): 1203–7.

Williamson, H. 1771. "An Attempt to Account for the Change Observed in the Middle Colonies in North America." *Transactions of the American Philosophical Society*. First Edition. (Cited in Ludlum 1987).

Willett Kempton is a research anthropologist at Princeton University's Center for Energy and Environmental Studies. He has done extensive research on the cognitive and behavioral aspects of residential energy use and energy efficiency. He received a Ph.D. in Anthropology from the University of Texas at Austin.

Chapter **3**

Environmental Improvement and Energy Efficiency in Buildings: Opportunities to Reduce CO_2 Emissions

Erich Unterwurzacher and Genevieve McInnes, *International Energy Agency*[1]

Introduction

Environmental considerations, including concerns about climate change, are playing a larger part in energy decisions. Policies are increasingly oriented towards environmental goals. Recent studies— such as the Brundtland Report, the IEA's Energy and Environment Policy Overview, and work carried out in the framework of the Inter-governmental Panel on Climate Change (IPCC)—recognize the close links between energy, the economy, and the environment. Energy use has a variety of significant environmental impacts, which reduced energy demand could diminish. In particular, the more efficient use of those fuels whose combustion emits greenhouse gases is a promising response strategy to combat climate change, in the absence of economically viable CO_2 abatement technologies.

Energy efficiency policy measures have in the past contributed to improved energy security in the member countries of the IEA, where oil dependence has been significantly reduced since the first oil shock,

---

[1] The views expressed in this chapter are those of the authors and do not necessarily reflect those of the International Energy Agency or its member countries. The International Energy Agency is an autonomous body within the Paris-based Organization for Economic Cooperation and Development (OECD). There are 21 member countries, listed in Table 3-4.

72 — Chapter Three

in 1973. Oil requirements between 1973 and 1988 decreased by 7.8%, or 137 million tons of oil equivalent (MMtoe), while economic activity increased substantially. During this period, the Gross Domestic Product (GDP) increased by 50% in real terms. At the same time, total primary energy requirements (TPER) increased by only 13.5%. Various studies have documented that a broad range of energy-efficient technologies is currently available (Schipper et al. 1987). Although the potential for cost-effective energy savings seems to be high, market barriers hinder the penetration of improved end-use technologies. Furthermore, softer energy prices since 1985 have reduced the economic incentive to carry out efficiency improvements and blurred awareness about the economic costs and longer-term security implications of energy use. The change in perspective due to growing environmental concern is providing renewed interest in potential energy savings and in their effects on emission levels, as well as a fresh impetus for the design of effective energy conservation programs.

The analysis presented here concentrates on the residential and commercial/public sector, which accounted in 1988 for about 30% of energy use in the IEA. It examines how the end use of energy contributes to emissions of CO_2 and how energy-efficient technologies and other measures, such as energy management, could be further developed in order to help reduce growth in energy demand and related emissions.

Pollutant Emissions and Energy Demand Trends

Recent studies (IPCC 1990a) show that CO_2 holds by far the largest share in both the greenhouse effect due to anthropogenic activities (71%) and the increase in the greenhouse effect due to these activities (about 50%). Though other human activities, such as deforestation and agriculture, are contributing to increases in the atmospheric concentration of CO_2, the energy sector is responsible for about 61% of anthropogenic CO_2 emissions (IPCC 1990b) and thus is the focus of much of the concern about the risk of global warming.

The important role of CO_2 is of particular interest to those involved in the development of improved energy efficiency: in the current absence of any economically justifiable CO_2 abatement technology, energy efficiency appears to be one of the most promising response strategies to limit greenhouse-gas emissions. The possibilities offered by carbon offsetting through reforestation and "carbon-neutral" biomass plantations are also considered to be promising, though the need to rationalize our use of energy clearly appears to be

Environmental Improvement and Energy Efficiency in Buildings — 73

the central priority. In addition, CO_2 is a fuel-dependent emission, decreasing proportionally with the use of a given fuel. SO_2 and particulate matter are also essentially fuel-dependent pollutants; however, abatement technologies such as flue gas desulfurization or electrostatic filters are available for these pollutants and are widely used. Although increased efficiency can contribute to a reduction in the quantity of emissions of other pollutants such as NO_x, VOC and CO, the relationship between energy use and emission levels is not linear, as these pollutants are essentially technology-dependent. As for other pollutants, such as CH_4 and CFCs, their generation is not directly related to energy combustion.

CO_2 emissions therefore appear to be a special case for energy efficiency efforts: for other fuel-dependent pollutants, abatement technologies are available and for technology-dependent pollutants, reduced energy use does not necessarily result in reduced emissions. As a result, improved end-use energy efficiency is not the "first-order" response that it is for CO_2 emissions. This paper therefore focuses mainly on the benefits of improved energy efficiency in terms of reductions in CO_2 emissions. It should nevertheless be emphasized that, where efficiency improvements displace fossil fuels, reduced energy use will reduce the need for expensive abatement technologies for fuel-dependent pollutants and may also decrease emissions of other pollutants. In addition, the cumulative effect of energy efficiency improvements will ultimately reduce other environmental impacts—land use and water quality problems, for instance—often associated with energy production, transformation, and transport.

Trends in Energy Demand

Energy consumption dropped substantially after the oil price hikes of 1973 and particularly of 1979. Since 1986, the price signals have weakened, as reflected in a gradual increase in energy demand. These developments appear clearly in the evolution of energy intensity between 1970 and 1988.[2] The period between 1980 and 1984 saw the greatest achievements in energy efficiency, and intensity declined by 2.6%/yr. But these improvements in energy productivity were largely driven by relatively high energy prices. The price-induced momentum dropped off during a second four-year period, between 1984 and 1988, when intensity declined by only 1.4%/yr.

But energy demand developments have not been uniform among different end-use sectors, as shown in Table 3-1. The strongest increase

---

[2] Energy intensity = TPER (in MMtoe) per unit of economic output (in constant U.S. dollars).

74 — Chapter Three

Table 3-1. Trends in Sectoral Energy Demand in IEA Member Countries						
	Energy Demand (Mtoe)			Annual Changes (%)		
	1973	1985	1988	73–85	85–88	73–88
Industry	965.90	842.79	894.75	−1.13	2.01	−0.51
Residential	492.45	489.13	502.67	−0.06	0.91	0.14
Comm/Publ.	209.92	250.78	273.63	1.49	2.95	1.78
Transport	635.35	735.33	823.30	1.22	3.84	1.74
Others	175.10	162.48	174.33	−0.62	2.37	−0.03
Total Final Consumption	2,478.72	2,480.51	2,668.68	0.01	2.47	0.49
Trans. Losses	834.11	1,029.00	1,094.16	1.76	2.07	1.83
TPER	3,312.82	3,509.51	3,762.84	0.48	2.35	0.85
Source: IEA 1990b.						

in demand between 1985 and 1988 was experienced in the transport sector, which is almost entirely oil dependent. The strong growth exhibited by the commercial and public-service sector, as well as industry, was driven largely by the economic expansion in the IEA region in recent years: compared to these trends, the growth of energy demand in the residential sector was rather modest.

In addition, the pattern of energy demand within the residential/commercial sector experienced significant shifts. Figure 3-1 depicts the changes in requirements for oil, solid fuels, gas, and electricity from 1973 to 1988. District heating is not shown in Figure 3-1, as it provided less than 1% of the energy requirements of the residential and commercial sectors in 1988. Demand for fuels that are primarily used to provide heat slightly declined between 1973 and 1988, while electricity demand increased significantly, and electricity increased its share in the fuel mix from 21% to 33%.

The trends described above reflect many, often related, developments—such as changes in energy prices, consumer behavior, and levels of disposable income—that affect the way individuals make investment decisions. These factors are significant in determining the scope for further energy efficiency improvements, as they have a strong influence on the likely achievable savings potential. For instance, in the case of electricity use, increases in income foster the market penetration of new household appliances, such as dishwashers, that improve the level of comfort and convenience. Furthermore, changes

Environmental Improvement and Energy Efficiency in Buildings — 75

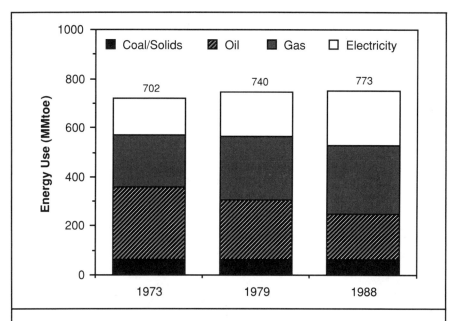

Figure 3-1. Residential/Commercial Energy Use by Fuel in IEA Member Countries

in lifestyle result in stronger demand for certain leisure or business services that require more electricity.

Carbon Dioxide Emissions from Energy End-Use Sectors

Table 3-2 shows estimates of CO_2 emissions from different end-use sectors in the IEA countries in 1988. The calculations are based on energy consumption data derived from IEA energy balances and on standard primary energy emission factors (Grubb 1989), which express CO_2 emissions in tons of carbon released by the combustion of fossil fuels. However, the emissions shown in Table 3-2 also take into account a range of upstream energy uses in order to provide estimates of emissions at end-use level. They rely on the calculation of delivered energy emission factors (EFde) that incorporate emissions that occur at other stages of the fuel cycles, particularly during transformation and electricity generation. These emissions are allocated to the different end-use sectors on a pro-rate basis. Compared to the methodology that applies emission factors only to primary energy requirements, the approach chosen here provides a more accurate picture of CO_2 emis-

76 — Chapter Three

Table 3-2. Carbon Emissions of the IEA Countries, 1988[a]		
	Carbon (MMt)	**Percent**
Industry	987.6	36.5
Transport	732.4	27.1
Road	596.6	22.1
Air	94.0	3.5
Other	983.7	36.4
Residential	566.5	21.0
Commercial/Public	349.0	12.9
Total Final Consumption	2,703.7	100.0

Note:
[a] Calculations are based on delivered energy
Source: IEA 1990b; Grubb 1989.

Table 3-3. Emission Factors on a Delivered Energy Basis for the IEA[a]	
Coal	1.13
Other solid fuels	0.89
Oil	0.88
Gas	0.73
Electricity	1.95

Note:
[a] Values show MMt of carbon emitted per MMtoe of final electricity demand (MMt carbon/MMtoe).

sions brought about by the various sectors of economic activity and is better suited to the investigation of response measures such as improved energy efficiency.

Electricity poses specific CO_2 accounting problems because emissions result from the combustion of fossil fuels in power stations, not from end uses. The emission factors calculated here are *averages* that can be related to 1 MMtoe of final electricity demand (Table 3-3). They reflect the fuel mix used to generate electricity in the IEA countries (or, in the case of national data, in the country concerned) but do not take account of the use of different generation sources for base load versus peak load. Application of these emissions factors to end-use electricity consumption figures is therefore an approximation to be

Environmental Improvement and Energy Efficiency in Buildings — 77

treated with caution when changes in electricity demand are being considered. The short-term effect of demand reduction is to reduce the load on the marginal power station, which is usually oil-fired in the daytime in winter and coal-fired or hydro at most other times. Over the longer term, the effect of electricity demand reduction on fuel use depends on the load profile. Where the load profile is not significantly altered, the savings are likely to fall approximately proportionally on each type of station, in which case the average emission factor calculated here is in fact appropriate. Reductions in electricity demand can also provide greater flexibility to reduce the operation of high-emitting power stations. CO_2 reductions could then be higher than the average emission factor used here suggests.

The industrial sector contributes the largest quantity—36.5%—of IEA-wide CO_2 emissions. Emissions from transport energy use cause approximately 27% of total CO_2 releases. The residential sector contributes about 21% and the commercial/public service sector almost 13%. Similar calculations carried out for IEA countries are summarized in Table 3-4 below. These calculations reveal that the carbon emitted by electricity generation plays a key role in the buildings sector's share of total CO_2 emissions. In countries that rely heavily on fossil fuels, particularly coal, for electricity production, the buildings sector typically represents over 40% of total emissions. This is the case for the U.K. (43%) and Denmark (48%). Where electricity is essentially produced by nonfossil fuels, the share of the buildings sector falls to 28% (in Sweden or Canada), or even as low as 14% (Norway).

Residential and Service Sector CO_2 Emissions by End-Use Categories

Although it is extremely difficult to allocate carbon emissions to the different end-use categories (heating, warm water, refrigeration, lighting, and so forth), doing so is necessary in order to evaluate different options to reduce emissions. The results of a preliminary analysis of end-use CO_2 releases are shown in Figure 3-2. These estimates, based on country-specific data where such data are available, represent about 80% of total service sector demand and over 75% of residential sector demand in IEA countries. These energy uses were extrapolated to countries where data are not available in order to obtain IEA-wide estimates of end-use demand; these estimates were then multiplied by fuel-specific EFde to give estimates of CO_2 emissions.

Taking the commercial and residential sectors together, about 54% of carbon releases are related to space conditioning, 13% to water-heating, 12% to lighting, and 6% to residential refrigeration. The

78 — Chapter Three

Table 3-4. Share of CO_2 Emissions by End-Use Sector

	% Share of CO_2 Emissions			Electricity Emission
	Industry	Transport	Other	Factor (MMt carbon/MMtoe)
IEA Countries	36.5	27.0	37.0	1.96
Australia	41.1	26.9	32.0	3.32
Austria	32.1	27.8	40.1	0.70
Belgium	43.1	22.6	34.3	1.15
Canada	40.1	28.9	31.0	0.77
Denmark	23.0	22.0	55.0	2.92
Germany	38.4	21.5	40.1	2.11
Greece	35.6	25.6	38.9	3.49
Ireland	33.5	20.0	46.4	2.87
Italy	39.6	26.6	33.6	1.82
Japan	51.9	20.9	27.2	1.49
Luxembourg	58.9	24.5	16.4	0.47
Netherlands	43.0	20.2	36.8	2.04
New Zealand	45.9	35.1	19.0	0.55
Norway	36.7	34.8	16.5	0.02
Portugal	49.7	27.9	22.3	1.32
Spain	43.2	34.0	22.8	1.43
Sweden	40.3	30.3	29.4	0.17
Switzerland	15.3	36.6	48.1	0.12
Turkey	35.7	20.1	44.2	1.72
U.K.	32.4	23.8	43.8	2.64
United States	31.8	29.2	39.0	2.34

Note: "Other" includes buildings (residential, commercial, and public sector) and agriculture.

remaining 15% are caused by other uses, such as cooking and various domestic appliances. These estimates, though only indicative, provide a more detailed picture of the origin of carbon emissions in the buildings sector, which need to be taken into account in any response strategy. The choice of measures to reduce energy requirements and emissions depends crucially on the characteristics of these different end-use categories. About 54% of emissions are related to heating, cooling, and ventilation, whereas the remaining 46% include a varied range of dispersed categories, from office automation and lighting to

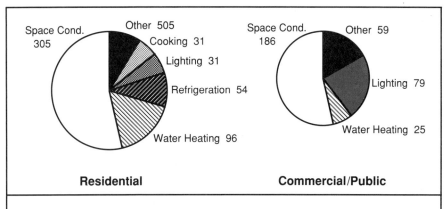

Figure 3-2. Estimated End-Use CO_2 Emissions for Buildings Sectors in IEA Countries (in MMt carbon)

residential uses such as dishwashers and refrigerators. Energy requirements of space conditioning can be lowered by improvements in the building shell, by other measures to reduce the heat load, such as the passive use of solar energy; and by efficiency improvements for space heating technologies, including boilers and energy management systems. In addition, the lengths of time required to translate the full scope of efficiency improvements into demand reductions differs significantly: appliances are replaced every 10 to 15 years, five times faster than buildings are replaced. Initiatives aimed at increasing energy efficiency, such as retrofit programs or replacement of boilers, have to take these differences into account.

Technology Options

Better insulation and appliance efficiency have significantly improved the efficiency of the buildings sector, largely in response to high energy prices in the late 1970s and early 1980s. Technology options can further improve the efficiency of both space conditioning and appliances.

Space Conditioning

Energy requirements for space heating are primarily influenced by the thermal efficiency of the building shell, by the conversion efficiency of burners and furnaces, and by distribution losses in the heating system. Other factors that influence energy requirements include climatic vari-

80 — Chapter Three

Table 3-5. Potential Improvements in Energy Efficiency for Space Heating in the Service Sector in the U.K.

Category	Examples	Technical	Cost-effective[a]
Building shell improvements	Insulation of roofs and walls, double glazing	40–50	20–40
Infiltration reduction	Draft proofing, door seals, etc.	10–15	5–10
Improved control systems	Time and temperature energy-use optimizers, energy management systems	20–25	10–20
Reduction in distribution losses, increased heat recovery		10–20	5–10
Overall for electric heating systems		55–65	30–55

Note:
[a] "Cost-effective" = measure pays back in less than six years or, in the case of minor changes, in less than three years.
Source: Energy Efficiency Office 1988.

ations, the amount of space to be heated or cooled, and the required comfort level.

The performance of the building shell is influenced by thermal insulation, air infiltration, window characteristics, and the orientation of the building (passive use of solar energy). Table 3-5 gives an overview of the scope for efficiency improvements for service sector space heating in the U.K., both in technically achievable and economically feasible terms. The largest potential is for measures that reduce heat losses through walls and roofs, for which improvements of 20% to 40% are regarded as being cost-effective (payback less than six years or, in the case of minor changes, less than three years). Heat losses can also be significantly reduced by replacing single-glazed windows with multiple glazing. A comparison of heat losses based on a survey of different types of windows can be found in Giovannini *et al.* 1989.

Improved technologies are also available to increase the efficiency of space heating and cooling. Figure 3-3 illustrates the scope of efficiency improvements based on efficiency data on current stock, new stock, and best available technology. The potential is particularly high for heating, ventilation, and air-conditioning systems (HVAC). For

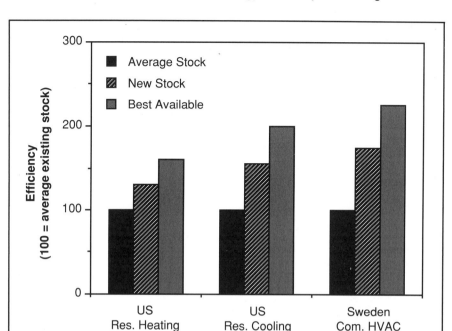

Figure 3-3. Efficiency Potential of Selected End Uses in the United States and Sweden

example, in Sweden, new systems are more than twice as efficient as the current stock.

Appliances

About 33% of carbon releases in the buildings sector of the IEA are caused by lighting, refrigeration, and electric appliances other than those used for space conditioning. As noted above, this IEA average figure is much higher in countries that use a large share of coal for their electricity generation. As a response to rising concerns about the energy, environmental, and economic costs of growing electricity consumption, the IEA has recently carried out a study to evaluate the economic potential for efficiency improvements in electric appliances and uses (IEA 1989) and analyzed end uses that together account for about 70% of total electricity use: lighting, residential space heating and water-heating, refrigeration, commercial/public building space conditioning, and industrial motors. Table 3-6 summarizes the results of this study.

82 — Chapter Three

Table 3-6. Economic Opportunities for Efficiency Improvements of Selected Electricity End Uses[a]

	(A) Share of total electricity final consumption[b]	(B) Total savings possible[c]	(C) Existing market/inst. barriers[d]	(D) Potential savings not likely to be achieved[e]	(E) Time-frame for savings (years)[f]
Residential space heating	4.5%	Medium/high	Some/many	Mixed	More than 20
Residential water-heating	5.4%	Mixed	Some/many	Mixed	10–20
Residential refrigeration	6.8%	High	Many	Medium	10–20
Lighting	16.7%	Very high	Many	High	10–20
Commercial space heating	9.9%	Mixed	Some/many	Mixed	20 or more

Notes:

[a] *How to read this table:* For example, for lighting, "very high" (more than 50% per unit) savings would result if the best available technology were used to replace the average lighting stock in use today over the next 10–20 years. Some of these savings would take place under existing market and policy conditions. But due to the "many" market and institutional barriers, there would remain a "high" (30–50%) economic potential for savings that would not be achieved.

[b] Average share for the six countries examined (United States, Japan, Germany, United Kingdom, Italy, Sweden). 27% of electricity is used by industrial motors in the six countries selected.

[c] Based on a comparison of the average efficiency of existing capital stocks to the efficiency of the best available new technology. This estimate includes the savings likely to be achieved in response to current market forces and government policies as well as those potential savings (column D) not likely to be achieved by current efforts: low (0–10% reductions per unit, on average); medium (10–30%/unit); high (30–50%/unit); very high (more than 50%/unit); and mixed (spanning at least three categories).

[d] Extent of existing market and institutional barriers to efficiency investments.

[e] Potential savings (reduction per unit) *not* likely to be achieved in response to current market forces and government policies (part of total indicated in column B).

[f] Required to achieve most of the economic potential for savings.

Source: IEA 1989.

Although economically justifiable improvements are available in all end uses, the largest economic saving potential was found in lighting. Electricity consumption in most lighting systems could be more than halved by the use of high-efficiency light bulbs, electronic ballasts, improved reflectors, and better controls. The efficiency of home refrigerators could also be significantly improved. An efficiency

Environmental Improvement and Energy Efficiency in Buildings — 83

improvement was assumed to be economic if savings can pay back the first cost in less than about five years, based on current prices.

Based on the shares of consumption and the potential savings not likely if current trends are not changed (column D, Table 3-6), the scope of overall efficiency improvements from the five end-use categories can be calculated. These estimates suggest that savings in the range of 10% to 20% per unit of service could result from economically viable improvements not likely to be realized by present efforts. The full achievement of this potential would require replacing most of the current capital stock and could only happen over about two decades. A 10–20% saving from improved efficiency, if achieved, would amount to a reduction of 0.5–1.1%/yr in the growth rate of total electricity use (IEA analyses of possible trends in electricity consumption to 2005 indicate that without efficiency improvements, electricity demand may grow by 2.7%/yr). But this amount of savings would require that the numerous barriers to efficiency investments, described in the following section, be overcome.

In the commercial sector, the most important uses for electricity are lighting, space conditioning, and office automation. The contribution to electricity demand of the last category is likely to increase in the future as our economies move to more service sector activities that require more electronic devices. In some commercial premises, office automation already has electricity requirements comparable to those of lighting (Harris *et al.* 1989). Growing electricity demand for these devices may outweigh the impact on energy demand of energy-efficiency improvements. There are nevertheless technologies that significantly reduce energy demand (microchips used for battery-driven lap-top computers, for example).

Barriers to the Efficient Use of Energy

One of the first measures likely to accelerate the market penetration of the energy-efficient technologies described above is to remove market distortions and institutional barriers that still hinder the economically efficient use of energy in the residential/commercial sector.

Market barriers in the residential sector are largely due to lack of information about energy use and related costs and about technologies available to reduce energy use. Furthermore, individual consumers often do not have access to information on means of financing investments in general and energy-efficiency technologies in particular. They make decisions to meet their day-to-day requirements, of which energy-related decisions represent only a minor part, and are usually

84 — Chapter Three

only moderately interested in their energy bills. Energy efficiency is not among the most important criteria for purchase decisions. This means that individual discount rates significantly exceed those usually applied in business (25% to 35%). For certain residential appliances, the purchasers use an implicit discount rate exceeding 60% (Meier 1983).

There are many examples of market imperfections. A building owner or developer may be interested in minimizing investment, and the tenant who will have to pay the energy bills is rarely in a position to influence these decisions. Furthermore, in multifamily houses where individual heat requirements are not separately metered, the single customer is not aware of the cost of providing heat and might regulate indoor air temperature by opening the windows instead of turning down the thermostat. Even if the dwelling is individually metered, billing and metering procedures often do not provide accurate information on energy use and costs. Such imperfections can interrupt the feedback loop that is required for the market to function correctly and that can be summarized as follows:

Energy use → energy costs → consumer action (behavioral changes, investment decisions) → cost reductions → money saved

Review of Policy and Program Options

Policy and program options within IEA member countries include information programs, regulatory instruments, and economic instruments. Policymakers have acquired considerable experience in these areas over the past 15 years.

Once a policy has been established, *information programs* can urge compliance with and continued support for the chosen policy and provide information about available benefits that might not otherwise be widely known to the general public or to specific target groups such as industrial managers or architects. Informing energy users about the energy consumption of equipment is a well-established policy in IEA countries. In addition to energy efficiency labeling, a number of member countries have introduced the "eco-label," which guides consumers to products that are environmentally friendly at a time when green consumerism is becoming a major force in the marketplace. Training programs by governments have also been used at times to improve the skills of personnel closely involved in energy efficiency developments. Testing and training can entail substantial costs, and one alternative to government involvement is to shift such tasks to industry.

Information programs alone tend to work best for actions that make good economic sense for an industry or for the consumer but that may not be widely known. In instances where the economic benefits are not so straightforward, direct market interventions, in addition to some form of information program, are probably necessary.

Regulatory instruments applied to energy demand policy include the broad array of standards and control mechanisms, such as restrictions on fuel use, building codes, and efficiency and emission requirements for burners. While standards and regulations can be fairly effective and easy to promulgate, their initial design often assumes and requires considerable technical knowledge. For this reason, standard setting is likely to be an iterative process, whereby standards are promulgated and revised to reflect current experience and technology as well as national situations. As a result, this process often produces quite different results in different countries (Boyle 1989).

Unless regularly revised or upgraded, standards can fail to encourage the adoption of new technologies to fulfill or improve upon existing minimum requirements. Though the use of energy-efficiency standards is well established as a policy instrument, the introduction and upgrading of such standards pose technical and economic problems. The mandatory use of available, cost-effective technologies is usually the center of a debate involving issues such as market choice (leaving its to users to decide the trade-offs between convenience features, purchase price, and operating costs) and the commercialization of new technologies (particularly in terms of reliable, low-cost mass production). These familiar issues are being revived by the need to reconsider new energy standards and technologies in terms of their potential for reducing polluting emissions.

Building standards usually apply only to new buildings and therefore are slow in producing an effect on energy demand. Typical rates of building for housing are 1–3% of the stock per year. Rapid improvements to the energy efficiency of the building stock would require improving existing buildings, in which case regulations could take the form of the procedure applied in Denmark, where energy inspections are required when houses are sold.

Economic instruments include taxes, pricing, charges, subsidies, and other financial inducements. These instruments have been widely applied in years of high energy prices. Most schemes have been gradually reduced in recent years because energy prices have declined and the political need to limit public spending has reduced the availability of funds. Moreover, governments are increasingly reluctant to grant financial incentives because individuals who would have invested in

86 — Chapter Three

energy-demand reducing measures anyway also benefit from the programs. Many policymakers contend that this so-called free-rider effect results in a nonoptimal allocation of public resources.

Developments in the area of environmental protection are fueling interest in market-based instruments and taxes, which, when implemented in the energy sector, are likely to affect energy-efficiency improvements and policy instruments designed to encourage their development. Market-based economic instruments are policies that use the power of the marketplace to achieve environmental goals. One such policy, emissions trading, is particularly interesting to energy-efficiency efforts because it was specifically developed to increase the cost-effectiveness of emission control and allow industry a large flexibility in choosing technical solutions. Emission reductions achieved through improved energy-efficiency actions can be fully credited in emissions trading, which is not always the case for approaches based on traditional emissions standards. If the market-based trading system is appropriately designed, it can provide much incentive to energy-efficiency actions. So far, experience with emission trading schemes has been limited to the United States and Germany and applied to the control of traditional pollutants such as SO_2 in the industrial and power-generation sectors. Within the IPCC and other forums, the use of similar schemes to introduce more flexibility and equity into any international agreement to limit CO_2 is being examined (IPCC 1990b).

Carbon taxes, proportional to the carbon content of fossil fuels, are also the subject of much attention. As is the case with any tax increasing energy prices, the effect on demand is a function of the price elasticity. Though few would question that higher oil prices have in the past prompted improved energy efficiency, the relative impacts of nonprice policy measures, structural change, and prices on energy demand are difficult to separate. A review of international literature carried out recently (Mills 1989) showed a bewildering degree of variation in elasticity estimates for a given country, fuel, and sector. Given the investment criteria applied by energy users in the residential and commercial sectors and the market barriers these investments need to overcome, any price increase would probably have to be more marked in these than in other sectors to produce a significant effect not only on fuel choice but also on absolute levels of demand.

There is nevertheless a range of measures that can be taken or encouraged to enhance the development of cost-effective energy-efficiency improvements even at present low energy prices. The achievements of demand-side management programs run by utilities and local authorities in North America are promising, providing this

Environmental Improvement and Energy Efficiency in Buildings — 87

experience can be transferred to other parts of the IEA where utilities function under different regulatory regimes and supply constraints.

Conclusions

The efficient use of energy—both in terms of economic soundness and rational allocation—can help reduce the anticipated growth rates of greenhouse-gas emissions. A variety of cost-effective measures can reduce energy consumption without sacrificing individual comfort. The buildings sector requires special attention from policymakers. Residential and commercial end users are important contributors to the anthropogenic CO_2 emissions. However, substantial market imperfections reduce or slow down the penetration of available emissions-reducing technology. Fortunately, an array of policy options can help reduce these market barriers. Careful selection of appropriate policies, flexibility, and strong commitment will be required to address the remaining potential for energy savings and translate it into emission reductions.

References

Boyle, S., and I. Brown 1989. "Reducing Greenhouse Gas Emissions—A Case Study of the Electricity in the UK." In *Energy Technologies for Reducing Emissions of Greenhouse Gases*. Paris: OECD.

Danish Ministry of Energy 1989. *Statement on Energy Policy*. Copenhagen: Danish Ministry of Energy.

Department of Energy 1990. *An Evaluation of Energy Related Greenhouse Gas Emissions and Measures to Ameliorate Them*. Energy Paper Number 58. London: HMSO.

Deutscher Bundestag, ed. 1990. *Schutz der Erdatmosphäre: Eine Internationale Herausforderung*. Interim Report of Enquiry Commission. Bonn.

Energy Efficiency Office 1988. *Energy Use and Energy Efficiency in UK Commercial and Public Buildings up to Year 2000*. London: HMSO.

Giovannini, B., B. Aebischer, and D. Pain 1989. *Scientific and Technical Arguments for the Optimal Use of Energy*. Geneva: University of Geneva.

Grubb, M. 1989. "On Coefficients for Determining Greenhouse Gas Emission Factors from Fossil Fuel Production and Consumption."

88 — Chapter Three

In *Energy Technologies for Reducing Emissions of Greenhouse Gases*. Paris: OECD.

Harris, J., L. Norford, A. Rabl, and J. Roturier 1989. "Electronic Office Equipment: The Impact of Market Trends and Technology on End-Use Demand for Electricity." In *Electricity—Efficient End Use and New Generation Technologies, and Their Planning Implications*. Lund: Lund University Press.

IEA 1987. *Energy Conservation in IEA Countries*. Paris: OECD.

———— 1989. *Electricity End-Use Efficiency*. Paris: OECD.

———— 1990a. *Energy and the Environment: Policy Overview*. Paris: OECD.

———— 1990b. *Energy Balances for OECD Countries: 1987/1988*. Paris: OECD.

IPCC 1990a. *Report of Working Group I*. Geneva: UNEP/WMO.

———— 1990b. *Report of Working Group II*. Geneva: UNEP/WMO.

Meier, A., and J. Whittier 1983. "Consumer Discount Rates Implied by Consumer Purchases of Energy-Efficient Refrigerators." In *IEnergy, The International Journal* 8 (12).

Mills, E. 1989. "An End-Use Perspective on Electricity Price Responsiveness." *Vattenfalls Updrag 2000 Report*. Stockholm: Voltenfall.

Schipper, L., A. Ketoff, S. Meyers, and P. Hawk 1987. "Residential Electricity Consumption in Industrialised Countries: Changes since 1973." In *Energy, The International Journal* 12(12).

Streb, A. 1989. "Energy Efficiency and Global Warming." In *Electricity—Efficient End-Use and New Generation Technologies, and Their Planning Implications*. Lund: Lund University Press.

Erich Unterwurzacher and Genevieve McInnes are both energy analysts in the Office of Long Term Co-operation and Policy Analysis of the International Energy Agency. They analyze energy demand and energy policies, the potential for more efficient electricity use and strategies to reduce environmental impacts of energy use.

Unterwurzacher holds a doctorate in energy economics and a master's in electrical engineering from the University of Technology in Vienna. McInnes received an M.S. in Environmental Technology (speciality Energy) from Imperial College of Science and Technology, University of London.

Chapter 4

Carbon Dioxide Emissions and Energy Efficiency in U.K. Buildings

George Henderson and Les Shorrock, *The Building Research Establishment, United Kingdom*

Introduction

Global industrial development is now widely recognized to be on a scale large enough to cause changes to the composition of the atmosphere and, consequently, changes to climate. The principal humanly caused agents of global climate change are increased concentrations of greenhouse gases, which raise average temperatures in the lower atmosphere by absorbing infrared radiation. A very important contribution arises from the carbon dioxide (CO_2), which is released to the atmosphere whenever fossil fuels are burned. Other significant greenhouse gases include methane, nitrous oxide, ozone, and the chlorofluorocarbons (CFCs).

Buildings require energy for space heating and water-heating, lighting, refrigeration, ventilation, and other services. These uses, together with domestic appliances and office equipment, account for about half of total U.K. demand for energy and a similar proportion of all energy-related CO_2 emissions. It follows that improvements to the efficiency with which energy is used in buildings could offer considerable opportunities for reducing those emissions. In the U.K., buildings provide many opportunities for improving energy efficiency cost-effectively and, thus, at no net cost. Doing so would reduce both energy requirements and greenhouse-gas emissions. The aim of the work described in this chapter was to estimate the extent to which greenhouse emissions could be reduced through applying energy-

90 — Chapter Four

efficiency measures that are both cost-effective and technically well proven.

Energy Use in Buildings

Information on the total quantities of energy consumed by various sectors of the economy is available from the Digest of U.K. Energy Statistics (Department of Energy 1989). The Digest does not have a single category for buildings but includes them instead in various sectors according to economic criteria. Energy use in residential buildings can be estimated readily, however, since it is reasonable to attribute all of residential energy consumption to buildings. How much nonresidential consumption can be attributed to buildings is less certain. For present purposes, we use a study by Hardcastle (1984). Figure 4-1 shows the energy consumed by various sectors of the U.K. economy in 1987 in units of delivered energy (the calorific value of energy delivered to consumers). This figure illustrates the importance of buildings, which account for over two-thirds of all electricity consumed and a similar fraction of natural gas.

Understanding of energy use in buildings requires knowing the amounts of energy and of different fuels consumed for various end uses. These data are needed to evaluate the potential effects of energy-efficiency improvements, which generally apply to specific end uses and, hence, to particular fuels.

Domestic buildings alone account for 28% of all energy consumed in the U.K. and are a particularly important market for natural gas, accounting for 55% of all consumption of that fuel. Domestic electricity consumption is also important, accounting for 36% of electricity supplied to all consumers. By contrast, the domestic market for oil is small and accounts for only 4% of oil supplied in units of delivered energy. A detailed study of energy use in the domestic sector has been undertaken by the Building Research Establishment (BRE) (Henderson and Shorrock 1989). Figure 4-2, showing a breakdown of residential energy use by fuel and end use in units of delivered energy, illustrates the dominance of natural gas as a heating fuel and the high proportion of total energy used for heating.

Much less detailed information is available on energy consumption in nonresidential buildings, which include many building types and variations of activity within buildings. A study of energy use in commercial and public buildings sponsored by the U.K. Department of Energy assembled information from many sources and provides a valuable compilation of existing data (Herring et al. 1988). However,

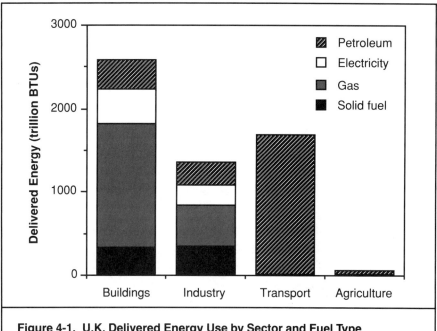

Figure 4-1. U.K. Delivered Energy Use by Sector and Fuel Type

this study also pointed out the lack of representative survey data for many sectors.

In general, space heating and water-heating account for a lower proportion, and lighting for a higher proportion, of consumption in nonresidential buildings than in dwellings. Also, in the U.K. air-conditioning is a significant end use in some types of nonresidential buildings but a negligible one in dwellings.

Carbon Dioxide Emissions from Fuel Consumption

The different fossil fuels contain varying proportions of carbon and have different caloric values. Accordingly, they emit varying amounts of CO_2 for each unit of heat they produce when burned. Typical emissions figures for the primary energy available from the main fuels used in the U.K. are shown in Table 4-1(a).

Electricity presents a much more complicated picture, since its CO_2 emissions depend entirely on how it is generated. For example, hydroelectric generation does not involve the burning of fossil fuels

92 — Chapter Four

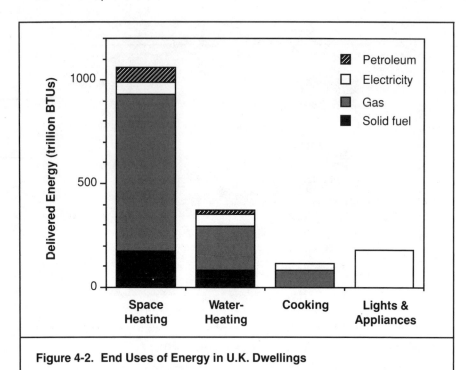

Figure 4-2. End Uses of Energy in U.K. Dwellings

Table 4-1. Carbon Dioxide Emissions from Various Fuels Used in the U.K. (lbs of CO_2/kWh)

Fuel	(a) Based on primary energy available	(b) Based on energy delivered to consumers[a]
Coal	0.72	0.73
Natural gas	0.40	0.44
Petroleum	0.55	0.67
Electricity	N/A	1.83

Note:
[a] Values for nonelectricity sources are higher in column (b) than in column (a) because of overheads associated with conversion and distribution.

Carbon Dioxide Emissions and Energy Efficiency in U.K. Buildings — 93

and emits no CO_2, while electricity generated from coal or oil emits relatively high levels. For the latter case, the efficiency of the generation process is also important, since the CO_2 emissions are related to the fuel consumed rather than the electricity produced. In the U.K., most electricity is generated from coal in large power stations with no means of utilizing the waste heat from the generation process. Despite the relatively high efficiency of many of the power stations, the overall thermal efficiency of the process is low, and about three units of energy are released by the combustion of the coal for each unit of electricity. It is possible to calculate the CO_2 emission associated with each unit of electricity used by consumers by taking account of efficiency and the mix of fuels used in generation. Thus, although electricity produces no CO_2 at the point of consumption, an average unit of electricity consumed in the U.K. produces large amounts at the point of generation. Table 4-1(b) shows the amounts of CO_2 associated with each unit of fuel delivered to the consumer, including the overheads associated with generation and distribution. They have been calculated using the values in Table 4-1(a) and data from the Digest of U.K. Energy Statistics (*op. cit.*). In the case of electricity, this reflects the present mix of generating fuels which included 67% coal, 17% nuclear and 8% oil in 1988.

The figures given in Table 4-1(b) have been used to convert energy consumption into CO_2 emissions throughout this report. It is important to note that they only apply to the U.K. energy economy at the present time and would not hold for other circumstances. In particular, the value for electricity applies only to that supplied by the U.K. national grid.

Carbon Dioxide Emissions from Building Energy Use

When energy use is converted into CO_2 using the factors in Table 4-1(b), the relative importance of the various fuels and end-uses changes considerably (see Figures 4-3 and 4-4). A comparison between Figure 4-3 and Figure 4-1 shows that electricity accounts for a larger proportion of the total CO_2 emissions than of delivered energy. Similarly, Figures 4-4 and 4-2 show that lighting and appliance consumption in dwellings become more significant as sources of CO_2 emissions. Conversely, the relative importance of space heating is diminished, although it is still the largest single contributor to CO_2 emissions. The reason for the changes in the relative importance of the different end uses is clear: those that rely heavily on electricity become more significant, while those that rely mainly on natural gas become

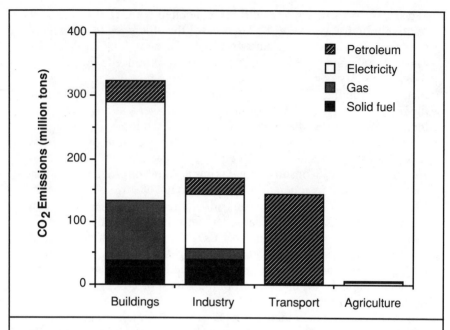

Figure 4-3. CO_2 Emissions in the U.K. by Sector and Fuel Type

less so. This fact has important consequences in considering the effects of energy-efficiency measures on CO_2 emissions.

In total, CO_2 emissions related to energy use in buildings are calculated to be 48% of all U.K. energy-related emissions. This percentage is very similar to the proportion of total delivered energy consumed by buildings.

Improvements to Energy Efficiency in Existing Buildings

The slow replacement rate of the building stock makes it necessary to improve existing buildings if a rapid overall improvement is to be achieved. For example, the rate at which new houses have been built in the U.K. over recent years is about 1% of the existing stock per year. The U.K. has many old buildings that are structurally sound but poorly insulated, because they were built before thermal insulation was required by the national building regulations or the local building by-laws that preceded the regulations. The first national regulations

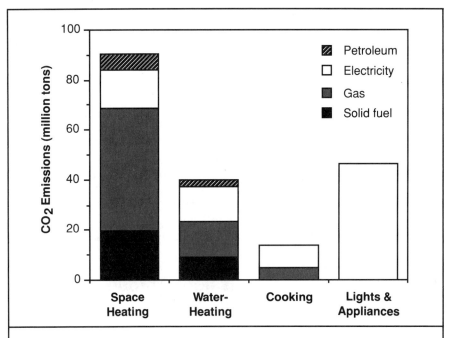

Figure 4-4. CO_2 Attributable to U.K. Dwellings, by End Use and by Delivered Fuel

were introduced in 1965, and higher standards were established in 1976, 1982, and 1990. Thus, many buildings erected as recently as a decade ago have thermal insulation at well below present standards, as do a large number of older buildings.

Although insulation is generally most economically installed during construction, it can be cost-effectively improved in existing buildings. For example, most U.K. houses have pitched roofs with accessible attics that can be insulated very simply and cost-effectively. Other cost-effective measures include weather stripping of windows and doors and insulation of hot-water storage tanks and external cavity walls. Further measures are cost-effective under certain circumstances, such as double glazing when windows need to be replaced.

Apart from improvements to insulation, energy efficiency can also be improved by installing more efficient heating systems and better controls. A particularly important opportunity exists for improving the efficiency of domestic boilers when they need to be replaced. About half of all households have gas central heating using a boiler to heat

96 — Chapter Four

water circulated to radiators. Many of those boilers were installed 10 to 20 years ago and are being replaced with new models. The replacements are generally more efficient than the old boilers, but it has been shown that using condensing boilers could increase efficiency even more. Field trials in occupied houses have shown that condensing boilers can achieve a yearly average efficiency of about 85% compared with about 70% for a modern conventional boiler operating under the same conditions (Trim 1988).

There are also many opportunities for improving energy efficiency in nonresidential buildings. For example, a recent study of energy-efficient office buildings highlighted the importance of lighting and of the energy loads imposed by large computer installations and their associated cooling requirements (Brownhill 1990). This work also showed that air-conditioning significantly increased demand for energy, largely through the power consumed by the system's pumps and ventilation fans.

Reductions in CO_2 Emissions Through Energy-Efficiency Improvements

Particular energy-efficiency measures affect specific uses of energy and, hence, reduce consumption of the particular fuels that serve those uses. When considering CO_2 emissions, it is important to attribute reductions in energy use to particular fuels because of the large fuel-specific differences in emissions per unit of energy (Table 4-1). Linking energy savings to fuels is particularly important in the case of electricity and those end-uses, such as lighting and appliances, that rely mostly on electricity.

BRE has analyzed the reductions in CO_2 that could be obtained from energy-efficiency improvements to the housing stock in some detail (Shorrock and Henderson 1990). Two cases were considered. The first case was confined to those measures that are considered cost-effective, while the second also included measures that are technically quite feasible but not generally found to be cost-effective at present prices and fuel costs.

The cost-effective category was based on results from a variety of existing studies rather than on new analysis. Accordingly, no single criterion was used to define "cost-effective." The booklets on cutting home energy cost published by the U.K. Department of Energy give examples of costs and savings for various house types (UK Energy Efficiency Office 1986). Most improvements, with the exception of double glazing, have a simple payback of less than seven years. Double glazing is included because reduced condensation on windows and

Carbon Dioxide Emissions and Energy Efficiency in U.K. Buildings — 97

Table 4-2. Residential Energy-Saving Improvements Capable of Reducing CO_2 Emissions	
Cost-Effective[a]	Technically Possible[b]
80% of all cavity walls insulated	All walls insulated
Attics with $< = 25$ mm insulated to 150 mm	All attics insulated to 150 mm
Full double glazing in all homes (as windows are replaced)	Full double glazing in all homes
Weather stripping in homes with <80% of rooms already treated	Weather stripping in all homes
Hot-water tanks with <50 mm insulation increased to 80 mm	All hot-water tanks insulated to 80 mm
Condensing boilers in all homes with gas central heating (as boilers are replaced)	Condensing boilers in all homes with gas central heating
13% efficiency improvements to gas and electric cookers	25% efficiency improvements to gas and electric cookers
38% efficiency improvements to lighting	75% efficiency improvements to lighting
25% efficiency improvements to refrigeration equipment	50% efficiency improvements to refrigeration equipment
20% efficiency improvements to washing machines	20% efficiency improvements to washing machines and tumble clothes dryers
25% efficiency improvements to televisions	25% efficiency improvements to televisions

Notes:
[a] "Cost-effective" is generally defined here as measures with simple paybacks of less than seven years.
[b] Improvements that can be carried out by well-tried methods and readily available materials but that are not necessarily cost-effective as here defined.
Source: Shorrock and Henderson 1990.

value added to the home, in addition to reduced energy expenditures, are considered to justify a longer simple payback.

The "technically possible" category has been defined very conservatively and is based on improvements that can be carried out now using well-tried methods and readily available materials. In the case of lighting and domestic appliances, this definition has been taken to mean equipment that can be bought in the U.K. at present.

Both cost-effective and technically feasible categories of improvements are partly based on judgments made by the authors and are intended to be indicative rather than definitive. The improvements included in each category are shown in Table 4-2.

The scope for applying the various improvements and the associ-

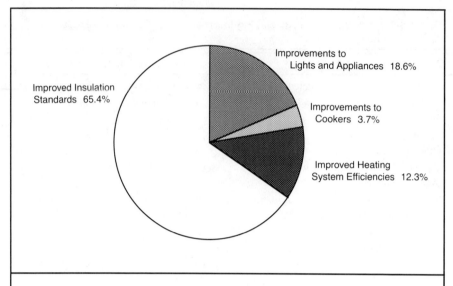

Figure 4-5. Potential Reductions in CO_2 Emissions Through the Application of Cost-Effective Energy Efficiency Measures in U.K. Dwellings

ated reductions in energy use were calculated using BREHOMES, a model of energy use in the U.K. residential sector (Henderson and Shorrock 1988). Reductions in CO_2 emissions were then calculated using the emission factors shown in Table 4-1(b). Ideally, the emission factor for electricity should have varied with the order in which power stations are dispatched onto the national grid (merit order). In the absence of detailed information on the merit order, the overall factor given in Table 4-1(b) was used. This decision is likely to cause a slight underestimate, since the avoided load is mostly served by coal-fired stations across a wide range of loads.

The calculated reductions amount to about 25% of present emissions for the cost-effective case and about 35% for the technically possible case. Figure 4-5 shows the estimated CO_2 reductions that could be derived from the cost-effective case. About two-thirds of the reductions are due to insulation measures, and the remainder are due to improvements in the efficiency of heating systems and appliances. We estimate the total capital cost of applying the cost-effective case to be $25–30 billion, equivalent to about one-and-a-half times current annual fuel expenditure by households.

No detailed calculations have yet been undertaken for nonresiden-

Carbon Dioxide Emissions and Energy Efficiency in U.K. Buildings — 99

tial buildings. However, Herring *et al.* 1988 identified cost-effective potential reductions of 40–55% of delivered energy in commercial and public-sector buildings. Taking account of the mix of fuels used, it is reasonable to assume that such reductions would also substantially reduce CO_2 emissions. In general, nonresidential buildings offer greater scope than do dwellings for making reductions through better plant maintenance and energy management. Also, the efficiency and control of lighting are of much greater importance than in dwellings.

Current Trends in CO_2 Emissions

It is important to note that the foregoing analysis is not a forecast of what is likely to happen over the next 20 years but rather an estimate of the likely effect of applying those measures to *present* patterns of use. Estimates of actual emissions at some point in the future require that trends in patterns of demand should also be considered, taking account of saturation effects where appropriate. A detailed analysis of this sort has not yet been undertaken by the authors, but preliminary work indicates that, given a continuation of present trends, emissions from the residential sector are likely to remain close to present levels over the next decade. A similar conclusion was reached by Leach and Nowak (1990). Emissions have actually fallen by about 5% since 1970 despite a growth of 15% in the number of households and greatly increased penetration of central heating during the intervening period. The reduction has been due to the retrofitting of insulation measures to existing stock and to natural gas displacing coal as the main fuel for space heating and water-heating (Henderson and Shorrock 1989). Central heating must begin to reach saturation within the present decade, and growth in the number of households is forecast to be slower than in the two preceding decades. Both trends will tend to slow the growth in underlying demand. On the other hand, switching to natural gas for heating must also reach saturation, as must the insulation of previously uninsulated roof spaces. Furthermore, average indoor temperatures in the U.K. are still well below North American and Scandinavian levels and could rise considerably.

Lights and appliances are of particular interest because these energy uses have shown the fastest growth of all domestic end uses over the past two decades. Table 4-3 shows the estimated consumption by type of appliance and level of ownership inn the U.K. Refrigeration equipment accounts for 35% of the total, lighting for 16%, and home laundry equipment for 12%. In many cases there is a reasonable prospect that increased efficiency could compensate for increased owner-

100 — Chapter Four

Table 4-3. Energy Use in U.K. Dwellings for Lights and Appliances in 1987			
Appliance(s)	Typical Consumption (kWh/year)	Ownership Level (%)	Aggregate Consumption (petajoules)
Washing machines	200	86	14.2
Clothes dryers	300	31	7.6
Dishwashers	500	7	2.9
Refrigerators	300	57	14.4
Refrigerator/freezers	740	43	26.1
Freezers	740	39	23.4
Electric kettles	250	86	17.9
Irons	75	98	6.1
Vacuum cleaners	25	98	2.0
Televisions	235	98	19.2
Lighting	360	100	30.1
Miscellaneous	240	100	20.0
TOTAL			184 PJ
Source: Data supplied by the U.K. electricity supply industry.			

ship level. One of the largest potential reductions would result from the replacement of incandescent light bulbs by low-energy units that are available but have a low market penetration at present. In other cases, considerable increases in ownership levels are likely following patterns already established in some other European countries and North America, resulting in increased energy consumption. For example, dishwashers are heavy energy users and are at present owned by less than 10% of households in the U.K.

An analysis of possible reductions in CO_2 emissions from electricity generation is beyond the scope of this paper, but it is important to note that all above estimates of potential reductions of CO_2 assume a continuation of present electricity generation patterns. Clearly, any reductions deriving from a reduced demand for electricity would be even greater if the CO_2 emission factor for electricity were itself to be reduced. This possibility would need to be taken into account when considering future levels of emissions in order to avoid exaggerating the reductions brought about by energy-efficiency improvements.

Other Greenhouse Gases

Greenhouse gases other than CO_2 are also associated with buildings. CFCs are widely used as working fluids in refrigeration equipment that is vital to the proper functioning of some types of building, both for cold storage and for air-conditioning. They are also used as foaming agents in some types of insulation used in buildings. Present-day buildings, therefore, are important sources of CFCs released to the atmosphere. CFCs are already the subject of international agreements that will greatly reduce their future use, so their significance as greenhouse gases should also decline. However, such restrictions will necessitate CFC substitutes, some of which could cause a small increase in the amount of energy required for refrigeration and, consequently, in CO_2 emissions. Some CFC substitutes are themselves significant greenhouse gases. In the U.K., building-related emissions are estimated to have accounted for about 8% of CFC emissions in 1986 (Butler 1989). This proportion will already have increased considerably, as CFCs were still widely used in aerosols sold in the U.K. in 1986, although these products have since been phased out.

Methane, being the main constituent of natural gas, is another greenhouse gas associated with energy in buildings. Any leakage of methane from the gas distribution network could be seen as an overhead to be added to the CO_2 that results from its combustion. This effect could be very significant if such leakage were more than a few percent, because methane is much more effective than CO_2 as a greenhouse gas.

Nitrous oxide is produced in low concentrations during the burning of fossil fuels. No reliable data are yet available on the quantities involved, but they are thought to be small, both in relation to other building emissions and to other emissions of nitrous oxide.

Conclusions

Energy use in buildings is responsible for about half of total U.K. emissions of CO_2 at present. In the residential sector, those emissions could be reduced by a quarter of present levels through applying energy-efficiency measures that are already considered to be cost-effective, if present levels of demand were maintained. Actual future emissions will depend both on the rate at which energy efficiency is improved and the rate of growth in demand. Trends over the last decade have shown CO_2 emissions from residential energy use staying at about the same level.

102 — Chapter Four

Acknowledgment

This work forms part of the research program of the Building Research Establishment and is published with permission of the director.

References

Brownhill, D. 1990. "Energy Efficient Offices: Case Histories," *Building Services*, July.

Butler, D. 1989. "CFCs and the Building Industry." BRE Information Paper IP 23/89. London: Building Research Establishment.

Department of Energy 1989. *Digest of U.K. Energy Statistics*. London: HMSO.

Hardcastle, R. 1984. *The Pattern of Energy Use in the U.K.—1980*. London: HMSO.

Henderson, G., and L. Shorrock 1988. "Energy Efficiency and Consumption in the U.K. Housing Stock." In *Proceedings of the ACEEE Summer Study on Energy Efficiency in Buildings*. Vol. 10, *Performance Measurement and Analysis*. Washington, D.C.: American Council for an Energy-Efficient Economy.

——— 1989. *Domestic Energy Factfile*. BRE Report BR151. London: Building Research Establishment.

Herring, H., R. Hardcastle, and R. Philipson 1988. *Energy Use and Efficiency in U.K. Commercial and Public Buildings up to the Year 2000*. London: HMSO.

Leach, G., and Z. Nowak 1990. "Cutting Carbon Dioxide Emissions from Poland and the United Kingdom." In *Proceedings of the Workshop on Energy and Environment—Sustainable Energy Use*. Stockholm: Stockholm Environment Institute.

Shorrock, L., and G. Henderson 1990. *Energy Use in Buildings and Carbon Dioxide Emissions*. BRE Report BR170. London: Building Research Establishment.

Trim, M. 1988. "The Performance of Gas-fired Condensing Boilers on Family Housing." BRE Information Paper IP 10/88. London: Building Research Establishment.

U.K. Energy Efficiency Office 1986. *Cutting Home Energy Costs—A Step-by-Step Monergy Guide*. London: Energy Efficiency Office.

George Henderson is head of the Energy Economics and Statistics Section of the Building Research Establishment. This section works on numerous aspects of energy efficiency in buildings, such as developing and testing energy models, building energy-use databases, analysing

Carbon Dioxide Emissions and Energy Efficiency in U.K. Buildings — 103

cost-effectiveness of energy efficiency measures, researching greenhouse-gas emissions, and energy labelling. He has a B.Sc. (Physics) and an M.Sc. (Instrumentation) from the University of Aberdeen.

Les Shorrock is a senior scientific officer in the Energy Economics Statistics Section of the Building Research Establishment. He has a B.A. (Physics) from the University of York and a Ph.D. (Applied Physics) from the University of Hull.

Chapter 5

Environmental Benefits of Energy Efficiency: Impact of Washington State Residential Energy Codes on Greenhouse-Gas Emissions

Richard Byers, *Washington State Energy Office*

Introduction

The United States has made major contributions to the research and scientific consensus concerning the significance and reality of the environmental threats posed by current energy policy. However, progressive decisions to implement policies and allocate budget resources to address these threats comprehensively have either been slow in coming (acid rain) or have yet to be made at all (greenhouse-gas reduction). The reluctance to act has been largely driven by concern over the impacts such policies might have on the U.S. economy. To the degree that policies aimed at reducing greenhouse-gas emissions, acid gas emissions, other air pollutants, or non-air-related environmental threats (for example, oil spills) are expensive, they are perceived to place a burden on the economy and lead to reduced productivity and competitiveness. Consequently, the last two administrations have taken the position that until a threat can be scientifically proven beyond controversy, the principal action taken should be further study.

Many policy analysts have been quick to point out, however, that even if a cautious, wait-and-see strategy is taken, those actions that achieve the environmental objectives while at the same time providing other economic benefits should be implemented now. Energy conser-

105

106 — Chapter Five

vation and efficiency measures offer such an opportunity. Because these measures improve the efficiency, both thermodynamic and economic, with which an energy service is delivered, they are often justified on conventional economic grounds, even before environmental costs and benefits are considered. In light of the fact that many of our energy resources are finite and that a sizable portion of our gross national product is devoted to energy costs, these are efficiencies that should be pursued anyway—even if they offered no environmental benefits.

The purpose of this chapter is to provide a real-world example of such a "win-win" energy efficiency policy action. The State of Washington enforces a residential building code that sets minimum levels of insulation for new buildings. This paper discusses the cost of these insulation levels, their energy savings, their economics from the perspectives of the individual home buyer and the state as a whole, and the magnitude of the environmental benefits achieved. Assessment of environmental benefits focuses on reduction in annual emissions of greenhouse gases.

The Washington State Residential Building Energy Code

In 1977 (effective 1978), the Washington State Legislature enacted the state's first energy code covering residential construction. This code established basic insulation requirements for ceilings, walls, and floors in new residential buildings. Before enactment of the 1978 code, Washington had no minimum specific insulation standards for residential structures.

The 1978 Washington State Energy Code (WSEC) was upgraded in 1980 and 1986 and was revised again by the legislature in 1990 to be upgraded effective in 1991. This analysis evaluates the cost, savings, and reductions in greenhouse-gas emissions associated with the 1980, 1986, and 1991 codes, taking the 1978 code levels as a base case.

Both climate severity and type of heating fuel are considered in the insulation specifications of the 1986 and 1991 codes. The 1980 code was fuel-blind and nearly uniform across the state's climate zones. Table 5-1 presents the insulation levels called for by the current code (1991), as well as levels required by previous codes in the two Washington climate zones. The levels for the 1986 and 1991 codes were established on the basis of consumer cost-effectiveness. The code is structured such that these insulation prescriptions establish a target level for space heating performance. Flexibility is permitted in actual

Environmental Benefits of Energy Efficiency — 107

Table 5-1. Insulation Levels Required by Washington State Residential Energy Codes Climate Zones

	Ceiling (R-value)	Walls (R-value)	Floors (R-value)	Windows[a] (U-value)	Doors (U-value)	Infil. (type)
1978	19/19	11/11	11/11	single/single	NR	NR[b]
1980	30/30	11/11	11/11	double/double	NR	clk/wthrstrp
1986	38/30	19/19	19/19[a] 25/19[b]	.60/.75	NR	clk/wthrstrp
1991	38/30[c]	19/19[a]	30/19[a]	.40/.65[c]	.19/NR	clk/wthrstrp
	38/38[d]	24/19[b]	30/25[b]	.40/.60[d]		

Notes:
[a] Window requirements by basic description in 1978 and 1980 codes. Tested U-value maximums (AAMA 1503.1) required by 1986 and 1991 codes.
[b] NR = no requirement
[c] For western Washington climate zone (<6,000 HDD 65°F).
[d] For eastern Washington climate zone (>6,000 HDD 65°F)
All other codes pertain to both climate zones.

building component insulation levels, so long as performance of the house as a whole measures up to the prescribed levels.

For comparison, the current draft of American Society of Heating, Refrigerating and Air-Conditioning Engineers (ASHRAE) Standard 90.2 recommends insulation levels for Washington that fall between the 1980 and 1986 WSEC levels.

Cost and Energy Savings for Energy Code Insulation Levels

The additional construction costs for new residences incorporating the insulation levels required by the energy codes have been documented in two ways. First, incremental construction costs were collected from the builders of 226 energy-efficient demonstration homes as a part of the state Residential Standards Demonstration Project (RSDP) (Tangora *et al.* 1986). Data were collected in 1985. Second, incremental construction costs were estimated by standard construction cost estimation procedures as a part of the University of Washington's (UW) Component Testing Project (Ossinger *et al.* 1989). These two data sources agree quite closely (Byers 1989). Because of their more extensive documentation, costs from the UW project have been used in this analysis. Table 5-2 presents the added costs (as 1989 $/sq ft of building

108 — Chapter Five

Table 5-2. Component Level Insulation Costs (1989$/sq ft)

Ceilings	Walls (net sq ft)	Floors	Windows	Doors
R19:Base	R11:Base	R11:Base	Single:Base	Wood:Base
R30:$.17	R19:$.56	R19:$.20	.75[a]:$1.75	R5:$1.12
R38:$.30	R24:$1.24	R25:$.29	.65[a]:$2.54	—
—	—	R30:$.46	.60[a]:$3.99	—
—	—	—	.40[a]:$6.87	—

Note: Costs are cumulative from indicated base and include 40% builder overhead and profit.
[a] Average U-values.

component) for insulation components included in the Washington energy codes. These costs include a 40% markup for builder overhead and profit.

The energy savings calculations derived from a simple thermal simulation model, called SUNDAY, assume that future household operating conditions will remain similar to current conditions: for example, internal temperatures of 67–68°F, internal heat gains from appliances and occupants of 3,000 Btu/hr (see appendix for further details of the SUNDAY model). This is a conservative assumption. As appliances become more efficient, internal heat gains will decrease and space heat requirements will increase; thus, savings due to improved insulation will increase as well. SUNDAY indicates that a reduction of 1,000 Btu/hr in available internal gains yields an increase in space heat of about 2,050 kWh/yr for a home insulated to the 1978 standards and only 1,600 kWh/yr for the 1991 standards, a savings of 450 kWh/yr.

While no homes using natural gas heat have been monitored, it is assumed in this analysis that insulation performance will be similar to performance in electrically heated homes and that assumptions need only be made concerning heating system efficiency. An annual fuel utilization index (AFUE) of 78% is assumed (minimum set by National Appliance Efficiency Act) with a distribution system efficiency of 70.5% (derived from engineering estimation of conductive and convective heat loss from ducts and from pressure differences, due to forced-air heating, across the building envelope). These assumptions lead to a total natural gas heat delivery efficiency of 55% (Harris and Maloney 1989).

Figure 5-1 compares the estimated typical annual space heat for each of the heating fuel types and energy codes. Implementation of

Environmental Benefits of Energy Efficiency — 109

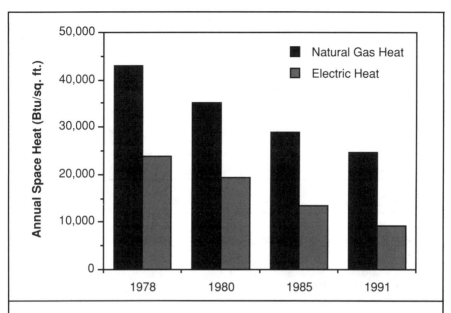

Figure 5-1. Annual Space Heat for Typical Home (1,650 sq ft) in Washington State

insulation standards has reduced typical space heating energy by approximately 60% in electrically heated homes and 43% in homes heated with natural gas. The wide disparity in performance levels between electric- and natural gas–heated homes reflects the difference in consumer cost for these two fuels. Natural gas costs only about 30% as much as electricity per Btu (53% if adjusted for end-use efficiency). Consequently, the insulation levels required in natural gas–heated homes are less stringent than those in electrically heated homes.

Recent analysis of space heat end-use-monitored data indicates that, for electrically heated homes, peak load or peak system capacity savings may be as much as five to eight times annual average capacity savings (Foley 1989). These figures are based on preliminary analysis of hourly end-use data and the observation that electrical system load factors for residential space heating are typically in the neighborhood of 20% (see Gillman *et al.* 1990 for a more detailed analysis of this issue).

The additional construction costs and energy savings attributable to the Washington State Energy Codes are summarized in Table 5-3.

110 — Chapter Five

Table 5-3. Costs and Annual Energy Savings for Washington State Residential Energy Codes

| | Western Washington | | | | Eastern Washington | | | |
| | Savings | | Cost | | Savings | | Cost | |
Code	Elec. (kWh/yr)	Gas (therms/yr)	Elec.	Gas	Elec. (kWh/yr)	Gas (therms/yr)	Elec.	Gas
1980	1,980	123	$973	$973	3,052	198	$973	$973
1986	4,785	225	$2491	$1864	6,995	330	$2573	$1864
1991	6,715	293	$3662	$2326	10,031	483	$4503	$2920

Note: Based on average 1,650 sq ft home. Base case equals 1978 WSEC. Costs and savings are cumulative from 1978 base case.

Figures are based on a typical 1,650-square-foot home. Costs are stated in 1989 dollars. Both costs and energy savings are cumulative from the 1978 code base case.

Estimation of Greenhouse-Gas Reductions

The energy savings achieved by the energy codes displace greenhouse-gas emissions of both carbon dioxide (CO_2) and methane (CH_4). These two gases are considered to contribute about 80% of the human-caused warming potential due to atmospheric radiative absorption (Lashof and Ahuja 1989).

Reduction of gas emissions due to energy savings in natural gas–heated homes assumes 118 pounds of CO_2 per million Btu of natural gas and methane distribution losses of 0.5% of volume (the lower end of the 0.5–1.6% range for lost/unaccounted-for gas reported by natural gas utilities in Washington). Current estimates indicate that methane is approximately 10 times (on a weight basis) more potent than CO_2 for greenhouse warming (Lashof and Ahuja 1989). Assuming 1,000 cu ft/MMBtu and a weight of 0.0424 lb/cu ft of natural gas, the methane leaks contribute another 2.1 lb/MMBtu of CO_2 equivalent.

Reduction of greenhouse-gas emissions due to energy savings in electrically heated homes is more difficult to estimate. The calculation depends on the fuel mix used to generate electricity. Electricity savings attributable to improved efficiency in new residential loads can be considered to displace the kind of electricity generation that is growing the

Environmental Benefits of Energy Efficiency — 111

fastest to serve increased electricity loads (that is, the marginal resource).[1]

As shown in Figure 5-2, in 1980, 72% of generation was hydro-electric. By 1988, this percentage had declined to 56%. Over the same period, the proportion of coal-fired generation increased from 22% to 37%. Because the share of regional generation provided by coal is growing, this analysis assumes that efficiency improvements are saving coal-fired electricity.

The large coal-fired plants in Washington, Oregon, Montana, Wyoming, and Nevada have an average heat rate of 10,685 Btu/kWh (PNUCC 1987). Considering this heat rate and an average CO_2 emission rate of 204 lb/MMBtu of coal (Marland and Rotty 1983), the plants produce an average of 2.18 lb/kWh of CO_2. Including the effect of electricity transmission losses averaging 8.9% (Edison 1987), each kilowatt-hour of end-use electricity saved reduces carbon dioxide emissions by 2.38 lb.

Current electricity planning in the Northwest appears to be leaning toward reliance on combined-cycle, gas-fired turbines for new electricity supplies. Consequently, the marginal electricity resource in the future may be a combined-cycle, gas-fired facility operating at a higher efficiency than the coal-fired resource assumed in this analysis. Carbon dioxide emissions from these facilities would be approximately halved and methane emissions increased approximately tenfold (Rosen 1990). These new generating facilities are not likely to be constructed before the late 1990s. Because of the uncertainty in this timing and the excess coal-fired capacity currently available in the PNW (the Boardman Plant is not being utilized to capacity), this analysis assumes that coal-fired electricity will continue to make up the margin over the period evaluated.

Table 5-4 presents the CO_2, methane, and total CO_2-equivalent emissions per million Btu of natural gas and electricity. These figures reflect only combustion for direct use or electricity generation; they do not reflect contributions of greenhouse gas over the fuels' life cycles,

---

[1] Washington is part of an electricity distribution grid that encompasses a large portion of the northwestern United States and Canada. The full range of electricity generation that supplies this grid must be considered in the identification of this marginal resource. Fully 30% of the electricity that serves the Pacific Northwest region comes from outside the region (as defined by the Pacific Northwest Electric Power Planning and Conservation Act, 1980) (PNUCC 1987). To reflect this fact, Figure 5-2 plots each generation source as a proportion of total generation mix throughout the 1980s in the states of Washington, Oregon, Idaho, Montana, Wyoming, and Nevada (EIA 1982–1988). The majority of electricity generation sources owned by Washington utilities is confined to these six states.

112 — Chapter Five

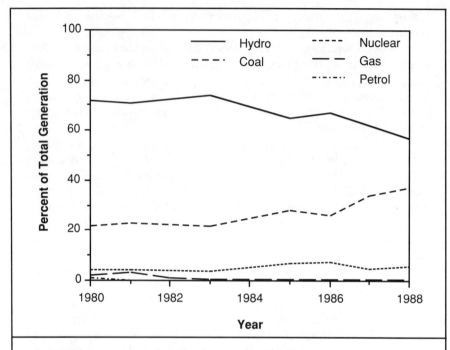

Figure 5-2. Northwestern U.S. Electricity Generation Sources as a Percent of Total

Note: The region covered includes the states of Washington, Oregon, Idaho, Montana, Wyoming, and Nevada.

which would include coal mining and natural gas extraction. Based on these assumptions, the annual greenhouse-gas savings for each of the building code levels are presented in Table 5-5. Again, these figures are reported as cumulative from the 1978 code base case.

Economics of Energy Code and Cost of Environmental Benefits Based on Individual Homes

Based on the cumulative energy savings and costs for all three building codes, the economics for homes built to the 1991 code are presented in Tables 5-6 and 5-7 for homes heated with electricity and natural gas, respectively. The net present value (Net PV) of energy savings to the home buyer assumes an economic life of 30 years and a nominal dis-

Table 5-4. Greenhouse-Gas Emission Rates per MMBTU of End-Use Electricity and Natural Gas (lb)

	CO_2	CH_4	CO_2 Equiv.
Natural Gas	118	2.1	120.1
Electricity	698	.005	698

Table 5-5. Greenhouse-Gas Savings for Typical-Sized Single-Family Home in Washington State (1,650 sq ft)

	Western		Eastern	
	Electric	Nat. Gas	Electric	Nat. Gas
1980	4,716	1,486	7,272	2,378
1986	11,398	2,715	16,666	3,963
1991	15,996	3,527	23,898	5,805

Note: Annual pounds of CO_2 equivalent, cumulative from 1978 base codes (lb/yr).

Table 5-6. Costs and Benefits of Building Energy Code for Typical Electric Resistance-Heated (1989$) Home[a]

	Savings (kWh/yr)	Cost	Net PV	B/C[b]	$/ton CO_2[c]	$/ton CO_2[d]
West Wash.	6,715	$3,662	$1,713	1.47	$6.54	($3.05)
East Wash.	10,031	$4,503	$3,526	1.78	$5.38	($4.21)

Notes:
[a] Based on 1991 building code and 1978 code base case.
[b] Benefit-cost ratio.
[c] Cost of lifetime displaced CO_2 before consideration of value of energy savings. Assumes 70-year physical life for average home.
[d] Cost of lifetime displaced CO_2 after consideration of value of energy savings. Assumes 70-year physical life for average home.

count rate equal to a 10.5% mortgage rate. Electricity prices are assumed to escalate at a real annual rate of 0.2%, and natural gas prices are assumed to escalate at a real annual rate of 1.6% (NPPC 1989).

In all cases presented in Tables 5-6 and 5-7, the benefit-cost ratio exceeds 1, indicating that the investment in the building code–required

114 — Chapter Five

Table 5-7. Costs and Benefits of Building Energy Code for Typical Natural Gas–Heated Home (1989$)[a]

	Savings (therms/yr)	Cost	Net PV	B/C[b]	$/ton CO_2[c]	$/ton CO_2[d]
West Wash.	293	$2,326	$18	1.01	$18.84	($0.14)
East Wash.	483	$2,920	$943	1.32	$14.37	($4.64)

Notes:
[a] Based on 1991 building code and 1978 code base case.
[b] Benefit-cost ratio.
[c] Cost of lifetime displaced CO_2 before consideration of value of energy savings. Assumes 70-year physical life for average home.
[d] Cost of lifetime displaced CO_2 after consideration of value of energy savings. Assumes 70-year physical life for average home.

---

insulation measures has positive value for the home buyer under the economic assumptions used. In simple terms, the measures pay for themselves.

Two costs per ton of displaced CO_2 are calculated. Both assume that new homes will last for 70 years. The first calculation ignores the value of saved energy. If the building code had been implemented for the sole purpose of saving CO_2, with no consideration of energy dollar savings, this figure would reflect the cost of CO_2 abatement. The costs range from $5.38 to $18.84 per ton. These figures compare favorably with estimates ranging from $4.95 to $33.00 per ton of CO_2 abatement (Bernow and Marron 1990; Nordhaus 1990).

The second CO_2 cost calculated does consider the value of the saved energy. Because all of the cases have benefit-cost ratios greater than unity, all of these costs are effectively negative. The CO_2 abatement is more than paid for by the value of the energy savings.

These two calculations provide a sensitivity test for the cost of CO_2 reduction achieved through the building codes. The most extreme economic assumptions might consider that consumer discount rates are so high that energy saved in the future has little or no value. Even under this case, the cost of CO_2 reduction is very comparable to other methods currently estimated or being considered in carbon tax proposals (Nordhaus 1990; Flavin 1990).

Characteristics of the Washington State Housing Stock

To estimate the cumulative impacts of Washington's residential energy codes, the construction cost, consumer benefits, and environmental

Environmental Benefits of Energy Efficiency — 115

benefits can be extrapolated to the housing stock. Currently, about 60% of the single-family housing stock and 90% of the multifamily housing stock in Washington is heated with electricity. While energy prices and the geographic availability of natural gas may influence these proportions in the future, the magnitude and even the direction of these changes are difficult to predict. This analysis assumes that the fuel mix will remain constant through 2005.

A similar assumption has been made concerning the distribution of housing starts between the eastern and western Washington climate zones. Currently, about 94% of new houses are built in western Washington. The analysis assumes this split will continue.

Housing start data for 1981 through 1989 were obtained from building permit records. Housing start projections for the period 1990 through 2005 are derived from economic forecasts done by the Bonneville Power Administration (BPA) and the Northwest Power Planning Council (NPPC) (BPA 1988). Projections are drawn from the medium-growth scenario in these forecasts.

Statewide Energy, Peak Electrical Load, and Equivalent Carbon Dioxide Savings

Slightly more than 300,000 single-family homes and apartment units were added to the state's housing stock from 1981 through 1989. Table 5-8 presents the annual housing starts, annual rate of electricity savings, estimated peak electricity capacity savings, annual rate of natural gas savings, and annual rate of CO_2 displacement.

Electricity savings from housing starts through 1989 amount to 589 GWh, enough energy to serve the needs of a city the size of Everett, Washington (population 64,170). The total cost to achieve these savings is estimated at $294 million (1989$). The present value of energy savings over the 30-year mortgage period of these homes is estimated at $462 million (benefit-cost ratio of 1.57). Taking a social discount rate of 3% real and assuming a 70-year physical life for the homes yields a levelized cost of $.017/kWh ($.041 in nominal terms if a 5% inflation rate is assumed). This amount is less than half the estimated cost of generating electricity from a new coal-fired plant (NPPC 1990).

Natural gas savings from housing starts through 1989 amount to 13.5 million therms per year, or an average of 3.7 million cu ft/day. These savings were achieved for a cost of $106 million. The present value of energy savings is estimated to be $124 million (benefit-cost ratio of 1.17). Again, taking a 3% social discount rate and considering the 70-year physical life for the structures yields a levelized real cost

116 — Chapter Five

Table 5-8. Statewide Aggregate Totals for Energy and CO_2 Savings from Residential Building Codes[a]

Year	Housing (units)	Elec. (10^6 kWh)	Peak MW (MW)[b]	Gas (10^6 thms)	CO_2 (2,000 lbs)
1981	23,853	27.6	15.7	.8	37,888
1982	17,586	47.6	27.2	1.5	65,591
1983	27,278	78.8	45.0	2.5	108,885
1984	30,944	113.8	64.9	3.5	156,597
1985	35,475	152.5	87.0	4.5	208,721
1986	36,428	248.2	141.6	6.5	335,321
1987	38,341	348.1	198.7	8.6	465,907
1988	44,553	463.4	264.5	10.9	616,758
1989	47,607	588.7	336.0	13.5	782,046
Forecast through 2005		2526.0	1442.0	57.9	3,353,889

Notes:
[a] Cumulative Annual Rates for Homes Built 1981–1989.
[b] Estimated at 20 percent load factor.

of \$.27/therm (1989\$). This cost is significantly less than the \$.41–.44/therm estimated as a long-term avoided gas cost by two major gas utilities serving Washington State and Oregon (Cascade 1990; Northwest NG 1990).

Carbon dioxide displacement reached 782,000 tons per year for the cohort of homes built between 1981 and 1989 (Table 5-8). If these tons are valued at the \$5.00/ton figure discussed by Nordhaus (1990), the value of the greenhouse-gas benefits from the homes built during the decade of the 1980s is \$3.9 million/yr. Cumulative tons of CO_2 displaced reached 2.8 million tons through 1989, worth \$14 million at \$5.00 per ton.

These figures only reflect the energy savings and CO_2 benefits through 1989. These homes will continue to save energy and CO_2 throughout their lifetimes, and additional savings will accrue as new homes are built under the 1991 building code. Table 5-8 also includes an estimate of energy savings and CO_2 benefits for housing starts forecasted to occur through 2005. By the year 2005, 822,900 homes will have been added to the state's housing stock, 2,526 GWh of electricity will have been saved, nearly 58 million therms/yr of natural gas will have been saved (an average of 15.8 million cu ft/day), and CO_2

emissions will have been reduced by 3.3 million tons/yr. The annual CO_2 savings amount to about 3.9% of the state's total energy-related CO_2 emissions (86.7 million tons in 1988) and 14.0% of energy-related emissions from residential and commercial buildings (WSEO 1990).

Conclusions

Washington state's residential energy code not only has met its original objective of achieving energy efficiency that was cost-effective to the new home buyer, the state, and the region but has also netted substantial environmental benefits. Because the energy-efficiency measures required by the code are cost-effective to the home buyer and state, the environmental benefits are obtained at no net cost. In simple terms, they are paid for by the value of the energy savings.

By the year 2005, CO_2 savings will reach 3.3 million tons/yr, an environmental bonus from energy efficiency worth an estimated $16.5 million/yr (at $5.00/ton). Similar policies aimed at achieving energy efficiency throughout the state and U.S. economies can reasonably be expected to achieve similar economic and environmental benefits.

Appendix: Accuracy of the SUNDAY Model

Energy savings have been both directly measured and estimated through use of computer models. As a part of the RSDP and later demonstration project, the Residential Construction Demonstration Project (RCDP), end uses in some 350 electrically heated homes were monitored. The results of the space heat monitoring for these homes were compared with predictions from a simple thermal simulation model, SUNDAY (Ecotope 1984). An analysis of the agreement between model predictions and actual use indicated that, on average, the model's predictions agree to within 7–10% of actual space heat use (Byers and Palmiter 1988; Downey 1989). In addition, UW compared SUNDAY predictions with those of other models (including DOE-2 and CALPAS) and with the monitored space heat use of four extensively instrumented test homes. In this test, SUNDAY actually underpredicted the energy savings from a package of insulation measures by 16% (Emery *et al.* 1989). As a consequence of these two tests, the Washington State Energy Office (WSEO) has concluded that the computer model SUNDAY is sufficiently accurate to estimate the typical

118 — Chapter Five

savings to be expected, on average, from packages of insulation measures.

References

Bernow, S., and D. Marron 1990. *Valuation of Environmental Externalities for Energy Planning and Operations.* Boston: Tellus Institute.

Bonneville Power Administration 1988. *Economic Forecast for the Pacific Northwest.* DOE/BP-1138. Portland: Bonneville Power Administration.

BPA. See Bonneville Power Administration.

Byers, R. 1989. *Cost-effectiveness of Residential Building Energy Codes. Results of the University of Washington Component Test and the Residential Standards Demonstration Program.* WAOENG-89–66. Olympia: Washington State Energy Office.

Byers, R., and L. Palmiter 1988. "Analysis of Agreement between Predicted and Monitored Annual Space Heat Use for a Large Sample of Homes in the Pacific Northwest." In *Proceedings from the ACEEE 1988 Summer Study on Energy Efficiency in Buildings.* Vol. 10, *Performance Measurement and Analysis.* Washington, D.C.: American Council for an Energy-Efficient Economy.

Cascade Natural Gas Corporation 1990. *Least Cost Plan.* Second draft.

Department of Energy 1989. *Energy and Climate Change.* Draft report of the DOE Multi-Laboratory Climate Change Committee. UCRL-102476. Washington: U.S. Department of Energy.

Downey, P. 1989. *An Analysis of Predicted Versus Monitored Space Heat Energy Use in 83 Homes.* WAOENG 89–67. Olympia: Washington State Energy Office.

Ecotope Incorporated 1984. *SUNDAY 2.0.* Seattle: Ecotope Incorporated.

Edison Electric Institute 1987. *Statistical Yearbook of the Electrical Utility Industry.* Number 56. Washington, D.C.: Edison Electric Institute.

Emery, A. *et al.* 1989. *Dynamic Response of Building Components in Residential Homes.* Final simulation report. Olympia: Washington State Energy Office.

Energy Information Administration 1982–1988. *Electric Power Annual.* DOE/EIA-0348(82–88). Washington, D.C.: U.S. Department of Energy.

Flavin, C. 1990. "Slowing Global Warming." Ch. 2 in *State of the*

Environmental Benefits of Energy Efficiency — 119

World. A Worldwatch Institute Report on Progress Toward a Sustainable Society. New York: W. W. Norton.

Foley, T. 1989. "MCS Performance During February 1989 Cold Snap." Memo to Northwest Power Planning Council, Nov. 1.

Gillman, R., R. Sands, and R. Lucas 1990. "Observations on Residential and Commercial Load Shapes During a Cold Snap." In *Proceedings of the ACEEE 1990 Summer Study on Energy Efficiency in Buildings.* Vol. 10, *Performance Measurement and Analysis.* Washington, D.C.: American Council for an Energy-Efficient Economy.

Harris, J., and J. Maloney 1989. "A Simplified Energy Analysis Model Incorporating Duct Loss Impacts on Heating System Efficiency." Unpublished. Available from the authors at Northwest Power Planning Council, Portland, Ore.

Krause, F. 1989. *Energy Policy in the Greenhouse. Warming Fate to Warming Limit: Benchmarks for a Global Climate Convention.* Vol. 1. El Cerrito, Calif.: International Project for Suitable Energy Paths.

Lashof, D., and D. Ahuja 1989. "Relative Global Warming Potentials of Greenhouse Gas Emissions." *Nature* 344 (April 5): 529–31.

Marland, G., and R. Rotty 1983. *CO_2 Emissions from Fossil Fuels: A Procedure for Estimation and Results for 1950–1982.* DOE/NBB-0036. Oak Ridge, Tenn.: Oak Ridge Associated Universities.

Nordhaus, W. 1990. "Economic Policy in the Face of Global Warming." Presented at Energy and the Environment in the 21st Century, Massachusetts Institute of Technology, Cambridge, March 26.

Northwest Natural Gas Company 1990. *Least Cost Plan.* Part 4. Portland: Northwest Natural Gas Company.

Northwest NG. See Northwest Natural Gas Company.

Northwest Power Planning Council 1989. *1989 Supplement to the 1986 Northwest Conservation and Electric Power Plan.* Vol. 2. Portland: Northwest Power Planning Council.

———— 1990. "New Resources: Supply Curves and Environmental Effects." Table 6, Staff Issue Paper 90-1. Portland: Northwest Power Planning Council.

NPPC. See Northwest Power Planning Council.

Ossinger, T., M. Morrison, J. Ristoff, and R. Bourg 1989. *University of Washington Component Test Study. Cost and Economic Analysis.* Olympia: Washington State Energy Office.

Pacific Northwest Utilities Conference Committee 1987. *Thermal*

120 — Chapter Five

Resources Data Base. Portland: Pacific Northwest Utilities Conference Committee.

PNUCC. See Pacific Northwest Utilities Conference Committee.

Rosen, R. 1990. *The Role of Hydro-Quebec Power in a Least-Cost Energy Resource Plan for Vermont.* Tellus Institute Study 89-078. Boston: Tellus Institute, Boston, Mass., January 1990.

Tangora, P., R. Byers, J. Douglass, and B. Whitney 1986. *Energy Conservation in New Residential Construction. Cost Analysis from the Thermabilt Program in Washington State.* WAOENG-86-17. Olympia: Washington State Energy Office.

Washington State Energy Office 1990. *Inventory of Washington State Greenhouse Gas Sources.* Unpublished draft. Olympia: Washington State Energy Office.

WSEO. See Washington State Energy Office.

Richard Byers serves as the senior energy policy analyst for the Washington State Energy Office. Before assuming this position, he served as the lead research analyst evaluating residential and institutional building energy usage data for WSEO. He received a B.S. in Natural Resources from the University of Michigan.

Chapter 6

The CO_2 Diet for a Greenhouse Planet: Assessing Individual Actions for Slowing Global Warming

John M. DeCicco, *American Council for an Energy-Efficient Economy (formerly with National Audubon Society)*

James H. Cook, Dorene Bolze, and Jan Beyea, *National Audubon Society*

Introduction

Because of uncontrolled population growth and a short-sighted choice of technologies, human activity is emitting enormous quantities of greenhouse gases. Reducing emissions of these gases that can disrupt the Earth's climate will require action by individuals as well as by governments and industries. Most energy use currently results in carbon dioxide (CO_2) emissions because fossil fuels dominate our energy supply. Increasing energy efficiency can therefore lower CO_2 emissions. Reducing emissions of other greenhouse gases, such as halocarbons, is also necessary. Following such a "low-CO_2 diet" will require prudent consumption and lifestyle choices by individuals. This chapter focuses on activities related to greenhouse-gas emissions in the United States over which individuals have the most control.

We present a method by which individuals or households can quantify their direct greenhouse-gas emissions. This CO_2 "calorie count" can then be used to devise a *CO_2 Diet*, a personal plan for reducing emissions. The *CO_2 Diet* goal is to cut greenhouse emissions by 20% within ten years. We examine the significance of such individual action by working out the *CO_2 Diet* for a sample household.

122 — Chapter Six

Finally, we discuss the policies needed to complement individual CO_2 *Diet* efforts so that the United States as a whole can achieve targeted reductions in its emissions of greenhouse gases.

Public awareness of the greenhouse effect has grown in recent years, but the connection between greenhouse emissions and energy consumption is not always understood (see Kempton, chapter 2). The CO_2 *Diet* concept evolved from our recognition of the need to better inform the public about the connection between their lifestyles, particularly energy use, and greenhouse-gas emissions. This paper is based on a National Audubon Society report (DeCicco *et al.* 1990), which was written for a lay audience and to which readers are referred for further background on what is covered here.

Our Greenhouse Planet

The greenhouse effect and the need for a substantial curtailment of fossil-fuel use are well documented (see, for example, Schneider 1989a, 1989b; Krause *et al.* 1989). Greenhouse gases—carbon dioxide is the most important one—enhance the heat-trapping ability of the Earth's atmosphere. In Figure 6-1, the upper curve shows past and projected atmospheric CO_2 levels as a percent increase over the preindustrial level, which was 290 parts per million (ppm), so that 100% implies a doubled concentration of 580 ppm. If the present trend continues, this doubling is likely to occur before the end of the next century. Such a rapid change of atmospheric composition is unprecedented in the history of the Earth's climate.

The man-made changes are almost certain to lead to an overall increase in the Earth's average surface temperature; however, the particular timing and geographic effects of the attendant climatic disruptions are less certain and are likely to be very uneven (IPCC 1990). The tropics may be relatively unaffected, whereas greatly rising temperatures in polar regions could melt ice caps and devastate ecosystems in higher latitudes. Rising sea levels could flood coastal areas on most continents. Changed patterns of rainfall could inundate arid regions; areas now important for growing food could become too dry for agriculture (Schneider 1989a). Plant and animal populations throughout many regions will be thrown out of balance; some species may become extinct because of the human-induced climate change (Lester and Myers 1989).

By examining worldwide sources of greenhouse-gas emissions and modeling their buildup in the atmosphere, Krause *et al.* developed an international emissions budget for stabilizing the atmospheric CO_2 concentration at about 400 ppm. This budget specifies the cumulative

The CO₂ Diet for a Greenhouse Planet — 123

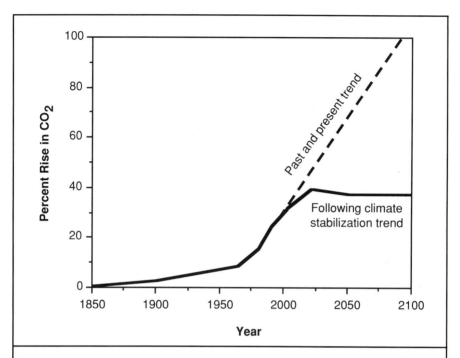

**Figure 6-1. Atmospheric CO₂ Levels, Present Trend vs. a Climate-Stabilizing Scenario, Shown as Percentage Rise above Preindustrial Level
Sources: Schneider 1989b; Krause et al. 1989**

total amount of fossil-source CO_2 emissions that cannot be exceeded without risking climate disruption. A feasible scenario entails a global budget of one trillion (10^{12}) tons of fossil-source CO_2 emissions through year 2100 as well as a sharp curtailment in the emissions of halocarbons and other greenhouse gases. The CO_2 levels involved in following this scenario are shown as the lower curve in Figure 6-1.

Meeting the greenhouse emissions budget will require a firm belt-tightening on fossil-fuel consumption by nations like the United States. The initial goal of the *CO₂ Diet* would be to cut global warming calories (greenhouse-gas emissions) by 20% in ten years or by an average annual reduction of 2% of one's current emissions. This reduction rate is consistent with the atmospheric CO_2 level shown in the lower curve in Figure 6-1. Note that cutting 20% in ten years is just a first step. Krause *et al.* suggest that reductions on the order of 75% will be required by the middle of the next century if the United States is to

124 — Chapter Six

achieve an economy that is sustainable from a global climatic perspective.

It is essential to start reducing greenhouse emissions immediately. Because CO_2 remains in the atmosphere for decades once it is emitted, it will be more difficult to achieve acceptably stable atmospheric concentrations with cuts in the future than with comparably sized cuts in the present.

Greenhouse Gases

The most important greenhouse gas is CO_2, which contributes to well over half of the greenhouse warming effect (Lashof and Ahuja 1990). Most of the human-caused CO_2 buildup in the atmosphere has come from the use of fossil fuels. Large amounts of CO_2 are also released when forests are burned (Houghton 1990).

Many halocarbons—compounds, mostly man-made, such as chlorofluorocarbons (CFCs) and halons—also have a significant greenhouse impact (Fisher *et al.* 1990; Lashof and Ahuja 1990). Although halocarbons are released in much smaller quantities than CO_2, the global warming impact of a halocarbon molecule can be thousands of times that of a CO_2 molecule. For the *CO_2 Diet* we count halocarbon emissions in pounds of CO_2 equivalence—that is, the amount of CO_2 that would produce the same global warming effect, as defined by Cook and Beyea (1990).

Other gases also contribute to global warming, notably methane (CH_4) and nitrous oxide (N_2O). However, because of uncertainties about the impact of human activities on these compounds' natural atmospheric cycles, we do not include methane and nitrous oxide emissions in the *CO_2 Diet*.

Man-Made Emissions

As a context for the *CO_2 Diet*, it is instructive to examine emissions globally and for the United States as a whole, which are given in Table 6-1. Although the United States has only 5% of the world's population (250 million out of 5.3 billion, Keyfitz 1989), its CO_2 emissions are about 20% of the world's total. This percentage amounts to 44,000 pounds of CO_2 per person each year. Breaking down U.S. CO_2 emissions by economic sector, 35% are from industry, 33% are from transportation, 18% are from the residential sector, and 14% are from the commercial sector.

The next largest contribution to the greenhouse effect is from halocarbons (Table 6-2). The annual global warming impact is equivalent to another 1.4 billion tons of CO_2 for the United States, or 11,000 pounds per capita. This amount brings the total U.S. annual emissions

The CO_2 Diet for a Greenhouse Planet — 125

Table 6-1. CO_2 Emissions Worldwide and in the United States During 1986

Source of Emissions	Annual CO_2 emissions	
	(10^9 tons)	**(lbs per capita)**
Global human activity[a]	30	12,000
United States[b]	5.5	44,000
U.S. Breakdown by economic sector:		
Industrial	1.9	15,400 (35%)
Transportation	1.8	14,500 (33%)
Residential	1.0	7,900 (18%)
Commercial	0.7	6,200 (14%)

Notes:
[a] Houghton 1990; conversion by 3.67 tons CO_2 per ton of carbon; using a world population estimate of 5 billion (Keyfitz 1989).
[b] Using energy statistics from EIA 1986 and emission rates of 26, 21, and 14 kg CO_2 per GJ of coal, oil, and natural gas, respectively, from SRI 1979.

of greenhouse gases to nearly 7 billion tons, or 55,000 pounds per capita, of CO_2 equivalent.

Individuals have the most control over emissions from residences and personal transportation. Summing all residential sector CO_2 emissions with the automobile, air, bus, and passenger rail portions of transportation-sector emissions and the mobile air-conditioning and residential refrigeration portions of halocarbon emissions (DeCicco *et al.* 1990) yields an estimate of 18,000 pounds of CO_2 equivalence per capita for the share of emissions over which individuals have direct control. This figure amounts to about one-third of the total emissions in the U.S. economy. Such aggregate statistics, however, do not reveal one's own emissions. For individuals to determine their contribution to—and their potential for reducing—greenhouse-gas emissions, an individualized analysis must be performed.

What Do You Emit?

We developed a worksheet, shown on the following page, for assessing greenhouse-gas emissions by individuals. The CO_2 *Diet* worksheet provides a personal greenhouse emissions audit, much like a household energy audit. In fact, conducting an actual energy audit and then carrying out its energy conservation recommendations is one of the best ways for individuals to reduce their greenhouse-gas emissions.

126 — Chapter Six

Table 6-2. Halocarbon Emissions in the United States			
	Annual emissions[a] Halocarbons	Global warming impact[b] CO_2 equivalence	
Source	(tons)	(lbs per capita)	(Fraction)
Mobile air conditioners	60,000	2,640	24%
Solvent cleaning	75,000	1,810	16%
Rigid plastic foam insulation	62,000	1,760	16%
Retail refrigeration	18,000	730	7%
Flexible plastic foam	16,000	370	3%
Air-conditioning	11,000	330	3%
Residential refrigeration	2,600	110	1%
Halon fire extinguishers	1,600	40	< 1%
Other	87,000	3,200	29%
TOTAL (rounded)	330,000	11,000	100%

Notes:
[a] EPA 1987, excluding methyl chloroform.
[b] Based on CO_2 equivalence conversions by Cook and Beyea 1990.

Instead of reporting the results in Btu or kWh, however, we use pounds of CO_2 equivalence. We have incorporated not only CO_2 emissions from residential activities but also greenhouse-related emissions of halocarbons and CO_2 from personal transportation.

There are three steps to filling out the *CO₂ Diet* worksheet. First, determine the annual consumption of fuels and other products that contribute to greenhouse-gas emissions and enter the estimates into the second column ("your use"). Then multiply the annual consumption by the CO_2 emissions factor given in the third column.[1] This calculation yields emissions estimates for each product or activity, which are entered into the fourth column ("annual emissions"). Finally, add up the numbers in the fourth column to determine the total annual greenhouse emissions, expressed as pounds of CO_2 equivalence. More specific guidelines for filling out the worksheet follow.

Residential Utilities. Households that save their utility bills can determine their annual consumption by tallying the amounts shown on

---

[1] Emissions are based on SRI (1979) for fossil fuels and the average U.S. generation mix (EIA 1987) for electricity. See DeCicco *et al.* 1990 for details.

The CO_2 Diet for a Greenhouse Planet — 127

Worksheet for the *CO_2 Diet*

Consumption or activity	Your use (units per year)	CO_2 factor (lbs CO_2/unit)	Annual emissions (lbs CO_2 equiv.)
RESIDENTIAL UTILITIES			
Electricity	_____ kWh	1.5 lbs/kWh	_____
Oil	_____ gal	22 lbs/gal	_____
Natural gas	_____ therms	11 lbs/therm	_____
Propane or bottled gas	_____ gal	13 lbs/gal	_____
TRANSPORTATION			
Automobile fuel use	_____ gal	22 lbs/gal	_____
Other motor fuel use	_____ gal	22 lbs/gal	_____
Air travel	_____ miles	0.9 lbs/mile	_____
Bus, urban	_____ miles	0.7 lbs/mile	_____
Bus, intercity	_____ miles	0.2 lbs/mile	_____
Railway or subway	_____ miles	0.6 lbs/mile	_____
Taxi or limousine	_____ miles	1.5 lbs/mile	_____
HOUSEHOLD WASTE			
Trash (anything discarded)	_____ lbs	3 lbs/lb	_____
Recycled items	_____ lbs	2 lbs/lb	_____
HALOCARBON PRODUCTS			
Refrigerators and freezers	_____ (number)	830 lbs each	_____
Automobile air conditioners	_____ (number)	4,800 lbs each	_____
Other halocarbon products (see Table 6-3 for equivalences)			_____
TOTAL ANNUAL GREENHOUSE-GAS EMISSIONS (pounds of CO_2 equivalent)			_____

the bills over a year. An ideal way to determine heating and cooling energy use is by using a weather-normalized annual consumption index, if one is provided on a utility statement. If a yearlong record is not available, you can divide the estimated dollar amount of annual payments by the unit price, which can be found on a recent bill or by calling the utility company. Tenants of a master-metered apartment building can ask their management for the same type of information.

128 — Chapter Six

Even though consumption may vary among apartments, a rough guess of your share can be obtained by dividing the building's total utility consumption by the number of apartments.

Transportation. For each motorized vehicle (car, truck, motorcycle, boat, snowmobile, and so on), divide the number of miles driven by the average miles per gallon (mpg) of the vehicle to estimate the number of gallons of motor fuel used in a year. (Treat diesel like gasoline.) Multiply by the CO_2 factor of 22 pounds per gallon to get the total yearly emissions. For a car that gets 20 mpg driven 10,000 miles a year, for example, the annual gasoline use is 500 gallons and the CO_2 emissions are 11,000 pounds. Individuals who often travel in a car with others can reduce their miles driven according to the average number of people in the car. For example, people who drive 10,000 miles per year with one other person on average should figure their emissions on the basis of 5,000 miles. One should also include distance traveled in rental cars.

Also make rough estimates of travel by air and other means of transportation.[2] Statements from frequent-flyer programs can be used to estimate annual air travel (deducting bonus miles that do not reflect actual travel). For public transportation, tally the annual mileage by multiplying the number of trips made by the average distance traveled each trip. Consider, for example, a commute involving a 12-mile bus ride followed by a 3-mile metro rail ride, twice a day on 240 workdays a year, for a total of 480 trips per year. The distances would then total $12 \times 480 = 5,760$ urban bus miles and $3 \times 480 = 1,440$ rail miles. For taxi rides, one may have to make very rough estimates of the distances involved. For shared trips, divide the per-trip mileage by the number of passengers.

Household Waste. Estimating the emissions associated with household wastes can be tricky, since we are not be used to thinking of such things in terms of their weight. A simple approach is to use representative CO_2 emissions factors for household garbage. Using information on the typical composition of municipal wastes (Franklin Assoc. 1988) and embodied energy (Gaines 1981), we estimated average CO_2 emissions factors for discarded (unrecycled) and recycled wastes.[3] Our calculations suggest an average CO_2 emissions reduction of one-third for recycling.

---

[2] Emissions factors for air and public transportation are based on energy statistics from Davis *et al.* 1989.

[3] The energy saved by recycling varies among different materials; our factors do not account for halocarbons. See DeCicco *et al.* 1990 for details.

The CO_2 Diet for a Greenhouse Planet — 129

For trash that is squashed down by hand, a full 30-gallon garbage bag (large size) holds about 40 pounds, a 10-gallon bag about 13 pounds, and a 5-gallon bag (about the same as a standard paper grocery bag) holds about 7 pounds. Estimate how many bags you generate each week and multiply by 52 weeks per year. For newspaper, a stack one foot high weighs about 40 pounds, and with daily delivery, the stack can reach 6 feet in a year, amounting to 240 pounds (this may double for a major city newspaper). Magazines and other paper products are usually heavier, at about 60 pounds per cubic foot, and can be tallied on a per-issue basis. A standard magazine weighs between 6 and 14 ounces. A lightweight magazine (for example, *Time* or *Newsweek*) averages about 8 ounces, so a weekly subscription (52 issues per year) adds up to 26 pounds. Heavier magazines, like *Scientific American* or *National Geographic*, weigh about 14 ounces apiece, so a monthly subscription to one of these works out to about 10 pounds per year.

Halocarbons. Estimating the primary source of halocarbon emissions is easy: simply note the number of automobile air conditioners. They emit an average of 4,800 pounds per year regardless of whether or not they are used. The emissions will be lower if the air conditioner is serviced at a shop that recycles CFCs.[4] Also note the number of separate refrigerators and freezers (a combined refrigerator/freezer counts as one). For other common halocarbon products, emissions can be estimated from the CO_2 equivalence factors provided in Table 6-3. In the case of other major appliances (such as room or central air conditioners), obtain an annual emission figure by dividing the total greenhouse-gas content (in CO_2 equivalence) by the number of years the item is kept.

Interpreting the Greenhouse Calorie Total

After filling in all of the entries that apply, a household's total annual CO_2 emissions are determined by summing the last column of the worksheet. This sum is the greenhouse calorie consumption, which needs to be reduced by personal action in residences and for transportation.

Cutting calories is an appropriate analogy for the need to reduce greenhouse-gas emissions. The calorie used to measure nutritional intake is, in fact, a measure of the energy value of the food. The lifestyle-related energy consumption that is such a large source of CO_2

---

[4] CFC recycling by air-conditioning and refrigeration repair shops may soon be mandatory as new provisions of the Clean Air Act are implemented (EPA 1990).

130 — Chapter Six

Table 6-3. Halocarbons in Consumer Items

Item	Halocarbon content	Pounds of CO_2 equivalence
Air conditioners		
Automobile	3.3 lbs CFC-12	17,500
Central	4.4 lbs CFC-22	6,200
Room	0.9 lbs CFC-22	1,200
Refrigerators or freezers		
in insulation	2.2 lbs CFC-11	8,300
in cooling system	0.4 lbs CFC-12	
Portable halon fire extinguisher	1.6 lbs halon-1211	2,700
Foam insulation ($4' \times 8' \times 1''$ sheet)	0.9 lbs CFC-11	2,400
CFCs in cans:		
"air" horn	1.0 lbs CFC-12	5,300
dust remover	0.8 lbs CFC-12	4,000
"air" brush cartridge	0.7 lbs CFC-12	3,500
foam party streamers	0.3 lbs CFC-12	1,800
electronic parts cleaner	0.3 lbs CFC-113	1,000

Source: Cook and Beyea 1990. Note that the halocarbons in some items may not be completely released for some time, so that annual release rates must be estimated according to lifetimes or leakage rates by item (see text).

emissions could also be measured in calories. A calorie (the so-called large or kilogram calorie) is equivalent to about 4 British thermal units (Btu) or 4.2 kilojoules (kJ). (In fact, some European dietary books list food values in kilojoules as well as calories.) A 2,000-calorie-per-day diet is equivalent to about 640 pounds of CO_2 per year.[5] However, because this amount is roughly equivalent to the amount of CO_2 removed from the atmosphere by plants at the base of the food chain, the CO_2 *Diet* need not consider emissions resulting from food consumed.

Mechanically and chemically intensive agriculture, processing, transportation, and marketing of food do, however, result in greenhouse-gas emissions. These activities result in an annual release of

_____

[5] Based on plant biomass, assuming 50% carbon by dry mass and using a woodlike energy content of 8,333 Btu/lb (EIA 1987).

The CO_2 Diet for a Greenhouse Planet — 131

approximately 330 million tons of CO_2 in the United States, or 2,700 pounds per capita.[6] Note that this amount is about four times the CO_2 directly associated with food consumed. We do not have sufficient data to estimate the food-related halocarbon, methane, or nitrous oxide emissions. Since these indirect food-related emissions are difficult to quantify and fall under the commercial and industrial sectors anyway, we do not include them in the CO_2 Diet.

For the residential and transportation-related emissions over which individuals do have control, the U.S. per capita average is, as noted earlier, about 18,000 pounds of CO_2 equivalent. Using this figure, an average four-person household would be responsible for 72,000 pounds per year. However, any particular household's emissions are likely to be quite a bit above or below the average depending on the region and other factors over which the household members may have little control. The best way to approach the issue is not by comparing your own consumption to the U.S. average but rather by using your current emissions, computed using the CO_2 Diet worksheet, as a basis for setting emissions reduction goals. Unless you already lead a particularly CO_2-lean lifestyle—for example, by living in a superinsulated house, making minimal use of motorized transportation, and relying on a high-mpg car when needed, you should be able to find plenty of fat among the calories listed in your worksheet.

Reducing Greenhouse Emissions

All of the residential energy conservation measures and efficient technologies developed over the past two decades apply to the CO_2 Diet. An appropriate substitution of energy sources that do not emit greenhouse gases (solar sources, for example) can also help. The best energy conservation measures for an individual or household depend very much on circumstances. One thing learned with certainty from energy conservation research is that the largest energy savings are reliably obtained only when the measures undertaken are carefully targeted to the inefficiencies particular to each end use. The same is true for reducing greenhouse-gas emissions, which is why a quantitative assessment like the CO_2 Diet worksheet is a good place to start.

Since the ways to reduce residential energy consumption are too numerous to cover here, the best we can do is cite some common examples, show the expected CO_2 reductions, and then point readers to the literature on energy efficiency. We also give examples of the

---

[6] Based on apportioning the values in Table 6-1 with statistics from Bureau of the Census 1990, Davis *et al.* 1989, and EIA 1985.

132 — Chapter Six

CO_2-equivalent emissions cuts from reducing or eliminating certain halocarbon uses. For measures not mentioned here, one can apply the emissions factors given in the *CO_2 Diet* worksheet to translate energy savings directly into CO_2 reductions. For example, assuming the average U.S. electricity-generating mix, each kilowatt-hour of electricity savings implies a 1.5-pound reduction in CO_2 emissions.

CO_2 Diet Calorie Cutters

(See the appendix for further comments on each item, as identified by the letters in parentheses.)

Boiler or furnace tune-up. Each percentage point gain in combustion efficiency from a tune-up will cut CO_2 emissions by approximately the same percentage. (a)

Buy high-efficiency equipment when replacing a heating system. For example, improving an oil-fired system from 65% to 85% efficiency would save an average of 3,600 lbs/yr of CO_2. (b)

Use night setback. The effect is climate dependent, but a rule of thumb is a 1% reduction in heating-related CO_2 emissions per °F of eight-hour setback: a 10°F setback would save about 10% of heating-related emissions.

Lower thermostat heating setpoint. Effect also varies with climate; in colder areas a 2°F change would typically yield a 6% reduction in heating-related CO_2 emissions. In mild climates, the percent reduction can be much greater. (c)

Use an automatic setback thermostat. The savings are the same as for night setback, but you make an investment to avoid forgetting. If you have both heating and air-conditioning, an automatic setback of 5°F in both winter and summer can reduce CO_2 emissions by 13%/yr. (d)

Water heater insulation. With gas systems, the annual reductions can be 20 therms and 220 lbs CO_2; for electric systems, 700 kWh and 1,100 lbs CO_2. (e)

Low-flow showerhead. For electric water-heating, the saving is about 200 kWh and 300 lbs CO_2 per year. For gas water-heating, the saving amounts to about 8 therms and 80 lbs CO_2 per year. (f)

Wash clothes in cold water when possible. If instead of always using hot water, you do four out of five washes in cold water, you can save up to 350 lbs/yr CO_2 (gas water-heating), 500 lbs/yr CO_2 (oil), or 1,000 lbs/yr CO_2 (electric). (e)

The CO_2 Diet for a Greenhouse Planet — 133

Choose efficient appliances. For a guide to efficient appliances, see Wilson 1991. When shopping, check the yellow *Energy Guide* labels to be sure the model you select is one of the more efficient in its class. The electricity cost savings will almost always more than offset any purchase price premium (which might even be zero), and you will reduce your CO_2 emissions, too (see text for an example of how to estimate the emissions reduction).

Use compact fluorescent light bulbs. Replacing a 100-W incandescent bulb that is used six hours a day with a compact fluorescent bulb cuts emissions by 260 lbs/yr CO_2. The savings increases in proportion to the number of hours of use. (g)

Turn off lights and appliances when they are not needed. The CO_2 savings will vary, of course, depending on your habits and needs, but do not forget about this old-fashioned, common-sense measure!

Automobile tune-up. On average, a well-tuned car emits 1,100 lbs/yr CO_2 less than a poorly tuned car. (h)

Buy a fuel-efficient model when you need a new car. For example, replacing a 20 mpg vehicle with a 35 mpg vehicle cuts that portion of CO_2 emissions by over 40%.

Carpool. Commuting with one other person rather than alone cuts that portion of your CO_2 emissions in half; a four-person carpool emits about one-fourth the CO_2 of four solo-driven cars.

Use transportation alternatives. Persons living in urban areas should also press for improved mass transit, ridesharing, and ways to encourage walking and bicycle use. If widely implemented, measures to reduce automobile use would significantly cut CO_2 emissions and reduce urban air pollution.

Avoid nonessential uses of halocarbons. Check Table 6-3 for the CO_2-equivalent emissions for items containing halocarbons.

Plant a tree. Growing trees remove CO_2 from the air—about 26 lbs/yr for a healthy tree. The CO_2 savings can be as much as 300 lbs/yr, counting the shade and windbreak benefits of a tree planted around a house. (i)

For More Information

A comprehensive practical source for saving energy in residences is Wilson 1991, which includes listings of the most efficient appliances, including furnaces, boilers, water heaters, and air conditioners, as well

134 — Chapter Six

as other information on reducing energy use. For new-home construction and major retrofits, see Nisson and Dutt 1985. For warmer climates, a guide for energy-efficient home building and many fact sheets on appropriate conservation measures and solar water-heating are available from the Florida Solar Energy Center (300 State Road 401, Cape Canaveral, FL 32920).

Many states have offices that provide energy-efficiency information particular to their regions. Utility companies are also a source of information, and some of them may offer audits that can help you identify the best ways to save energy in your own situation.

CO_2 *Diet* Example

By way of example, the CO_2 *Diet for a Greenhouse Planet* was worked out for the household of one of the authors of this chapter. We report the results here in an anecdotal style. The household is a two-person, two-career family, with two cats, two cars, and one old truck; this family lives in a rented house in the mid-Atlantic region. The filled-out sample worksheet is shown in Table 6-4. Our household's annual emissions are 68,120 lbs of CO_2 equivalent. The CO_2 *Diet* therefore involves cutting the emissions by an average of 1,360 lbs/yr (2%), working toward a cumulative reduction by 13,600 pounds over ten years.

The actions taken, planned, and possible for reducing greenhouse gas emissions by the sample household are listed in Table 6-5. Listed first are measures that the household has implemented this year ("current actions"). The second category lists measures planned over the next ten years. Only measures that most individuals would be able to undertake at their own discretion were considered. Finally, we list the cuts that would be possible with new governmental policies to support and complement individual CO_2 dieting.

The best way to look for ways to cut calories is by examining the items having the largest emissions. In our worksheet, the largest emissions come from oil use for heating—space conditioning is typically the main energy use in residences. The boiler had not been recently serviced; it in fact needed a tune-up. The serviceman measured the efficiency before and after the tune-up, identifying a gain of 9%. We therefore reduced our space heating CO_2 emissions by at least 9%, for a cut of 1,200 lbs/yr CO_2. We already use night setback and cannot reduce the thermostat setting any further without sacrificing comfort. We did not include some common energy conservation measures, such as low-flow showerheads, faucet aerators, water heater insulation, and cold-water washing in our plan because we were already doing them to save energy.

The CO_2 Diet for a Greenhouse Planet — 135

Table 6-4. Worksheet for a Sample Household on the CO_2 Diet			
Consumption or activity	**Your use (units per year)**	**CO_2 factor (lbs CO_2/unit)**	**Annual emissions (lbs CO_2 equiv.)**
RESIDENTIAL UTILITIES			
Electricity	_7,200_ kWh	1.5 lbs/kWh	_10,800_
Oil	_625_ gal	22 lbs/gal	_13,800_
Natural gas	_0_ therms	11 lbs/therm	_0_
Propane or bottled gas	_70_ gal	13 lbs/gal	_910_
TRANSPORTATION			
Automobile fuel use	_550_ gal	22 lbs/gal	_12,100_
Other motor fuel use	_0_ gal	22 lbs/gal	_0_
Air travel	_23,000_ miles	0.9 lbs/mile	_20,700_
Bus, urban	_0_ miles	0.7 lbs/mile	_0_
Bus, intercity	_0_ miles	0.2 lbs/mile	_0_
Railway or subway	_1,500_ miles	0.6 lbs/mile	_900_
Taxi or limousine	_400_ miles	1.5 lbs/mile	_600_
HOUSEHOLD WASTE			
Trash (anything discarded)	_2,200_ lbs	3 lbs/lb	_6,600_
Recycled items	_320_ lbs	2 lbs/lb	_640_
HALOCARBON PRODUCTS			
Refrigerators and freezers	_1_ (number)	830 lbs each	_830_
Automobile air conditioners	_0_ (number)	4,800 lbs each	_0_
Other halocarbon products (see Table 3 for equivalences)			_240_
TOTAL ANNUAL GREENHOUSE-GAS EMISSIONS (pounds of CO_2 equivalent)			_68,120_

The second largest source of emissions is air travel, which is mostly work-related. Since we do not have much discretion here, we are stuck with this portion of our greenhouse-gas emissions. Hopefully, commercial airlines will continue to improve their fuel efficiency, which has already increased by over 20% since 1980 and has doubled since 1970 (Davis *et al.* 1989).

Filling out the worksheet made us notice how much more often we use the larger car than we use the smaller one. If we save the larger car

136 — Chapter Six

Table 6-5. *CO$_2$ Diet* Reducing Plan for a Sample Household	Pounds of CO$_2$ equivalence
ASSESSMENT AND GOALS	
Current annual emissions	68,120
Average annual reduction needed (2%)	1,360
Ten-year reduction goal	13,600
	CO$_2$ calories cut:
CURRENT ACTIONS	
Boiler tune-up	1,200
Fluorescent lighting	1,030
Using smaller car more often	880
SUBTOTAL (rounded)	3,100
ACTIONS OVER NEXT TEN YEARS	
Storm windows	3,200
More efficient refrigerator	1,500
Replace 33 mpg car with 40 mpg car	900
Bicycling for errands	600
SUBTOTAL	6,200
POSSIBLE WITH POLICY SUPPORT	
Improved airline fuel efficiency	2,100
Elimination of global impact CFCs	1,070
High-efficiency furnace	1,000
Car that gets 50 instead of 40 mpg	900
40% community recycling	900
Further lighting efficiency	500
SUBTOTAL (rounded)	6,500
TOTAL REDUCTION	15,800

for two-person, long trips, cutting its annual miles from 12,000 to 8,000 (it gets about 33 mpg), and use the smaller car for 8,000 miles (it gets about 36 mpg), we'll save 40 gal/yr, a reduction of 880 lbs of CO$_2$.

To reduce electricity-related emissions, we used three 13-W compact fluorescent bulbs (replacing 60-W bulbs) in fixtures that are on 8 hrs/day, cutting 630 lbs/yr of CO$_2$. We put in a 22-W (100-W replace-

The CO$_2$ Diet for a Greenhouse Planet — 137

ment) bulb in a lamp used about 12 hrs/day and another in one used 4 hrs/day, saving another 400 lbs/yr of CO$_2$.

The sum of our current actions cuts 3,100 lbs/yr, which covers about 2 years' worth of the 10-yr *CO$_2$ Diet* goal.

Looking ahead on what we can do over the next 10 years, we find that installing storm windows provides a big opportunity for saving energy, since most of our windows are now single glazed. Fortunately, our landlords were willing to put in storm windows and have started installing some of them. Converting the estimated reduction in oil heating load to CO$_2$ emissions yields a predicted cut of annual emissions by 3,200 lbs. As for electricity, our biggest use is the refrigerator. (The air conditioner is used for only a few weeks in midsummer, and the hot water is heated by oil.) Our refrigerator is a rather inefficient, secondhand model. Although we are not ready to replace it as present, we hope to afford a newer, more efficient one in a few years, at which time we can obtain a significant reduction in both electricity use and CO$_2$ emissions. When the time comes to replace a car, we will, we hope, be able to buy one that gets at least 40 mpg; using it to replace the less efficient of our present cars would cut 900 lbs/yr of CO$_2$. Finally, if we reduce car use by bicycling for errands, we can cut another 600 lbs/yr. The total of these expected actions results in an annual reduction of 6,200 lbs of greenhouse-gas emissions.

Adding the current and expected actions that our sample household can take on its own yields an expected cumulative cut of 9,300 lbs/yr within 10 years. This reduction is quite short of our goal of cutting annual emissions 13,600 lbs: we realistically expect to achieve only a 14% cut in CO$_2$ calories by ourselves, rather than the 20% cut that is needed. This shortfall brings us to the need for public- and private-sector actions to complement the *CO$_2$ Diet* of individuals.

Policy Complements to the *CO$_2$ Diet*

Further CO$_2$ calorie cutting is quite possible, including many energy-efficiency measures that are also very cost-effective. Using the sample household, we explored the emissions reductions possible from efforts by industry to build more efficient products and from public policies that would make it easier for individuals to obtain energy-efficient and environmentally benign products and services.

In our sample household, the oil furnace, while being kept tuned up, is nonetheless an older model, much less efficient than what is now available. However, since we are renting, it is not economic to invest in a new furnace by ourselves. Moreover, since the landlords do not pay the oil bill, they are not likely to replace the furnace unless it totally breaks down, which it may not do for many years. If business

138 — Chapter Six

or government were to provide incentives, such as a way to share the fuel savings, we could cut another 1,000 lbs/yr by installing a high-efficiency furnace.

If improved automobile fuel economy standards are legislated, we would be able to replace our 33 mpg car with a 50 mpg model of comparable size, rather than a 40 mpg model as we assumed above. This would cut an additional 900 lb/yr of CO_2. If airline fuel efficiency improves by 10% over the next 10 years (this is about half the rate of improvement over the past 10 years) and our average air travel remains the same, we would cut another 2,100 lbs/yr.

If safe replacements and recovery methods are found for halocarbons and other chemicals that now contribute to the greenhouse effect, our household's halocarbon emissions can be eliminated, for a reduction of 1,070 lbs/yr CO_2 equivalent.

Our garbage-related emissions are sizable. We are already recycling what we can, given the recycling options available in the community. To reduce garbage-related emissions further, we need to push for policies that reduce the source of waste and provide for more comprehensive recycling programs. It is very likely that a decent program could easily achieve a 40% recycling rate, which would cut our trash-related CO_2 emissions by about 900 lbs/yr.

If new lamps and fixtures become available, we could make further improvements in household lighting efficiency, which would cut electricity-related emissions by another 500 lbs/yr.

The progress of the sample household's CO_2 Diet is illustrated in Figure 6-2, which shows household greenhouse emissions plotted over the next 10 years. The dashed line is the CO_2 Diet target: calorie cutting at 2%/yr, decreasing the annual emissions from the present value of 68,120 lbs to 54,500 lbs of CO_2 equivalent over 10 years. The upper line shows the step-by-step reductions the household can make on its own. It starts with a drop in the first year by 3,100 lbs (the first group of cuts discussed above). Over the following 9 years, another 6,200 lbs are cut, shown by the upper line of steps. As noted earlier, this reduction was not enough to meet the CO_2 Diet target.

The lower line in Figure 6-2 shows the calorie-cutting path that can be followed if government and industry also make commitments to reducing greenhouse-gas emissions. The steady decrease of emissions every year represents steady gradual changes, such as improving airline efficiency, phasing out of harmful halocarbons, and increasing recycling rates. Policies that help individuals make energy-efficiency investments they are not likely to make on their own, such as an early furnace replacement, are represented by steps down in emissions that did not appear on the upper line. Finally, some of the household's own

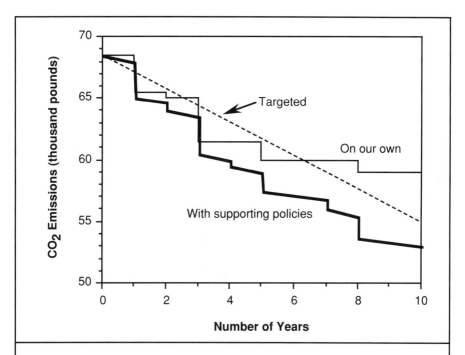

Figure 6-2. *CO₂ Diet* **for a Sample Household**
Note: Dashed line shows the targeted 2%/yr reduction plan. Upper line shows the household's likely progress on its own, without new policies for reducing nationwide greenhouse emissions. Lower line shows the progress that can be made with supporting policies by government and industry.

actions will result in bigger cuts if there is policy support—for example, better fuel economy standards that enable purchase of an even more efficient car. We see that, with collective action by individuals as well as by government and industry, a 20% cut in 10 years is now achievable. With such a concerted effort, we see that it may be possible to do even better than just meet the target and to gain a lead on the greater cuts in greenhouse-gas emissions that are ultimately needed.

Although we make no claim that the sample household is average in any statistical sense, our analysis does suggest the relative scales of personal action versus business- and public-sector action needed to significantly reduce greenhouse-gas emissions in the United States. The sample household is able to cut greenhouse-gas emissions by about 14% over 10 years. Since the residential and personal transportation portion is about one-third of the nation's total greenhouse-gas

140 — Chapter Six

emissions, this reduction suggests that individual action alone cannot be expected to achieve more than a 5% cut. To attain a nationwide reduction of 20%, therefore, the private sector and all levels of government must be responsible for the remaining 15%. That is to say, for every 1,000 CO_2 calories cut by individuals, 3,000 should be cut by business, industry, and government. On the other hand, we do see that individual CO_2 dieting is a necessary response to whatever policy steps might be taken to cut greenhouse emissions, since a significant fraction (one-fourth by our example) of the reduction must be implemented through decisions made by households.

The *CO₂ Diet* example certainly highlights the importance of energy conservation in reducing greenhouse-gas emissions. It is instructive to think of energy conservation measures as falling into three categories: management, investment, and sacrifice (Kempton *et al.* 1984). The same view is useful when considering ways to reduce greenhouse-gas emissions. Tuning up an oil burner or using night setback involves management. Purchasing efficient appliances or a new, more efficient car are investments. Choosing to use a smaller car more frequently may be viewed as a sacrifice. Most of the best opportunities for individual CO_2 dieters involve management and investment. Some emissions can be cut by sacrificing amenity or luxury, but people can only change so much without seriously disrupting their standard of living. Appeals to sacrifice, moreover, are not likely to engender the ongoing political support needed for a national commitment to reducing greenhouse-gas emissions.

Ultimately, the biggest reductions in greenhouse emissions will come through private and public investments in technologies that are more efficient and less dependent on fossil fuels. If, as a society, the United States makes the right investments, environmental impacts will be easier to manage and a high quality of life can be maintained without the need for sacrifices. Needed, therefore, are policy changes to induce the manufacture of more efficient cars, housing, and appliances, benign replacements for halocarbons, and better access to energy conservation services and products. Decision makers can help the nation's *CO₂ Diet* by redirecting capital investments and public subsidies away from fossil fuel and other nonrenewable energy supplies, pursuing management and investment strategies that foster energy efficiency, and promoting renewable sources of energy.

Conclusion

In writing *The CO₂ Diet for a Greenhouse Planet*, we wished not only to inform individuals about the relation between their lifestyles and the

The CO_2 Diet for a Greenhouse Planet — 141

greenhouse effect but also to empower them to take the steps, through both personal action and political activism, needed to reduce greenhouse-gas emissions in the United States. We hope that the CO_2 Diet will be useful in educating the public about the aspects of their lifestyles, particularly energy use, that relate to the greenhouse effect and climate change. The worksheet highlights the importance of energy efficiency in reducing greenhouse-gas emissions. It may therefore serve as a motivational tool for organizations wishing to promote energy efficiency for its environmental and economic benefits. We also reveal the importance of reducing the emissions of halocarbons, which occur in a number of items used by individuals. Finally, the CO_2 Diet shows how individual actions to reduce greenhouse emissions are a necessary part of a transition to an environmentally sustainable economy.

Appendix

Notes on the CO_2 Diet Calorie Cutters

(a) Strictly speaking, a percentage gain in heating system efficiency translates to a larger percentage reduction in fuel use (and CO_2 emissions) because of the reciprocal relation between efficiency and fuel use. For example, increasing efficiency by 10%, from 65% to 75%, reduces fuel use by 13%. The one-to-one ratio therefore understates the savings but simplifies the estimation for most people, and the error is small. Note also that tune-up savings are not additive with savings from other measures, such as setback. If setback is implemented first, the effect of tune-up must be based on the lowered fuel use due to setback rather than the higher, original fuel use.

(b) The national average annual space heating–related CO_2 emissions from a household with oil heat are 15,000 lbs (based on RECS data, EIA 1989). The stated efficiency change reduces consumption by 24%.

(c) Vieira and Sheinkopf (1988: p. A-2) estimate a 27% reduction in Florida heating energy use for a 2°F reduction in thermostat setting.

(d) Hunn et al. (1986: Table A-5), assuming a midsized, electrically heated house in a Texas climate and with the U.S. average CO_2 emission factor for electricity.

(e) Meier, Wright, and Rosenfeld (1983).

(f) Using a showerhead having a flow of approximately 2 gpm, energy savings for electric systems are based on Brown, White, and Purucker 1987; for gas, we assumed energy savings for the same site and converted to gas units.

142 — Chapter Six

(g) Authors' calculations based on average U.S. electricity generation.

(h) Based on the national averages of 10,000 miles/yr and 20 mpg (Davis *et al.* 1989) and assuming the tune-up gives a 9% improvement in fuel economy.

(i) Pers. comm., Deborah Gangloff, Global ReLeaf Program, American Forestry Association; the type of tree planted matters less than that it be well placed, kept healthy, and not crowded by other trees or plants. The review by Meier in this volume suggests a savings of 25% to 50% in air-conditioning energy use with the use of shade vegetation.

References

Brown, M., D. White, and S. Purucker 1987. *Impact of the Hood River Conservation Project on Electricity Use for Residential Water Heating*. Report ORNL/CON-238. Oak Ridge, Tenn.: Oak Ridge National Laboratory.

Bureau of the Census 1990. *1987 Census of Transportation, Truck Inventory and Use Survey*. Washington, D.C.: U.S. Department of Commerce.

Cook, J., and J. Beyea 1990. *Relative Halocarbon Contributions to Global Climate Change*. New York: National Audubon Society.

Davis, S., *et al.* 1989. *Transportation Energy Data Book, Edition 10*. Report ORNL-6565. Oak Ridge, Tenn.: Oak Ridge National Laboratory.

DeCicco, J., J. Cook, D. Bolze, and J. Beyea 1990. *The CO_2 Diet for a Greenhouse Planet: A Citizen's Guide to Slowing Global Warming*. Audubon Policy Report No. 1. New York: National Audubon Society.

EIA. See Energy Information Administration.

Energy Information Administration 1985. *Nonresidential Buildings Energy Consumption Survey, Characteristics of Commercial Buildings 1983*. Report DOE/EIA-0246(83). Washington, D.C.: U.S. Department of Energy.

———— 1986. *Energy Conservation Indicators 1986*, Report DOE/EIA-0441(86). Washington, D.C.: U.S. Department of Energy.

———— 1987. *Energy Facts 1987*. Publication DOE/EIA-0469(87). Washington, D.C.: U.S. Department of Energy.

———— 1989. *Household Energy Consumption and Expenditures 1987 (RECS), Parts 1 and 2*. Report DOE/EIA-0321(87). Washington, D.C.: U.S. Department of Energy.

Environmental Protection Agency 1987. *Regulatory Impact Assess-*

ment: Protection of Stratospheric Ozone. Vols. 1 and 3. Washington, D.C.: EPA.

EPA. See Environmental Protection Agency.

———— 1990. *The 1990 Clean Air Act Amendments, Summary Document.* Washington, D.C.: Office of Air and Radiation, EPA.

Fisher, D., *et al.* 1990. "Relative Effects on Global Warming of Halogenated Methanes and Ethanes of Social and Industrial Interest." In *Scientific Assessment of Stratospheric Ozone: 1989.* Vol. 2. Cambridge, England: Cambridge University Press.

Franklin Associates 1988. *Characterization of the Municipal Solid Waste Stream in the United States, 1960–2000* (update 1988). Prepared for the U.S. Environmental Protection Agency, Office of Solid Waste and Emergency Response. Prairie Village, Kans.: Franklin Associates.

Gaines, L. 1981. *Energy and Materials Use in Packaging.* ANL/CNSV-TM-58. Argonne, Ill.: Argonne National Laboratory.

Houghton, R. 1990. "The Global Effects of Tropical Deforestation." *Environmental Science and Technology* 24(4): 414–22.

Hunn, B., *et al.* 1986. *Technical Potential for Electrical Energy Conservation and Peak Demand Reduction in Texas Buildings.* Report to the Texas Public Utilities Commission. Austin: Center for Energy Studies, University of Texas.

Intergovernmental Panel on Climate Change 1990. *First Assessment Report.* Geneva, Switz.: World Meteorological Organization.

IPCC. See Intergovernmental Panel on Climate Change.

Kempton, W., *et al.* 1984. "Do Consumers Know 'What Works' in Energy Conservation?" In *What Works: Documenting Energy Conservation in Buildings.* J. Harris and C. Blumstein, eds. Washington, D.C.: American Council for an Energy-Efficient Economy.

Keyfitz, N. 1989. "The Growing Human Population." *Scientific American* 261(3): 119–26.

Krause, F., W. Bach, and J. Koomey 1989. *Energy Policy in the Greenhouse.* Warming Fate to Warming Limit: Benchmarks for a Global Climate Convention, Vol. 1. El Cerrito, Calif.: International Project for Sustainable Energy Paths.

Lashof, D., and D. Ahuja 1990. "Relative Contributions of Greenhouse Gas Emissions to Global Warming." *Nature* 344 (5 April): 529–31.

Lester, R., and J. Myers 1989. *Global Warming, Climate Disruption, and Biological Diversity.* Audubon Wildlife Report 1989/1990. New York: Academic Press.

Meier, A., J. Wright, and A. Rosenfeld 1983. *Supplying Energy*

144 — Chapter Six

through Greater Efficiency. Berkeley: University of California Press.

Nisson, J., and G. Dutt 1985. *The Superinsulated Home Book.* New York: John Wiley & Sons.

Schneider, S. 1989a. *Global Warming: Are We Entering the Greenhouse Century?* San Francisco: Sierra Club Books.

————— 1989b. "The Changing Climate." *Scientific American* 261(3): 70–79.

SRI 1979. *The Long-term Impact of Atmospheric Carbon Dioxide on Climate.* JASON Report JSR-78-07. Arlington, Va.: SRI International.

Vieira, R., and K. Sheinkopf 1988. *Energy-Efficient Florida Home Buiding.* Cape Canaveral: Florida Solar Energy Center.

Wilson, A. *1991 Consumer Guide to Home Energy Savings.* Washington, D.C.: American Council for an Energy-Efficient Economy.

John M. DeCicco is a research associate with the American Council for an Energy-Efficient Economy where he focuses on energy efficiency in the transportation sector. In his prior position with the National Audubon Society, he analyzed the environmental impacts of energy use. He received his Ph.D. in Mechanical Engineering from Princeton University.

James H. Cook is a senior environmental policy analyst with the National Audubon Society. Prior to his current position, he was an aide with the Wisconsin Legislature. He holds a B.S. from the California Institute of Technology and a Ph.D. from the University of Wisconsin.

Dorene Bolze is an environmental policy analyst with the National Audubon Society. She directs a new program on the international trade in wildlife and works on marine conservation issues. She has a B.S. from Duke University and an M.E.S. from the Yale School of Forestry and Environmental Studies.

Jan Beyea is chief scientist at the National Audubon Society. He has studied, written and testified about the environmental impacts of nearly all existing and proposed energy sources. He serves on the Secretary of Energy's Task Force on Economic Modeling. He received his Ph.D. in Physics from Columbia University.

Chapter 7

Emissions Impacts of Demand-Side Programs: What Have We Achieved So Far and How Will Recent Policy Decisions Change Program Choices?

Richard S. Tempchin and A. Joseph Van den Berg, *Edison Electric Institute*

Vera B. Geba, Curtis S. Felix, and Marc W. Goldsmith, *Energy Research Group, Inc.*

Introduction

Demand-side management (DSM) programs have been actively pursued by electric utilities nationwide because these programs can increase capacity, save energy, and satisfy customer needs while reducing customer costs. In addition to providing these benefits, present DSM efforts have also reduced emissions. Little has been done to evaluate and quantify these benefits nationwide, although several studies have quantified the *potential* impact of *possible* DSM program impacts on acid rain by selected region (Centolella *et al.* 1988, Geller *et al.* 1987; Nixon and Neme 1989) and nationally (Chupka *et al.* 1989).

Additionally, several studies that estimate the national energy and capacity impacts of utility programs that promote efficient electricity use have been completed. These studies have been based on surveys (Cogan and Williams 1987) or computer modeling (Keelin and Gell-

146 — Chapter Seven

ings 1986; Scheer 1990; Carlsmith *et al.* 1990; Faruqui *et al.* 1990). However, if energy efficiency is to be considered an option for reducing emissions, public utility commissions and utilities must develop methods for incorporating benefits and costs of this source of emissions reduction into strategic planning decisions. Studies are necessary to answer a number of questions on performance, financial and environmental impacts, and potential accomplishments in improving efficiencies and reducing emissions.

We have recently completed two studies (Geba *et al.* 1989, 1990) and have under way an additional study, which begins to develop preliminary information on how existing DSM programs have affected power plant emissions and how selected substitution of efficient electric technologies can reduce CO_2 (Felix *et al.* 1990).

Steps Toward Estimating the Impacts of Electric Utility Demand-Side Management Programs on Power Plant Emissions

Electric utility promotion of energy-efficient end-use technologies is routinely recommended by federal and state policymakers, utilities, and environmentalists as the primary tool for controlling emissions. However, very little is known about how the impacts of these programs compare with those of traditional clean air technologies, such as scrubbers. DSM programs have been implemented because they satisfy customer needs, reduce customer costs, increase revenues and profitability, improve market share, defer capacity additions, and promote the more efficient use of existing capacity. These objectives may or may not be consistent with controlling emissions. This determination requires a systematic evaluation of energy and capacity impacts and fuel mixes.

Background

Over the past 25 years, electric utilities have helped customers become more energy-efficient. Through information and education programs, financial incentives, energy audits, research and development, and innovative partnerships among utilities, customers, and state agencies, electric utilities have accumulated numerous success stories. Today, all utilities combined have sponsored more than 1,300 energy-efficiency programs in their service territories, up from about 130 in 1977. Customers have more opportunities than ever to collaborate with their local utilities to obtain greater electric value through enhanced efficiency.

These programs involve over 13 million customers and have resulted in almost 20 gigiwatts in capacity savings since 1977 (Faruqui *et al.* 1990). In 1988, electric utilities spent over $1 billion dollars on efficiency programs, and this amount continues to rise (Felix *et al.* 1990).

Methodology

Quantifying air-emissions benefits of DSM required gathering and analyzing a large number of data. The following is a simplified description of the process. We first collected information on utility DSM impacts from utilities, power pools, public utility commissions, and other state and federal agencies by means of mail and telephone surveys. We developed a mail survey and telephone script to ascertain actual DSM efforts over a two-year period ending in 1989 and to project efforts for 1990. To date, only the 1988 data have been analyzed. Several types of reference data were also requested to allow cross-verification of reported data with other utilities and sources. We made follow-up calls to responding and nonresponding utilities and to the other above sources in order to reduce biases and obtain the best possible data set. These data were aggregated state by state using both statistical methods and some engineering judgment to estimate kilowatts reduced, kilowatt-hours saved, and costs incurred.

The mail survey to investor-owned utilities (IOUs) provided data on the effect of DSM on over 50% of IOU kilowatt-hour sales. This percentage equals a little under 40% of the U.S. total kilowatt-hour sales. Phone calls to public utility commissions, public utilities, and power pools brought the total survey coverage to approximately 50% of U.S. kilowatt-hour sales. This percentage permitted statistical extrapolation of the DSM savings to 87% of the U.S. kilowatt-hour sales. We extrapolated the sample to national estimates by combining subjective data on state efforts—from interviews with public utility commission staff—with the national pool of data. States were stratified as having a low, medium, or high level of DSM program development. Strata were also developed from the national pool, and these estimates were applied using the closest proxy in terms of geography and peak characteristics. The remaining 13% was estimated from the existing data pool, interviews with utility commission staff, and engineering judgment.

Most savings data reported are engineering estimates, not direct measurements; data on national actual measured savings are relatively

148 — Chapter Seven

meager. Also, utilities and public utility commissions have focused on capacity savings in order to avoid constructing new power plants; thus, capacity estimates are more readily available and were more often reported than energy savings.

Next, we used a set of state-by-state generation profiles and fuel mixes to estimate the fuel displaced by DSM savings. Other factors, such as estimated DSM program load profiles and estimated dispatch orders were also integrated into the analysis. We did not develop a detailed state-by-state fuel dispatch model because the DSM data did not support this level of accuracy.[1]

Summary of Results

This major survey effort collected data on the summer and winter capacity (kW) and energy (kWh) saved by end-use energy efficiency. Table 7-1 shows the results. The estimated summer capacity saved was over 8,100 MW, and estimated winter capacity saved was over 5,200 MW; energy saved was over 8.9 billion kWh. The utility-only investment cost was estimated at $1.283 billion, or $158/kW, in 1988. These data are ballpark estimates.

Many factors influence the data: reporting is not standardized, measurement is just beginning, and determination of the specific causes of action is difficult, to name a few of those factors. This paper thus represents a preliminary attempt to obtain field data by utility and state. For reasons mentioned earlier, capacity data appear to be more accurate than energy data. We expect increased utility and regulatory attention to energy savings. As DSM reduces energy generated and capacity factor, emissions will decrease. Therefore, measuring emission reduction has become significantly more important. Improvements are needed in monitoring and evaluating the effects on emissions and the other energy impacts of DSM programs; such efforts are already under way.

To develop a statewide estimate for total demand-side management activity, the Edison Electric Institute survey of only investor-owned utilities needed to be aggregated to include the activities of non-investor-owned utilities. PUC data in most states also included only investor-owned utilities and had the same aggregation problem. Using the limited data available from non-investor-owned utilities, we assumed that their activity would compare with investor-owned utility

---

[1] CO_2, SO_2, and NO_2 emissions are presented in a separate paper by assuming heat rates and pounds of emissions per Btu input for each of the fuels displaced (Felix *et al.* 1990). The result will be a state-by-state preliminary estimate of the emissions reduced by DSM measures.

Emissions Impacts of Demand-Side Programs — 149

Table 7-1. Summary of Demand-Side Management Savings by State in 1988

State	Capacity Savings Summer Peak (MW)	Capacity Savings Winter Peak (MW)	Energy Savings (MWH)	Dollars (millions)	Savings as a Percent of Capacity	Savings as a Percent of Sales
Alabama	210.2		261,715	8.5	1.1	0.5
Alaska		17.5	7,420	2.0	1.0	0.2
Arizona	472.6		21,102	2.3	3.0	0.1
Arkansas	99.5		48,566	10.0	1.0	0.2
California	2,776.6		731,834	360.3	6.0	0.4
Colorado	280.8		99,576	12.0	3.9	0.3
Connecticut	157.0		449,772	17.0	2.1	1.6
District of Columbia	19.3		14,262	2.0	1.0	0.2
Delaware	8.8		17,919	1.0	1.0	0.2
Florida		1,741.0	2,294,000	114.0	4.8	1.8
Georgia	442.9		283,365	42.0	2.0	0.4
Hawaii	30.6		29,491	3.0	2.0	0.4
Idaho	21.3		32,552	2.0	1.0	0.2
Illinois	375.5		209,355	36.0	1.0	0.2
Indiana	47.9		11,506	3.8	0.2	0.0
Iowa	239.3		89,906	21.7	2.8	0.3
Kansas		5.0	1,834	2.2	0.0	0.0
Kentucky		338.2	206,234	32.0	2.0	0.4
Louisiana	185.6		112,772	18.0	1.0	0.2
Maine	22.0		168,745	10.0	0.9	1.5
Maryland	106.1		90,784	10.0	1.0	0.2
Massachusetts	135.9		287,207	19.0	1.4	0.6
Michigan	35.2		19,816	13.6	0.1	0.0
Minnesota	26.6		97,872	10.0	0.3	0.2
Mississippi	72.9		54,050	7.0	1.0	0.2
Missouri	236.9		26,420	23.0	1.4	0.1

150 — Chapter Seven

Table 7-1. (continued)

State	Capacity Savings		Energy Savings (MWH)	Dollars (millions)	Savings as a Percent of	
	Summer Peak (MW)	Winter Peak (MW)			Capacity	Sales
Montana		100.6	48,128	10.0	2.0	0.4
Nebraska	58.4		32,535	6.0	1.0	0.2
Nevada		31.0	22,555	1.7	0.6	0.2
New Hampshire		7.0	(8,670)	2.0	0.5	NA
New Jersey	276.8		165,180	35.2	1.9	0.3
New Mexico	57.0		24,158	5.0	1.0	0.2
New York	267.8		115,361	63.1	0.8	0.1
North Carolina	266.8		237,377	38.6	1.3	0.3
North Dakota		47.7	12,673	5.0	1.0	0.2
Ohio		1,282.2	18,185	15.4	4.6	0.0
Oklahoma	387.4		7,452	7.3	2.9	0.0
Oregon		21.7	16,936	0.4	0.2	0.0
Pennsylvania		749.2	846,301	42.5	2.1	0.8
Rhode Island	30.0		20,021	4.9	11.0	0.3
South Carolina		165.7	100,601	16.0	1.0	0.2
South Dakota		1.8	1,407	0.3	0.1	0.0
Tennessee		371.0	276,075	36.0	2.0	0.4
Texas	349.3		672,411	28.0	0.6	0.3
Utah		53.2	28,540	5.0	1.0	0.2
Vermont		13.0	666	3.0	1.2	0.0
Virginia	237.7		106,293	5.7	1.7	0.2
Washington		189.9	148,478	49.9	0.8	0.2
West Virginia		43.4	19,724	1.1	0.3	0.1
Wisconsin	230.8		333,321	114.9	2.1	0.7
Wyoming		59.6	21,149	6.0	1.0	0.2
Country Totals	8,165.6	5,238.6	8,934,932	1,283	1.9	0.3

Source: Felix *et al*. 1990.

activities, as both are driven primarily by capacity and regulatory pressures within a state. The survey data were treated as follows to derive the *capacity savings* estimates.

survey capacity savings $\times$ (1 – DSM activity rating)
 $\div$ (DSM activity rating $\times$ state coverage ratio)
 = state capacity savings

The DSM activity rating expression used a subjective rating of the level of uniformity among DSM activities throughout the state as the basis for developing the statewide estimate. For each state, the rating was developed from PUC staff telephone interview data and conversations with utility staff responsible for DSM efforts. A high rating indicates that DSM efforts are occurring throughout 80% of the state, a medium rating indicates a 50% level of activity, and a low rating indicates an activity level of 20% or less.

A coverage ratio was then calculated to reflect the percentage of the state's total energy sales under PUC jurisdiction. For example, if only two investor-owned utilities comprising 40% of a state's electricity consumption actually report to the PUC, the coverage ratio would be 40%.

This study developed an aggregate program load shape at the state level. States were also categorized as either summer or winter peaking based on the predominant features of the utilities within the state. In states where there are both summer- and winter-peaking utilities, a system-by-system peak load–weighted approach was used to determine the overall load shape within that state. DSM program impacts then were categorized as either summer or winter peak so as to avoid double counting in summing the results. Although this approach could somewhat skew the results, it was felt that most of the potential bias would be compensated for through offsetting errors.

Carbon Dioxide Reduction Through Electrification

The electric utility industry is perceived to be the largest potential generator of global warming gases in the United States (Streeb 1988). This perception has led to the assumption that substituting electricity for direct use of fossil fuels will increase CO_2. However, recent research indicates that in selected applications this assumption is incorrect: using highly efficient electric technologies in several traditionally fossil-fueled applications may significantly reduce CO_2 emissions (Geba *et al.* 1989, 1990).

152 — Chapter Seven

Background

CO_2, produced when fossil fuels are burned, is estimated to represent approximately 50% of the gases contributing to potential global warming trends. Other major gases contributing include chlorofluorocarbons, methane, and nitrous oxide. According to the U.S. Department of Energy (DOE), the electric utility industry emits approximately 35% of the CO_2 generated by human activities in the United States. The transportation sector produces 32% and the industrial sector 20%. The remaining 13% is emitted from other sources such as residential and commercial use of fossil fuels (Streeb 1988).

The higher efficiencies generally associated with electric over fossil-fueled end uses could reduce CO_2 in several of these sectors by decreasing overall energy use. Reductions in combustion products depend upon, first, the relative technology efficiencies, including end-use and generating efficiencies,[2] and, second, the mix of fossil and lower-polluting nonfossil fuels used to generate electricity.

In two studies we have shown that a number of commercially available and competitive electrotechnologies, when substituted for fossil-fueled processes, result in substantial national reductions of CO_2 emissions (Geba et al. 1989, 1990). Reductions of other gases, such as methane, and of additional pollutants were also cited as possible but were not analyzed.

Methodology

Using fuel burned per kilowatt-hour generated, and CO_2 produced per unit of fuel burned, we calculated the total amount of CO_2 produced by the utility sector in 1988 and the average pounds of CO_2 per kilowatt-hour generated. In 1988, based on the national average fuel mix, 1.51 pounds of CO_2 were produced for every kilowatt-hour generated. Only direct CO_2 emissions from combustion were considered in the analysis. Indirect CO_2 emissions during fuel extraction and refining processes for coal, natural gas, and petroleum were not included, since these would apply to both electricity generation from, and direct use of, these fossil fuels.[3]

To illustrate the importance of utility fuel mix variations on the amount of CO_2 generated per kilowatt-hour, CO_2 emissions from two

---

[2] As an approximation, due to the lack of data for comparison, the electrical transmission and distribution losses were assumed to have emissions impacts similar to fuel distribution losses.

[3] The exclusion of primary energy used in mining, extraction, and so on should not substantially influence the overall outcome, since it applies to energy used in utilities as well as in other sectors.

Emissions Impacts of Demand-Side Programs — 153

different regions were also calculated. The Pacific Northwest region (including California, Oregon, and Washington) was selected as the "low CO_2 per kilowatt-hour" case region because it has an electric fuel mix consisting predominantly of hydroelectric and nuclear power. In this region, on average, only 0.41 pounds of CO_2 are emitted for every kilowatt-hour generated. The Southeast Central region (including Alabama, Kentucky, Mississippi, and Tennessee) was selected as the "high CO_2 per kilowatt-hour" case region, having an electric fuel mix mainly of coal. In this region, on average, 1.80 pounds of CO_2 are emitted for every kilowatt-hour generated.

With many different ways of measuring energy, it is important to evaluate alternatives on a comparable basis. Since the Btu is a U.S. standard quantity of energy, it is convenient to convert energy consumption to Btus to make comparisons. An initial screening process for substitution opportunities based on the number of Btus consumed identified whether CO_2 savings would result. Table 7-2 shows the CO_2 produced per Btu at the end-use level for several energy sources. In most cases, electricity appears more CO_2 intensive than the others. This result is due to the fact that power plant conversion and losses have been taken into account to provide a fair comparison with other end-use energy sources.

The total CO_2 released is a function not only of the fuel or electricity used but also of the efficiency of the end use. This effort evaluated end uses to determine where savings in CO_2 production may exist.

One way of sorting through technology choices for reducing CO_2 is to look at the ratio of CO_2 generated per Btu using electricity compared to CO_2 generated per Btu using fossil fuels. As shown below, CO_2 reduction will occur through electric technology substitution only when the ratio of electric work per unit of energy input is higher than the fossil fuel work per unit of energy input by a specific amount. In other words, Table 7-2 shows how much more efficient an electric technology must be in order to reduce CO_2.

We estimated CO_2 emissions associated with comparable electrotechnologies and fossil-fired systems. These estimates were used to compare the CO_2-reducing potential of electrotechnologies in the residential and commercial, industrial, and transportation sectors. Even after accounting for power plant conversion losses (which contribute to CO_2), we found that many electric end uses are significantly more energy-efficient than comparable fossil-fueled processes and thus lower total Btu requirements considerably.

154 — Chapter Seven

Table 7-2. CO_2 Generated per End-Use Btu for Various Fuel Types	
Electricity	
• Low-Case	1.2×10^{-4} lbs CO_2/Btu
• Base-Case	4.4×10^{-4} lbs CO_2/Btu
• High-Case	5.3×10^{-4} lbs CO_2/Btu
Coal	2.2×10^{-4} lbs CO_2/Btu
Oil	1.7×10^{-4} lbs CO_2/Btu
Natural Gas	1.2×10^{-4} lbs CO_2/Btu

Summary of Results

Residential Sector. The technologies that were evaluated included the electric heat pump, electric heat pump water heater, and the microwave oven. CO_2 was reduced by substituting efficient electric heat pumps for average heat pumps, electric baseboard heating, natural-gas-fired furnaces of average efficiency, and oil-fired furnaces of average and high efficiency. These findings assumed that identical electric central air-conditioning systems were matched with each heating system. CO_2 was not reduced by substituting efficient heat pumps for high-efficiency (95%) natural gas furnaces in the base- and high-case regional fuel mix scenarios, which depict areas of the country where significant amounts of coal are used to generate electricity. In areas where non-CO_2-emitting fuels predominate for the production of electricity, efficient heat pumps do offer CO_2 savings compared to efficient gas-fired furnaces.

Residential heat pump water heaters (HPWHs) generate considerably less CO_2 than efficient electric resistance, natural gas, and oil-fired water-heating systems. Based on the national average fuel mix, about 220 billion pounds of CO_2 could be saved per year if all of the existing electric resistance water heaters and 50% of all gas- and oil-fired heaters were converted to efficient HPWHs. The microwave oven generates between 2 and 14 times less CO_2 than standard electric ranges and gas ranges in several cooking applications.

Table 7-3 summarizes the key findings of the residential analysis. Additional research is needed to further quantify the total CO_2 reduction potential from these technologies. Other CO_2-reducing applications of electrification are also likely (for example, dual-fuel systems). Our analysis evaluated only the most likely CO_2-reducing candidates.

Commercial Sector. The commercial analysis focused on several end uses, including space cooling, water-heating, commercial cook-

Emissions Impacts of Demand-Side Programs — 155

Table 7-3. Summary of CO_2 Reduction from Efficient Residential Electro-technologies Versus Comparable Systems

Electrotechnology/ Substitute	Estimated Annual CO_2 Savings (lbs.) per Unit			Percent Base-Case Reduction
	Low-Case	Base-Case	High-Case	
Electric Heat Pumps (most efficient)				
• Average Electric Heat Pump	2,835	9,497	11,057	30%
• Electric Resistance	3,781	12,666	14,747	36%
• Natural Gas (Electric A/C)				
—Average (62%)	9,084	2,776	1,300	11%
—High-Efficient (95%)	4,103	(4,305)	(6,272)	—
• Oil (Electric A/C)				
—Average (60%)	16,322	10,028	8,556	31%
—High-Efficient (80%)	10,696	2,343	389	10%
Heat Pump Water Heaters (HPWHs) (most efficient)				
• Average Electric Heat Pump	232	778	905	29%
• Electric Resistance	1,492	4,997	5,818	72%
• Natural Gas				
—Average (50%)	3,578	2,215	1,896	53%
—High-Efficient (83%)	1,925	562	243	22%
• Oil				
—Average (50%)	5,947	3,947	3,628	67%
—High-Efficient (63%)	4,095	2,732	2,413	58%
Microwave Oven				
• 4 Baked Potatoes				
—Electric Range	0.29	0.97	1.14	64%
—Gas Range	2.23	1.82	1.73	76%
• Cake				
—Electric Range	0.17	0.58	0.67	62%
—Gas Range	1.62	1.44	1.38	80%
• Roast				
—Electric Range	0.37	1.26	1.47	56%
—Gas Range	3.70	3.01	2.85	75%

Source: Geba *et al.* 1990.

156 — Chapter Seven

ing, and peripheral office activities with equipment such as facsimile machines.

Electric chillers, and possibly cool-storage technologies, generate significantly less CO_2 than do gas-cooling technologies under all three regional fuel mix scenarios. An exception is the use of small electric chiller systems in areas of the country where significant amounts of coal are used to generate electricity. In those cases, gas cooling generates less CO_2.

Similar to residential heat pump water heaters (HPWHs), commercial HPWHs offer substantial CO_2 reducing potential. An analysis of three case studies involving use of HPWHs in restaurant, hotel, and school applications revealed that annual CO_2-savings could range from about 100,000 pounds to over 4 million pounds per facility (EPRI 1990). Because savings are very facility-specific and because market penetration data on the use of commercial HPWHs were not evaluated, national CO_2 reduction potential using HPWHs was not estimated.

Electric commercial cooking technologies were also evaluated for their CO_2-reducing potential. Efficient electric fryers generate less CO_2 than average gas-fired models under all three regional fuel mix scenarios, but compared to the efficient gas fryer, the electric fryer generates less CO_2 only under the low-case scenario. Electric griddles were not found to produce CO_2 savings except in the low-case fuel mix region, where significant amounts of non-CO_2-producing fuels are used to generate electricity.

Lastly, the fax machine has the potential to save between two and seven times the amount of CO_2 per document compared to transporting documents via overnight delivery services. Nationally, increased use of the fax machine could save about 14 million pounds of CO_2 annually if 30% of the letters currently transported by express mail were delivered via fax machines.

Table 7-4 summarizes the key findings of the commercial analysis. For most of the technologies, aggregate national CO_2 reduction impacts of electrotechnology substitution were not estimated as part of this study. Additional research is needed to further quantify national impacts of these technologies and identify other CO_2-reducing applications.

Industrial Sector. By substituting the following electrotechnologies for comparable industrial fossil-fueled processes, a minimum reduction of 17% in annual CO_2 generation from the industrial sector may be realized: electric arc furnace; induction heating; electric glass melting, annealing, and conditioning; infrared heating; and freeze concentration.

Emissions Impacts of Demand-Side Programs — 157

Table 7-4. Summary of CO_2 Reduction from Efficient Commercial Electrotechnologies Versus Comparable Systems

Electrotechnology/ Substitute	Estimated Annual CO_2 Savings (lbs.) per Unit			Percent Base-Case Reduction
	Low-Case	Base-Case	High-Case	
Electric Chillers vs. Gas-Fired Chillers/Heaters				
• Northeast Region				
—30-Ton Unit	33,560	5,548	(1,008)	11%
—250-Ton Unit	292,000	151,940	119,160	37%
—500-Ton Unit	585,640	307,118	241,932	38%
• Southeast Region				
—30-Ton Unit	90,576	14,944	(2,758)	11%
—250-Ton Unit	788,400	410,238	321,732	37%
—500-Ton Unit	1,581,276	829,267	653,264	38%
Document Transfer: Fax Machines				
Conventional Overnight Document Delivery Service	0.4	0.25	0.20	54%
Chilled Storage	Since chilled storage units use about the same amount of energy as conventional electric chillers, the CO_2 savings may be similar to above. However, actual savings would be dependent upon off-peak utility fuel mix.			
Cooking: Electric Fryers (most efficient)	493	1,653	1,925	10%
• Natural Gas				
—High-Efficient	4,541	(3,394)	(5,251)	—
—Average	8,792	1,834	204	10%
Cooking: Electric Griddles				
• Natural Gas Model	9	(5)	(8)	—
Heat Pump Water Heaters				
• Restaurant (converted from electric resistance	—	106,128	—	35% (estimate)
• Hotel (converted from gas water-heating)	—	4,141,642	—	79% (estimate)
• School (converted from oil-fired boiler)	—	170,175	—	—

Source: Geba *et al.* 1990.

158 — Chapter Seven

Electrotechnologies are already gaining widespread acceptance within the industrial sector because of their high efficiencies, precise energy control capabilities, and high processing and production rates. The fact that they are CO_2 reducing as well should provide an added incentive for their development and use.

Transportation Sector. Electric modes of transportation also have CO_2-reducing potential. An 8.2% reduction in the total amount of CO_2 currently produced in the transportation sector could be realized from direct substitution of electric cars and buses for gas or diesel cars or buses, and from substitution of more efficient electric modes of transit for less efficient fossil-fueled modes.

Additional research is needed to further quantify the national CO_2 reductions of these and other potential CO_2-reducing technologies. The preliminary findings, however, provide a basis for future R&D and legislative efforts to reduce CO_2 through the use of highly efficient electric transport.

Conclusions

DSM has a proven potential to reduce both capacity and energy requirements and in 1988 caused capacity reduction of over 8,000 MW and energy reduction of almost 9 billion kWh. This reduction will reduce CO_2, SO_2, and NO_x emissions.

These reductions will increase and accelerate with DSM experience and expenditures as well as state and federal regulatory environmental concerns. In addition, we believe that energy savings have been underreported by utilities and regulators and, therefore, underestimated in our study. Thus, we have probably underestimated emissions savings. In addition, regulatory incentives are changing to promote greater emissions savings.

Efficiency improvements can greatly reduce CO_2 emissions from both electric-powered equipment and advanced gas-fired equipment. Legislation and regulation that requires more efficient equipment will need an improved data base on emissions. This data base should reflect the present study's finding that highly efficient electric equipment reduces CO_2 emissions compared to gas-fired or oil-fired alternatives.

References

Carlsmith, R., *et al.* 1990. *Energy Efficiency: How Far Can We Go?* ORNL/TM-11441. Oak Ridge, Tenn.: Oak Ridge National Laboratory.

Centolella, P., *et al.* 1988. *Clearing the Air: Using Energy Conservation to Reduce Acid Rain Compliance Costs in Ohio.* Columbus: Ohio Office of Consumers' Counsel.

Chupka, M., *et al.* 1989. *Staff Working Paper: The Role of Technology and Conservation in Controlling Acid Rain.* Washington, D.C.: Congressional Budget Office.

Cogan, D., and S. Williams 1987. *Generating Energy Alternatives.* Washington, D.C.: Investor Responsibility Research Center.

Electric Power Research Institute 1990. *Commercial Heat Pump Water Heaters Applications Handbook.* EPRI CU-6666. Palo Alto, Calif.: Electric Power Research Institute.

EPRI. See Electric Power Research Institute.

Faruqui, A., *et al.* 1990. *Impact of Demand-Side Management on Future Electricity Demand: An Update.* Edison Electric Institute, Washington, D.C. (07-90-26) and Electric Power Research Institute, Palo Alto, Calif. (CU-6953).

Felix, C., *et al.* 1990. "Impacts of Electric Utility Demand-Side Management Programs on Power Plant Emissions." Draft. Washington, D.C.: Edison Electric Institute.

Geba, V., *et al.* 1989. *Carbon Dioxide Reduction Through Electrification of the Industrial and Transportation Sectors.* Washington, D.C.: Edison Electric Institute.

———— 1990. *Carbon Dioxide Reduction Through Electrification of the Residential and Commercial Sectors.* Draft. Washington, D.C.: Edison Electric Institute.

Geller, H., *et al.* 1987. *Acid Rain and Electricity Conservation.* Washington, D.C.: American Council for an Energy-Efficient Economy.

Keelin, T., and C. Gellings 1986. *Impact of Demand-Side Management on Future Customer Electricity Demand.* EM-4815-SR. Palo Alto, Calif.: Electric Power Research Institute.

Nixon, E., and C. Neme 1989. *An Efficiency Approach to Reducing Acid Rain: The Environmental Benefits of Energy Conservation.* Washington, D.C.: Center for Clean Air Policy.

Scheer, R. 1990. *Electricity Conservation and Load Management Potential in the Year 2000: A Review of Ten Studies.* Draft. Washington, D.C.: U.S. Department of Energy, Office of Policy, Planning, and Analysis.

Streeb, A. 1988. "Technical Feasibility and Implications of Reducing U.S. CO_2 Emissions in the Period from 1995 to 2010." Briefing by Al Streeb, Deputy Assistant Secretary for Energy Conservation, 4 March.

160 — Chapter Seven

Richard S. Tempchin is the manager of Energy Efficiency Policy and Programs at the Edison Electric Institute. He previously worked as a policy analyst at the U.S. Department of Energy and as a utility demand-side management consultant. He has a B.A. from the University of Maryland and an M.B.A. from the University of Denver.

A. Joseph Van den Berg is the director of Technical Services at the Edison Electric Institute. Prior to joining EEI, he worked for the Potomac Electric Power Company. He received a B.S. in Mechanical Engineering from Virginia Polytechnical Institute and State University and an M.B.A. in Finance and Investments from George Washington University.

Vera B. Geba is vice president of Energy Research Group, Inc. She specializes in the regulatory, environmental and technical aspects of energy technologies. She has a degree in Chemical Engineering from the University of Lowell.

Curtis S. Felix is a senior analyst at the Energy Research Group, Inc. He has conducted analyses on the environmental benefits of demand-side management, energy externalities and the environmental impacts of fuel switching. He holds a B.A. in Economics and a B.A. in Political Science from the University of Vermont.

Marc W. Goldsmith is president of Energy Research Group, Inc. He provides strategic planning, technology assessment, regulatory, and environmental impact support to the energy industry. He has engineering degrees from the Massachusetts Institute of Technology and the State University of New York Maritime College.

Chapter 8

The Global Climate Change Issue—What It Is, Where It Is Going, and How It Will Impact Utility DSM

Bonnie B. Jacobson and David W. Kathan, *ICF Resources Incorporated*

Introduction

The possibility of global climate change is regarded by many U.S. scientists and policymakers as a serious environmental problem whose consequences could include rising sea levels, higher global temperatures, and significant changes in local climate patterns. Emissions of greenhouse gases from human activities are viewed as the major cause of global warming, and U.S. public policy and regulatory attention has focused on reducing the release of greenhouse gases, including carbon dioxide (CO_2), chlorofluorohydrocarbons (CFCs), methane (CH_4), and other trace gases.

We will discuss the issues associated with global climate change, review research and policy initiatives, and discuss the implications for utilities of climate change and of resulting environmental policy actions. We will particularly emphasize the roles of energy efficiency and of utility DSM programs in solving these problems. Energy efficiency will be an important and cost-effective tool to reduce greenhouse-gas emissions, particularly for the electric utility sector.

Summary of Global Climate Change

The greenhouse gases associated with global warming have natural as well as anthropogenic (human-caused) sources. The preindustrial con-

162 — Chapter Eight

centration of CO_2 in the atmosphere is thought to have been about 275 parts per million (ppm), with some fluctuations associated with the spread and retreat of ice ages (EPA 1989). Industrial activities have spurred a 30% increase in CO_2 concentrations to about 350 ppm (*ibid.*) since the beginning of the Industrial Revolution. The difference between preindustrial and current levels of greenhouse gases appears to have resulted from CO_2 released by the burning of carbon-based fuels, CH_4 released from increased rice and cattle cultivation, and the development of CFCs.

Approximately 50% of the global warming potential of green-house-gas concentrations worldwide can be attributed to CO_2 (*ibid.*), and CO_2 reduction is thus the aim of many proposed control policies. Of this 50%, about two-thirds comes from the combustion of fossil fuels used by the utility, transportation, and industrial sectors (*ibid.*). The remaining one-third has biological origins, such as agricultural and land-clearing activities that permanently reduce standing biomass. Methane and CFC levels, estimated to contribute approximately 18% and 14%, respectively, to the increase in warming potential, have been growing faster than CO_2, and are expected to account for a relatively larger amount of future greenhouse-gas concentrations (*ibid.*). We will focus primarily on CO_2 reduction in this chapter.

Policy Options

The primary means of reducing greenhouse-gas emissions are to utilize both supply and demand-side technologies that reduce primary energy requirements through improved energy efficiency, to use less carbon-intensive fuel (either from fossil or renewable sources), to increase the stock of stored carbon in the form of new forests, and to directly reduce the use or release of greenhouse gases like CFCs and CH_4. Legislators and regulators have proposed a number of possible options for encouraging speedy adoption of the strategies and technologies discussed above. The following section describes three major categories of global climate change policy: (1) regulatory "push" mechanisms, (2) market "pull" mechanisms, and (3) research, demonstration, and development activities.

Global climate change will ultimately be determined by how quickly global warming does occur, if at all, and eventual U.S. governmental policy may be a combination of all three of the policy types listed above. As a recent Office of Technology Assessment report (OTA 1991) stated:

> Whatever the CO_2 reduction goal, Congress will have to use a variety of policy instruments to stimulate a diverse set of decision-makers to use

The Global Climate Change Issue — 163

the appropriate fuels, technologies, and forestry and agricultural practices and to adopt energy use patterns that conserve energy.

Each of the three categories of policy options are reviewed below.

Regulatory "Push" Mechanisms

Past environmental regulation has relied heavily on regulatory "push" policies such as the New Source Performance Standards for sulfur dioxide. In most instances, current regulatory practice does not internalize the environmental costs by including those costs in the decision-making process but rather forces recognition of these costs through regulations and emission standards.[1] These "command and control" regulations control the rate of emission of key sources of pollution, require installation of control equipment, and so on.

The primary advantage of command and control regulatory policy is uniformity—all affected companies are subject to the same emission standards. However, regulatory approaches have two main drawbacks.

First, regulatory standards are generally not efficient because they do not give society the maximum value for its money. The least expensive way to control emissions from a group of generating plants is to concentrate efforts where the cost of control (dollar per ton removed) is cheapest, rather than to impose an arbitrary reduction to all plants, at some of which the reduction may be very expensive to obtain. Requiring reduction only where cost-effective, however, may not be perceived as equitable if all control costs fall on only a portion of emissions sources. Additional problems with standards include the substantial resources required for designing, implementing, and monitoring emission controls and the difficulties in enforcement and achieving compliance.

Second, while regulations control the rate of emission, the primary issue associated with greenhouse gases is not their rate of emission or their effect on ambient air quality, but the total and cumulative amount of gases emitted and remaining in the atmosphere. The objective of emission standards is to stay below threshold levels that affect health or welfare, and this objective will require a different regulatory approach.

Regulatory tools that can be used to reduce future greenhouse emissions include emission standards, efficiency standards, and utility regulatory reforms. Future global climate change policy could combine maximum total emission levels (or, equivalently, set total emis-

---

[1] A fundamental exception to these "command and control" policies is the tradable emission allowances recently enacted in the 1990 Clean Air Act Amendments.

164 — Chapter Eight

sion reductions percentages) with selected regulatory "push" tools such as building codes and efficiency standards. These efficiency standards will probably be used to remove the least efficient equipment, appliances, and buildings from the market (*ibid.*).

Emissions Standards. Emissions standards typically prescribe maximum levels of pollutants that may be released into the air. For greenhouse gases, standards may be developed for total national greenhouse-gas emission reduction in tons of carbon, with specific reductions targeted at individual sectors or regions. Implementation of these reduction targets would probably emphasize flexibility and would not specify technologies used to meet the target. These total emission reductions can be combined with market "pull" policies.

Alternatively, and less likely, specific emission rate limits in pounds of carbon per kWh (lbs C/kWh) may be developed. This form of standard would be extremely technology-forcing. For example, OTA estimates that a .55 lbs C/kWh emission rate would only allow utilities to build coal plants utilizing efficient integrated coal gasification combined-cycle technologies (*ibid.*). In addition, the effects of emission rate limits will differ across utilities. A utility depending heavily on coal generation could be affected immediately by an emission rate limit, while a utility depending more on natural gas will not be affected until later, if at all.

Allowance and Offset Policies. A more flexible form of emission reduction would be to require utilities that open new plants, or expand capacity at existing plants, to offset the additional CO_2 emissions by reducing emissions at other facilities. Reduced emissions could also be derived from improved power plant or customer efficiency. These offsets could also be tradable, and sold like marketable permits.

Energy Efficiency Standards. A major future role for regulation may be the development of efficiency standards and codes for energy-using equipment, technologies, and buildings. For example, in 1987, Congress adopted minimum efficiency standards for refrigeration, air conditioners, and other residential appliances. Other possible forms of standards include:

- implementing energy-efficiency standards for other appliances and/or higher standards for appliances currently subject to standards;

- implementing mandatory building energy codes;

- establishing new standards for energy labeling of light bulbs, windows, and insulation materials; and

- setting power plant efficiency standards.

These approaches could also serve as bases for performance standards for utilities. Utilities would be given incentives in the form of higher rates of return for performance exceeding the efficiency standards set by state regulatory commissions. Specific requirements might include minimum power-plant or system heat rates or reductions in operation and maintenance costs.

Least-Cost Utility Planning. Least-cost utility planning combines market-based policy mechanisms with utility regulation. Least-cost plans typically outline utility programs for conserving fossil energy sources and expanding the use of energy efficiency and renewable energy sources in order to meet short- and long-term energy demands. In a scenario of global warming, least-cost planning could be expanded to minimizing utility or societal costs subject to global warming reduction or stabilization goals.

Other utility regulatory actions that could contribute to reductions in CO_2 emissions include requiring utilities to consider the cost of externalities in evaluation of new resource options, either utility-built or solicited through bidding programs. According to the Ottinger paper in this volume, 22 states incorporate or are in the process of incorporating environmental externality costs in utility planning. Wisconsin provides a good example of this incorporation of externality costs in long-range planning. The Wisconsin Public Service Commission recently published rules that create a 15% noncombustion credit for nonfossil sources such as renewables or energy conservation (Wisconsin Public Service Commission, *Re Advanced Plans for Construction of Facilities*, 05-EP-5, 102 PUR 4th 245). Based on these rules, the technical cost of noncombustion options will be reduced by 15% to account for the environmental benefits of avoiding combustion. The Wisconsin Commission implemented this policy to ensure that some of the costs of combustion are explicitly accounted for and reflected in utility plans.

Market "Pull" Mechanisms

Mechanisms and policies that provide incentives for reducing greenhouse gas emissions are alternatives to regulatory "push" policies. These mechanisms could include incentives (such as rebates and tax credits that reward actions that reduce emissions) and penalties (such as carbon taxes on fuels or technologies that are associated with high greenhouse gas emissions). The rationale for these market "pull" mechanisms is that they are economically and administratively efficient. Specifically, market-based approaches require considerably less centralized information collection; minimize waste by allowing a con-

166 — Chapter Eight

tinuous balancing of costs and benefits; and send proper signals to the industry and its consumers about the costs of global warming (Breyer 1979).

The nature of the global climate change problem may be particularly amenable to market-based policies. Emission of greenhouse gases, particularly CO_2, is pervasive and is a by-product of economic activity. Adjustment of the relative economics of carbon-based fuels may provide sufficient incentives for emission reductions. Nevertheless, there are often limits to the effectiveness of market-based policies, either due to inelastic demand, the absence of alternatives, or political or practical considerations.

Specific market "pull" mechanisms that could be used to reduce greenhouse emissions are reviewed below.

Emissions Taxes. Taxes offer a way to make technologies that emit high levels of greenhouse gases more expensive than lower-emitting technologies. Emissions taxes, of which a carbon tax is the best example, would also stimulate greater energy efficiency by increasing fuel costs. A carbon tax would also tend to cause a shift in energy production and use from high-carbon-emitting fossil fuels to low-carbon fuels or nonfossil fuels. In addition to encouraging the utilities to minimize the emission of pollutants, emissions taxes would also provide a new source of revenue that could be used for research and development of alternative energy sources or for reducing the federal budget deficit.

Energy Taxes. Energy taxes, less specific than emissions taxes, would target all fuels and thus discourage demand for all energy. For utilities, an energy tax would raise fuel costs, depending on generation mix, by increasing the relative cost of the taxed fuel. An energy tax, like the emissions tax, also would generate revenues that could be used for research and development projects designed to slow the pace of global warming or for reducing the federal deficit. However, a broad energy tax may be regressive and could hurt low-income individuals or smaller firms disproportionately.

Marketable Permits. Under a marketable permit system, environmental policymakers would fix the total amount of greenhouse gases that could be emitted and issue permits equal to this total amount. Like fuels, these permits could be bought and sold by energy users. If demand for energy rises, the price of a permit may also rise, reflecting the additional cost of lowering emissions. Permit holders would also have an incentive to seek cost-effective emission reduction techniques that would allow them to sell their permits (for a profit) to others. Marketable permits are the current U.S. method for enforcing the Montreal Protocol (see chapter 16) and controlling CFC emissions. Such permits are also a regulatory mechanism used to control acid rain in the 1990 Clean Air Act Amendments.

Tax Credits. A primary constraint to reduced greenhouse-gas emissions in the utility sector is that the industry is very capital-intensive and has a very low capital turnover rate. Given the high costs of new capital and the financial and regulatory uncertainty associated with the construction of new plant, utilities have few incentives to invest in new, environmentally benign plants. Market-based incentives could be used to increase the capital turnover rate in the utility sector. These might include tax incentives and regulatory revisions that encourage the earlier retirements of older, dirtier equipment. Further fiscal incentives could be given for technologies that are low-emitting or nonemitting. These incentives could include tax incentives or higher rates of return for these technologies.

Incentives for Efficiency and Conservation. Efficiency and conservation are key policies for reducing emissions from the combustion of fossil fuels. The following are examples of incentives that can be used to promote energy efficiency and conservation:

- consumer tax credits and other incentives such as rebate programs for energy-efficient equipment;

- federal and state funding for home weatherization programs; and

- hook-up fees that vary according to energy efficiency of the shell and equipment in new buildings.

Research, Development, and Demonstration

Research and development funding may be increased for renewable and nonfossil energy sources to initiate a movement away from energy sources that contribute significantly to global warming. Since large reductions in greenhouse gases will require substantial modifications in energy use and production, low-emitting technologies will need to be available in the future. Increased R&D activity may provide new sources or information on which technologies can be used within stringent emission reduction targets. Specific examples of R&D activity include:

- research and development on energy efficiency and renewable energy sources such as wind, photovoltaic, solar thermal, biofuels, ocean thermal, geothermal, and hydrogen fuel cells;

- promoting natural gas as a transition fuel for power generation from coal and oil and supporting research and development for intercooled, steam-injected gas turbines;

168 — Chapter Eight

- providing for research and development of a new generation of "inherently safe" nuclear reactors and promoting nuclear power; and

- promoting efficient coal repowering technologies such as pressurized fluidized-bed combustion or integrated coal gasification combined cycle.

These R&D programs would not be cheap. The 1990 combined energy technology R&D budgets (in 1990 dollars) for renewable energy, conservation, and nuclear fission were 83% lower than they were in 1980. An increase of $2.6 billion would be required to reinstate this research to only 1980 levels (OTA 1991).

Legislative Proposals

A number of legislative proposals addressing global warming have been introduced in the U.S. Congress in the 101st and 102nd sessions. The majority of these proposals have focused on across-the-board limits on total CO_2 emissions and the promotion of energy efficiency. One research-oriented bill, the Global Change Research Act, did pass in the final days of the 101st session. This bill establishes a program aimed at studying the cumulative effects of human actions on the environment. As of this writing, three of the major legislative proposals from the 101st Congress have been reintroduced in the 102nd Congress. These three proposals are:

Wirth Bill (S. 201)   Stabilizes greenhouse emissions (base period and target date are not specified). It calls for least cost utility planning. The new version includes CO_2 allowance and offset provisions. The old version of this bill passed the Senate in 1990. Most of S. 201 has been incorporated in the National Energy Strategy, or will become part of NES amendments.

Stark Bill (H.R. 1086)   Would impose a $28/ton tax on the carbon content of fossil fuels. This bill is pending in committee.

Cooper-Synar Bill (H.R. 2663)   Introduced in June 1991 to amend the Clean Air Act, this bill would, over time, stabilize and ultimately reduce CO_2 emissions. It requires electric utilities and other new major stationary sources of CO_2 to offset emissions. Mechanisms for offset include cogeneration, fuel switching, and conservation investments.

Additional global warming bills are expected to be introduced in the 102nd Congress.

States have also proposed initiatives and laws to reduce CO_2 emis-

The Global Climate Change Issue — 169

sions. The best example of proposed state action was the Big Green Initiative (Proposition 128), which would require state CO_2 emissions to drop by 20% by the year 2000 and 40% by 2010. This proposition was not passed by California's voters in November 1990.

Research Activities Now in Progress

Along with the above legislative proposals, a multibillion-dollar federal research effort that will supply critical data needed to ameliorate climate change problems is under way. Under the aegis of the U.S. Global Change Research Program, directed by the White House Science Office, 13 different agencies are investigating all aspects of global change. Key research activities currently under way are the Mission to Planet Earth and EPA's Global Climate Change Program.

A major research effort to study the extent of climate change is the Mission to Planet Earth. Mission to Planet Earth is a multibillion-dollar program using satellite information to monitor and detect changes in our environment. The program involves coordinating Earth-orbiting satellites to study oceans, land, and atmosphere and provide comprehensive, long-term "planetary data sets" for the study of ozone depletion, oceanic weather effects, and deforestation. A major goal of this effort is to develop the scientific data and understanding necessary to determine the existence and magnitude of global climate change.

EPA's current global climate change program stems from a congressional request that EPA conduct two studies. The first program (EPA 1988) examined the health and environmental effects of climate change; developed climate change scenarios; and studied the potential impacts on human health, agriculture, forests, wetlands, rivers, lakes, and estuaries as well as on other ecosystems and societal activities. A second program (EPA 1989) examined the policy options that, if implemented, would stabilize current levels of greenhouse-gas concentrations. These studies were very general, though, so others were initiated to quantify the effects of policy implementation. These additional EPA projects are discussed below.

CO_2 Costing Studies

The purpose of a major research effort (ICF 1990), performed by ICF and others, was to collect available information on the costs of reducing emissions of CO_2 and other greenhouse gases. The overall methodology included projecting CO_2 and CO_2-equivalent emissions to 2010, identifying technologies and programs that could significantly

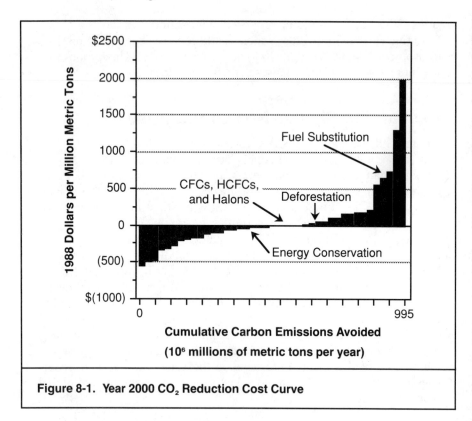

Figure 8-1. Year 2000 CO$_2$ Reduction Cost Curve

reduce these emissions, and estimating CO$_2$ reduction potential and the associated unit costs for these technologies and programs.

The principal programs investigated were energy conservation, fuel substitution, reforestation, methane control options such as coal bed methane recovery and landfill gas recovery, and the phase-out of CFCs. The cost curve shown in Figure 8-1 illustrates the results of the study effort. As shown in Figure 8-1, the least expensive policies that can reduce year 2000 emissions are energy conservation and the phase-out of CFCs. The low-cost energy conservation measures that have been identified in this effort include (listed in order of increasing cost): automobile and truck fuel efficiency, residential shell retrofits on electrically heated homes, residential appliance efficiency, and commercial energy conservation.

According to the study, for the year 2000, 15% of the 995 million metric tons of carbon potentially displaceable could be accomplished

Table 8-1. Summary of Potential Carbon Displacement for Year 2000		
Sector	Total Displaced (10^6MT)	Cost* (MM 88$)
Residential	107.585	$1,927
Commercial	71.780	-$4,461
Industrial	57.977	$17,940
Transportation	73.672	-$17,624
Electric Utility	82.163	$12,594
Non-Energy	601.970	$1,737
Total	995.147	$12,113

Note:
*A negative cost indicates that options would yield a net cost savings to the nation at a 7% discount rate.

cost-effectively through transportation and utility conservation alone. Table 8-1 indicates the project findings.

In another study examining the implications and costs of policies to reduce CO_2 emissions, ICF has developed a case study of CO_2 reduction strategies for the state of Michigan. The purpose of the project is to use existing state data and models from which to begin assessing the implications of alternative technical measures and strategies for reducing CO_2 emissions in the state. The work focuses on electric utilities and transportation, two important energy-using sectors in the state. Specific objectives include:

- identification of existing and emerging technologies and resource options that potentially can reduce CO_2 emissions;

- identification of combinations of technologies and resource options likely to achieve alternative, specified reductions in CO_2 emissions (for example, 20% by 2000);

- assessment of the technical and achievable emissions impacts, costs, and uncertainties associated with alternative strategies;

- estimation of the costs and resource implications of these strategies for Michigan; and

- assessment of state and federal policy options that could be used to implement preferred emissions reduction strategies.

172 — Chapter Eight

Results from this project indicate that significant reductions in CO_2 emissions can be reduced through cost-effective energy-efficiency measures.

Impact on Electric Utilities

Electric utilities and their customers would be substantially affected both by global climate change and by the policies—market-based or regulatory—to mitigate or forestall climate change. Utility impacts will be in the areas of electricity demand, electric utility operation, and utility DSM activities. This section briefly describes these impacts.

Electricity Demand

Climate change can affect the level and pattern of electricity demand. Changes in temperature and humidity can affect heating and air-conditioning needs, which influence peak demand and the shape of the load curve. Utilities serving agricultural areas with sizable pumped irrigation loads might also experience sharp demand fluctuations due to climate-induced changes in soil moisture. In addition, the overall demand in a region could be affected through population and demographic shifts that result from climate changes. A recent report (ICF 1989) examined these demand shifts. The study investigated how changes in temperature might affect energy demand through impacts on heating and air conditioning needs. Figure 8-2 shows the impact of a moderate temperature rise on patterns of demand in New York State in 2015. As can be seen, electricity demands would increase in the summer and decrease in the winter. These changes are greater downstate because of higher air-conditioning loads.

Electric Utility Operation

On the supply side, changes in temperature and precipitation can affect stream flow, which, in turn, affects the availability of hydro power. Delivery systems can also be affected by shifts in the frequency and intensity of extreme events predicted to accompany climate change, such as tornadoes, hurricanes, and severe storms. Finally, a rise in sea level could seriously affect utility operations in vulnerable coastal areas. The results of a recent ICF study (ICF 1989) indicated that temperature-induced reductions in stream flow could lower hydro generation by nearly 10% by the year 2015 in New York. Given hydropower's importance and status as New York's least expensive source of power, a decrease in this source was considered especially significant.

In addition, natural gas may become a transition fuel used by

The Global Climate Change Issue — 173

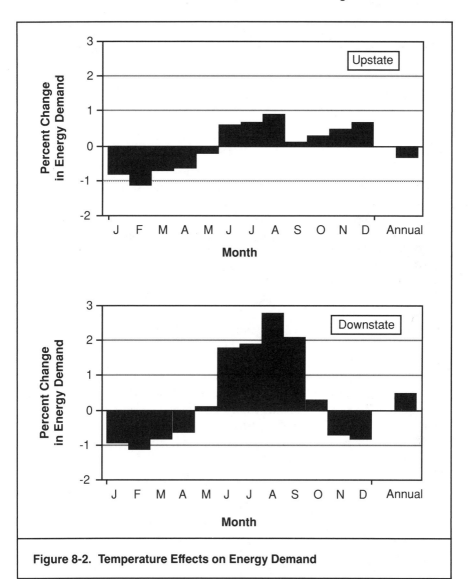

Figure 8-2. Temperature Effects on Energy Demand

174 — Chapter Eight

some utilities to help meet emission reduction targets while the utility explores alternative energy sources. For utilities with gas generation, hydro, and nuclear generating capacity, the transition will be smoother, although standards are likely to eventually impact all fossil-fuel plants.[2]

Impacts on Utility Demand-Side Management

A common thread in most of the proposed policies, legislation, and utility planning strategies to reduce CO_2 is greater reliance on energy efficiency. The results of our investigation of the costs to reduce CO_2 emissions support this position. Most of the legislative proposals responding to the climate change problem include provisions specifically geared towards the promotion of electric utility energy efficiency. In addition, energy efficiency can also be an important contributor to solving other environmental problems such as acid rain.

The impetus for utility DSM activity will come from two directions: (1) the direct need to reduce CO_2 emissions, and (2) the need to economically supply electric power for the future within the constraints associated with emission standards and/or increased costs of combustion technologies.

In order to achieve sizable demand and energy savings to meet the CO_2 goals, a much expanded level of energy efficiency will be required. This activity could result from either more stringent appliance efficiency and building standards or greater levels of utility investment in energy conservation programs. To meet this expanded level of effort, greater research will be needed on customer behavior, end-use patterns, and new energy-efficient technologies. Existing utility conservation programs will need to be enlarged and enhanced to encourage greater savings, and new programs will need to be initiated.

Restrictions in the use of carbon-intensive fossil fuels will create a substantial role for cost-effective conservation as a utility resource option. If trading of emissions permits, emission taxes, or fuel taxes are implemented, utility generation costs will increase, particularly for utilities with substantial coal- and oil-fired capacity. Utility avoided costs and customer rates will tend to increase at the affected utilities and will thus increase the amount of cost-effective DSM. Increased customer rates will encourage price-driven customer conservation

---

[2] Note that increased use of natural gas also could increase atmospheric emissions of methane, an important greenhouse gas. Methane is leaked into the atmosphere in varying amounts during extraction, transportation, and storage of natural gas.

activities. The interaction between this price-driven conservation and utility DSM programs needs to be examined in detail.

Finally, as utility DSM activity increases, utility managers, state regulators, and federal policy officials should be aware of the following issues:

- The forecasted increases in summer and winter temperatures could lead to a worsening of utility system load factors for some utilities. Increased air conditioner usage under these conditions will drive summer peaks and cause them to occur more frequently (ICF Resources 1989). This increase in usage will counteract improvements in customer end-use energy efficiency. Energy-efficient air conditioners may need to be emphasized to at least partially offset future increases in summer peaks.

- Actions to reduce or eliminate the usage of CFCs may increase the cost of residential refrigerators and freezers and of commercial and industrial chillers.

- DSM programs associated with the objectives of valley filling, load-shifting, or strategic load growth may not be good candidates for use as CO_2 reduction strategies. Depending on the base-load generation mix of a utility, promotion of these DSM programs could lead to an increase in CO_2 emissions from base-loaded coal and oil-fired generation.

- Federal and state environmental policy should explicitly account for the contribution of DSM to emission reductions. When a utility's customers reduce energy usage and thus emissions, emission offsets or credits should be given.

Summary

Global climate change, and especially global warming, are major concerns that will ultimately have a significant impact on the utility industry and its customers. Among these impacts is the need for more comprehensive efforts in energy conservation promotion. Energy conservation will reduce the need for fossil-fuel generation and would thus reduce the level of greenhouse gases emitted to the atmosphere. In addition, energy conservation and utility DSM seem to be the most cost-effective CO_2 abatement measures. So, beyond the new generation activities by utilities, we will see a marked and renewed interest in new utility conservation and demand-side activity.

176 — Chapter Eight

References

Breyer, S. 1979. "Analyzing Regulatory Failure: Mismatches, Less Restrictive Alternatives, and Reform," *Harvard Law Review* 92(1979): 552–53.

EPA. See U.S. Environmental Protection Agency.

ICF, Inc. 1989. *Potential Impacts of Climate Change on Electric Utilities*. EPRI EN-6249. Palo Alto, Calif.: Electric Power Research Institute.

——— 1990. *Preliminary Technology Cost Estimates of Measures Available to Reduce U.S. Greenhouse Gas Emissions by 2010*. Final Report to the U.S. Environmental Protection Agency. Fairfax, Va.: ICF, Inc.

ICF Resources, Inc. 1989. *Potential Impact of Global Warming on Pacific Gas and Electric Company*. PGE Report 009.5-88.6. San Francisco: Pacific Gas and Electric.

OTA. See U.S. Congress, Office of Technology Assessment.

U.S. Congress, Office of Technology Assessment 1991. *Changing by Degrees: Steps to Reduce Greenhouse Gases*. OTA-O-482. Washington, D.C.: U.S. Government Printing Office.

U.S. Environmental Protection Agency 1988. "The Potential Effects of Global Climate Change on the United States." Draft report to Congress.

U.S. Environmental Protection Agency 1989. "Policy Options for Stabilizing Global Climate." Draft report to Congress. Office of Policy, Planning and Evaluation.

Bonnie B. Jacobson is a vice president of ICF Resources, Inc. She heads the demand-side management and integrated resource planning practices. She holds a B.S. in Mathematics and Biology from LaRoche College in Pittsburgh and an M.Sc. in Biostatistics from the University of Pittsburgh Graduate School of Public Health.

David W. Kathan is a senior associate at ICF Resources, Inc. He specializes in demand-side management and integrated resource planning. He received a B.A. in Economics from Grinnell College and an M.A. in Public Policy from the University of Pennsylvania.

Chapter 9

Integrated Resource Planning and the Clean Air Act

Daniel M. Violette and Carolyn M. Lang, *RCG/Hagler, Bailly, Inc.*

Introduction

New environmental regulations, such as the Clean Air Act Amendments of 1990, will add a new dimension to utility resource planning. Under these amendments, utilities will have to account for each ton of sulfur dioxide (SO_2) emissions, not just the emissions that exceed allowable limits. This new emissions constraint was created by the national SO2 emissions cap and allowance system established by the acid rain title of the Clean Air Act. With an emissions cap, the control options for a utility are no longer mandated by the best available control technology (BACT). The utility is free to choose the control options that will be applied at each unit. Demand-side options, like any strategy that will reduce generation at affected units, are now eligible as emissions reduction strategies. In addition, the allowance system provides for exchange of emissions allowances between affected sources. This means that each allowance or ton will have a value within this market and that there is incentive to allocate allowances efficiently within a utility system to maximize allowance "profits."

The emissions allowance system is the first among a number of environmental regulations that could fundamentally change the integrated resource planning (IRP) process at utilities. Other proposals include a carbon tax and environmental externality adjustments.[1] By

---

[1] A carbon tax has been recently been introduced into Congress as part of global climate change omnibus bills. Several states have established orders to include environmental externalities in utility resource planning. The Massachusetts Department of Public Utilities in Order DP 89-239 established price adders for emissions of criteria air pollutants and greenhouse gases for new resources. The New York Public Service Commission has created a scoring system for utility resource planning decisions that can add up to 1.405 cents/kWh

178 — Chapter Nine

using an emissions cap rather than a command and control technology mandate, the Environmental Protection Agency (EPA) has allowed utilities flexibility in their compliance options. These compliance options will need to be considered in a utility plan. Emissions allowances will become another resource within the utility system that will need to be allocated efficiently to achieve the least-cost plan. Risk of emissions exceedance will also require consideration in utility planning to determine the optimal level of allowance reserves to hold to ensure a secure level of emissions allowances. The process of choosing the lowest-cost compliance options while considering risk and reliability could become another step in the integrated planning process, with environmental limitations becoming a new planning constraint.

This chapter will explore the effects of the Clean Air Act Amendments on utility resource planning. Under IRP, utilities will be expected by their regulators to manage risk on behalf of their ratepayers. Risk management evaluates risk mitigation strategies against their costs. With the large number of compliance options available to utilities under the Clean Air Act Amendments, utilities will need to weigh the cost of guaranteed reduction strategies against the potential risk of exceedance with some of the lower-cost compliance options. In addition, the new allowance market will bring opportunities for risks and rewards. The trade-off between compliance cost and risk, in combination with the uncertainty of a new market will add a new dimension to integrated resource planning.

The New Rules

The new rules of the Clean Air Act give some utilities a new asset to manage and, in some cases, a new card to play in the least-cost planning game. The acid rain titles in the Clean Air Act Amendments are based on an emissions cap rather than an emissions rate compliance system. An emissions cap is the total annual tons of pollutant, in this case SO_2, that an individual unit can emit. Current SO_2 regulations are based on an emissions rate that restricts each unit to a maximum rate, usually lbs/MMBtu, determined by the best available control technology: flue gas desulfurization. The primary difference between these two systems is that under an emissions rate or a technology requirement, a unit can run at any utilization level during the year as long as it stays below the allowable emissions rate in lbs SO_2 per MMBtu of

---

to the planning cost of a coal-fired unit. See chapter 10 for a detailed list of actions by state regulatory commissions to include consideration of externalities in the utility planning process.

fuel burned. An emissions cap, however, focuses on total emissions over the course of a year. Even if a particular unit operates below its maximum emissions rate, if it runs too long, it could exceed its emissions cap for that year.

The emissions cap for each unit is determined by calculations outlined in the Clean Air Act. The calculations use a baseline fuel consumption level, which is generally the average fuel usage for a unit in the years 1985–1987. This baseline is multiplied times an emissions rate, such as 1.2 lbs/MMBtu, to determine the total tons in a year that the unit can emit. The emissions rate used in the calculation depends on the size, age, and fuel type of the unit. Emissions caps for the largest SO_2 emitters are specified in Table A of the Clean Air Act Amendments of 1990. Other units can estimate their emissions caps by using the data in the National Allowance Data Base (version 1 released in 1990, version 2 to be released by EPA in 1991). This data file will be the basis for EPA's allowance calculation.

The emissions cap will be allocated to affected units through an emissions allowance system. An emissions allowance is permission to emit one ton of SO_2 in a specified year. For example, a unit with an emissions cap of 100 tons will be issued 100 allowances each year. It will be unlawful for any unit to emit SO_2 above the amount of allowances that it holds in any year. In this way, the allowance system is essentially the same as an emissions rate system with allowances serving as permits.

A new twist has been added to the emissions allowance system: allowance trading. Utilities will be allowed to trade allowances with units in their own system or with utilities across state lines. Since a finite number of allowances will be issued each year (currently estimated at nine million after 2000), a market will be created for this scarce resource. Under this market, utilities with excess emissions allowances can sell them to others who need more allowances to accommodate future growth or increased operating plans.

Allowance trading could reflect many types of compliance strategies. One type of intra-utility trade could consist of redispatching a utility system to reduce the electricity generated at selected units with high SO_2 emissions. This is a form of allowance trading because once the high emitting unit is redispatched, the utility must use another unit to generate electricity, and this alternate unit's emissions, in turn, will be increased. Other forms of trading could include either adding emissions controls to existing units to reduce emissions rates or shutting down units and transferring the allowances to a new power plant. New units are not eligible for allowances from EPA and will need to either purchase them on the market or receive them from other system units.

180 — Chapter Nine

Interstate trades will probably be transacted on a commodity market, where allowances are bought and sold between utilities at a price negotiated between the parties. Two different trading markets may develop: a long-term market selling streams of allowances and a short-term market selling single-year leases. The long-term market participants will probably be new power plant owners or nonutility generators (NUGs) who need allowances each year for the operating life of the new plant. These buyers might be interested in purchasing a lifetime stream of allowances that guarantee them access to the number of allowances required for a single year of operation. For example, a new plant emitting ten tons per year will need 200 allowances over a 20-year lifetime. Rather than buying all 200 initially, it may be easier to purchase a 20-year stream of allowances at a prenegotiated rate. The short-term market participants could be utilities exchanging allowances to meet annual fluctuations in demand and system operation. For example, a utility experiencing an outage at a nuclear plant may need to run coal plants more to compensate for the lost generating capacity. This utility may need to purchase allowances to allow its coal plants to emit more SO_2 during the nuclear outage. Utilities may even pool emissions allowances to ease the transaction process and allow for quick exchanges in emergencies. Until the market develops, it is not clear the types of transactions that will take place, but EPA is committed to creating a free allowance market with as few trade barriers as possible (EPA 1991).

Compliance with the New Rules

Compliance Options

Compliance planning is no longer just a matter of adding pollution control technology to new units. The emissions cap adds great flexibility to compliance, allowing many new options to become feasible. Several technology-based compliance options include flue gas desulfurization (FGD) (also known as scrubbing), switching to lower-sulfur coal, coal cleaning, and gas co-firing. The costs of an individual technology or compliance strategy can vary widely for different units because of site constraints and operating limitations.

In addition, because every ton of SO_2 has a cost under an SO_2 emissions cap, any measure that reduces SO_2 emissions in the system is a viable compliance option. Some of these new options include environmental dispatch, purchase power, allowance trading, and demand-side management. Environmental dispatch involves adjusting the dispatch order so that load is met with least cost *and* least emissions.

Given this new constraint, some utilities may find it cost-effective to invest in measures that increase the availability of low-emitting units, such as nuclear units, to reduce utilization of high-emitting coal units.

Purchase power becomes a compliance option because it is a way to meet load without any emissions. There are several risks with this option, especially the uncertainty of finding reliable, economical energy for sale. The acid rain emissions constraints could make some coal-based utilities unwilling to assume the risk of excess emissions produced while generating power for sale. The bulk power market will probably remain robust under the Clean Air Act Amendments, but there may be some new conditions added to sales, such as allowance leases or price adjustments, to compensate sellers for the additional emissions risk.

Emissions allowance trading and banking are new endeavors for utilities; as such, they carry with them a new set of risks and planning considerations. It is assumed that in the interstate allowance market, utilities with low marginal costs of compliance will have an incentive to over-comply and sell their extra allowances to utilities with high compliance costs or high growth rates. Using the marginal cost of pollution control as an indicator of allowance market potential, recent analysis shows a market clearing price around $750 per allowance (RCG 1990). Given this price, utilities whose cost to over-comply is less than $750 per ton will have an opportunity to gain by selling allowances. Utilities with compliance costs higher than $750 per ton will have incentive to purchase allowances rather than install controls. Utilities building new power plants will also be likely buyers, unless the new plant replaces an existing one and the allowance can be transferred internally.

Much speculation surrounds the allowance market and the future price of allowances. Some regulators fear that the past risk-averse behavior of utilities will cause them to retain allowances for their own use, thereby creating barriers for entry for nonutility generators and independent power producers who rely on the market to obtain allowances for new plants. This fear motivated legislators to withhold a portion of the allowance allocation for distribution to these new owners through a federal allowance auction and fixed price sale. Other utilities predict a drop in allowance prices after 2010 as the existing large coal units are retired and replaced by new clean units that will require only a fraction of the allowances made available by retirements. In fact, one objective of the emissions cap was to encourage construction of clean replacement units and allow for demand growth. The large number of retirements between 2010–2020, however, could cause a glut of allowances in the market and a corresponding price drop. Whatever the

182 — Chapter Nine

viewpoint, it is clear that the price and availability of allowances in this new market will be uncertain, and utilities may want to establish some strategies that reduce their risk related to these uncertainties.

Demand-Side Management as a Compliance Strategy

Any scheme that reduces generation at high-emitting plants will make it easier for utilities to meet their emissions target. DSM could play an important role in a compliance strategy as a way to reduce the demand for high-emitting plants or to shift the load to lower-emitting units. In addition, DSM could give utilities flexibility to meet load with a variety of resources and loosen the tight emissions reduction constraint. DSM is even receiving some preferential treatment as a compliance option with bonus allowance opportunities.

However, using DSM is not foolproof. For example, if one utility pursues an aggressive DSM program as part of its compliance plan and also is a member of a centrally dispatched power pool, there is no guarantee that this utility will receive credit for the conservation achieved. Instead, the reduced load may result in lower utilization of a second utility's plants, and the first utility would not see any reduction in emissions from its plants. Load shifting can prove to be perilous as well. If a utility implements a peak-shaving program and the majority of its peak units are low emitters, a reduction from these units will have little effect on overall emissions. However, if this utility shifts peak load to a low-emitting unit, such as a nuclear plant, overall emissions could be lower.

From these examples, it is clear that DSM alone does not ensure the emissions reduction, because the saved energy or generation needs to be taken from high-emitting units. Studies have shown that DSM is most effective in combination with environmental dispatch. A utility that implements a DSM program to target its high-emitting units' use patterns, or redispatches the system to maximize the DSM savings at high-emitting units, could reap substantial emissions savings. One study noted that energy conservation alone could reduce the cost of achieving one utility's share of a ten-million-ton SO_2 reduction by 40% (Center 1987).

In the Clean Air Act Amendments of 1990 (Section 404[f] and [g]), DSM is given preferential treatment as a compliance option with the Conservation and Renewable Energy Reserve of allowances, a two-for-one program for utilities that choose energy conervation as a compliance option. Utilities that pursue qualified conservation programs are eligible for bonus allowances related to their avoided emissions. A utility can bank extra allowances by reducing generation requirements at high-emitting units through DSM programs and receive bonus allowances for the avoided emissions. The provisions of the Conservation and Renewable Energy Reserve allocate up to 300,000 allowances to utilities that adopt new conservation programs. The provisions require that utilities adopt the conservation programs as part of a state-approved least-cost energy plan and create energy savings with qualified conservation programs between January 1, 1992, and December 31, 2000. Qualified programs will be determined jointly by EPA and the U.S. Department of Energy. The demonstrated energy savings from these programs will earn emissions allowances based on emissions that would be produced by a combustion turbine in generating the amount of energy saved. Although an individual program may only be expected to save enough energy to generate hundreds of allowances, this saving could be equivalent to hundreds of thousands of dollars on the market that could be used to offset the cost of the program. For utilities already planning to invest in DSM programs as part of a least-cost plan, these bonus funds could prove to make marginal programs big winners.

The DSM traps and the conflicting views of the allowance market highlight the uncertainty surrounding the new compliance options. This uncertainty may make some utilities unwilling to invest in them. However, the cost-effectiveness of these new options could reduce the total cost of compliance by as much as half (ICF Resources 1989). DSM, in particular, could give utilities flexibility in compliance by making more dispatch choices available or by reducing the need for new generation resources that could require precious allowances. Moreover, the energy conservation allowances could make more DSM programs cost-effective.

With all of these new options, there will be risks and rewards. Some utilities will be forced to pay more for allowances or controls than others. The stakes for utilities are high: utilities must choose their regulatory compliance options wisely, they must remain financially healthy in order to meet compliance costs, and they must plan carefully for these uncertainties if their systems are to grow.

Integrated Resource Planning and Compliance

Risk management will be the key to successful compliance planning. It implies more than costs of mitigating risks by strategies such as maintaining reserve margins. It involves the trade-off between mitigation costs and the acceptable level of risk. In order to fully evaluate all of the options, not just the technological fixes, emissions constraints should be included in IRP. There are several ways to incorporate compliance plans into IRP. Some utilities have chosen to evaluate acid rain

184 — Chapter Nine

compliance options separately and propose a plan to be incorporated into the IRP. Other utilities add an emissions constraint or externality cost to their dispatch plan and review the lowest-cost plans. Whatever the modeling technique, the key to a good plan will be fully evaluating *all* feasible compliance options and weighing them equally against the risks. For example, a key variable in the decision to scrub or switch coals at a coal unit is the future price of low-sulfur coal. If a utility believes that low-sulfur coal will be available at a low price in both the short and long term, switching will be the cost-effective option. Sensitivity to coal prices might be explored to determine their effect on the optimal plan. Our recent modeling experiences have uncovered several key variables in compliance planning.[2] These variables look remarkably familiar: fuel prices, demand growth, economic growth, and system constraints. The only new variable that we have found is the market price of allowances.

Risk Mitigation—Environmental Reserve Margin

Given that compliance planning is similar to IRP, it is not surprising that some of the expected risk-mitigating measures parallel existing provisions. Two areas that could develop in response to the risk of exceeding an annual emission cap, either because of system problems or market constraints, are an environmental reserve margin and emissions pooling. An environmental reserve margin has more dimensions than a reserve of emissions allowances. It can be a reserve of clean units that could be made available in case of an outage at another clean unit, such as a nuclear unit. This measure is similar in concept to a capacity reserve margin, but in addition to the size of the units, the type of fuel or emissions rate must also be considered. An environmental reserve margin can also be a bank of emissions allowances. With emissions banking, there is great incentive for utilities to hedge the emissions target. Rather than just meet the target every year, utilities with low-emitting base load plants (that is, nuclear units) might choose to reduce generation from their high-emitting units and bank allowances to compensate for outages at the low-emitting unit. The optimal level of banked allowances will depend on the fuel mix of the utility generation system and the availability of allowances in the market.

This environmental reserve margin could become a new requirement in least-cost planning. A reserve of emissions allowances or

---

[2] Based on Advanced Utility Simulation Model (AUSM) analyses for EPA and the National Acid Precipitation Assessment Program.

available nonemitting capacity may be required to stay under the annual emissions target in case of an emergency (such as an outage at a nuclear plant) that requires high-emitting plants to run longer. With a penalty of $2,000 per ton for every ton emitted in excess of the annual target *and* a reduction in the following year's target by tons exceeded, noncompliance under the acid rain provisions will have severe financial consequences for utilities. This risk penalty under noncompliance will motivate utilities to include an environmental reserve margin in their planning.

Risk Mitigation—Allowance Pooling

The economics of allowance pooling closely parallels that of power pooling and is based on the common goal of reducing risk. In the case of allowance pooling, the risk is that of exceeding the emissions target. Allowance pooling might look like this: in one year, a given utility may have emissions allowances to spare. This could result from high availability of nuclear plants. Another utility, partway through the year, may find that it is unlikely to meet the cap. An emissions pool could be established that automatically contains contract provisions for short-term lease of allowances. This practice would allow utilities to trade allowances on short notice without protracted negotiation. The negotiation would have been completed earlier as part of the pool agreement. Because emissions allowances are not bound by transmissions lines, pooling agreements could be transacted between utilities throughout the country, not just between neighboring utilities. Pooling partners could be utilities with similar allowance needs or utilities with widely different generation mixes to ensure that outages and weather do not increase allowance needs across the pool.

Overall, developing an optimal compliance strategy will probably follow the same process as IRP, with utilities evaluating strategies to minimize the risk of emissions exceedance with limited resources. New uncertainties such as the price of allowances and emissions constraints will be added to the analysis. In addition, new risk-mitigating measures may appear to reduce the likelihood of noncompliance. Utilities will be facing difficult risk management decisions in the trade-offs between the cost of various compliance options and the risk of not holding enough allowances to meet electric demand.

Winning Approval

A key player in the clean air debate is the public utility commission (PUC). Regulators will have significant influence over compliance

186 — Chapter Nine

plans through approval of capital projects and prudence reviews of trading strategies. Utilities will be required to justify their plans before commissions in order to win approval of capital investments and improvements.

The National Association of Regulatory Utility Commissioners (NARUC) and other regulatory agencies have been active in the legislative process but have made no final indication of their approval policy. Some preliminary issues, though, have risen to the surface. One especially thorny issue is treatment of allowance sales. It is not clear exactly what the value of an allowance should be and how the revenue from an allowance sale should be treated. Since allowances are given to utilities at no cost by EPA, the value of the asset is not easily measured. For example, the market value of an allowance may not truly reflect its value to an individual utility. A trade between a NUG and a utility for a stream of allowances may not reflect the value of spot allowances needed by a utiity in the current year. Although some precedents set by sales of real estate or other property may give some guidance, treatment of revenue from allowance sales, because it comes from the sale of an asset, not of electricity, represents new and difficult ground for commissions. If all of the revenue is returned to the ratepayers, there will be little incentive for utilities to overcomply and make trades. However, the ratepayers are financing the capital investment required to make the allowances available for trade.

The prudence of a trade may come under PUC review if at a later date a utility cannot meet demand growth because of an emissions constraint. Interstate operating companies may be especially vulnerable because they will have to coordinate their plans across state lines and commissions. Some commissions have suggested using a "hierarchy of needs" test before an allowance trade will be approved. Under this test, an interstate allowance trade will be approved if the needs of the following power producers are first met: (1) the individual utility (present and foreseeable future needs); (2) the other members of the power pool (if applicable); and (3) other utilities in the state, including municipals and cooperatives. Evaluating all these needs may be a difficult task. Commissions could even require utilities to coordinate their compliance plans and allowances trades to reach the overall lowest compliance cost for the state rather than just for each individual utility.

Given the strong influence that PUCs will have on the final approval of compliance plans, it may be prudent for utilities to work with commissions early on to understand their needs and concerns. Although the objective of the emissions cap was to move the burden of compliance planning off of the EPA and onto the utilities, the commissions may be caught in the middle.

Conclusion

IRP could be significantly changed as utilities consider their compliance options under the Clean Air Act Amendments. The newer compliance options hold new uncertainties and will create new risks for utilities. These new compliance options will require that compliance planning become part of IRP. Given that IRP is designed to choose optimal strategies in the context of limited resources, price uncertainties, and risks, some surprising combinations may result. Demand-side programs targeting base load may become more cost-effective for both capacity savings and emissions reductions options. Load-shifting programs may become too costly in emissions for some nuclear-based utilities. Environmental dispatch may become the least-cost dispatch pattern in a utility with emissions reduction goals. Strategies to build allowance reserves and contribute to allowance pools may alter the generation growth plans for utilities, with clean options gaining favor or retrofit control becoming cost-effective. DSM could play an important role in giving utilities access to the full range of compliance options by reducing the need for high-emitting units and allowing the system some slack in new generation requirements. Whatever the strategy, the Clean Air Act will redefine "least-cost" planning to include emissions allowances in the integrated resource plan.

References

Center. See Center for Clean Air Policy.

Center for Clean Air Policy 1987. *Acid Rain: Road to a Middleground Solution*. Washington, D.C.

EPA. See U.S. Environmental Protection Agency.

ICF Resources, Inc. 1989. *Economic Analysis of Title V of the Administration's Proposed Clean Air Act Amendments*. Prepared for the EPA. Fairfax, Va.: ICF Resources, Inc.

RCG. See RCG/Hagler, Bailly, Inc.

RCG/Hagler, Bailly, Inc. 1990. Memo to U.S. EPA, Air and Energy Engineering Laboratory, "Advanced Utility Simulation Model Analysis of National Acid Precipitation Assessment Program Scenarios." Boulder, Colo.: RCG/Hagler, Bailly, Inc.

U.S. Environmental Protection Agency 1991. Acid Rain Advisory Committee Issue Papers, Allowance Trading and Tracking Subcommittee. Washington, D.C.: Environmental Protection Agency.

Daniel M. Violette, Ph.D., is senior vice president of RCG/Hagler, Bailly, Inc. He is responsible for the utility planning division where he

188 — Chapter Nine

manages studies in the areas of utility economics, technical analyses and simulation modeling. He is also involved in integrating utility planning with environmental regulations.

Carolyn M. Lang is an associate with RCG/Hagler, Bailly, Inc. She concentrates on environmental and resource policy analyses, particularly in the energy industry. She holds a B.S. in Chemical Engineering from the University of Colorado and an M.S. from Stanford University in engineering-economic systems.

Chapter **10**

Consideration of Environmental Externality Costs in Electric Utility Resource Selections and Regulation

Richard L. Ottinger, *Pace University Law School Center for Environmental Legal Studies*

Approximately half the state electric utility regulatory commissions in the United States have started to require utilities to consider environmental externality costs in planning and resource selection. The principal rationale for doing so is that electric utility operations impose very real and substantial damage to human health and the environment, damage not taken into account either by traditional utility least-cost planning and resource selection procedures or by government pollution regulations. It is becoming clear that, given the high likelihood of more stringent controls in the future, failing to consider environmental externality costs when selecting resources is imprudent.

Most regulatory commissions requiring utilities to consider environmental externalities have left it to the utilities to compute the societal costs of residual environmental effects. Some commissions, however, have either set those costs themselves or used a proxy adder for the costs of polluting resources or a bonus for nonpolluting resources. These commissions have used control or pollution mitigation costs, rather than societal damage costs, in their regulatory computations.

This chapter recommends using damage costs where adequate

189

190 — Chapter Ten

studies exist to permit quantification; it also discusses methodologies for measuring damage costs and describes the means for incorporating such costs into the resource selection process.

Background

Internalizing the environmental costs imposed on society by polluters is the wave of the future in addressing environmental pollution. Governments are just starting to consider supplementing pollution regulations with pollution taxes or fees that would add to market prices the social costs of damages inflicted by polluting resources. The Organization for Economic Cooperation and Development (OECD) has recently published a review of pollution levies, indicating a total of 85 pollution taxes in six of its principal countries (Economist 1989). Germany, France, and Holland impose wastewater effluent charges ($2, $9, and $39 per capita respectively); Switzerland imposes extra landing fees on noisy aircraft; and Sweden and Norway require returnable deposits on automobile bodies to prevent their being dumped. West Germany is considering an auto tax based on tail pipe exhausts (ibid.). The U.S. House of Representatives Ways and Means Committee has held hearings on pollution taxes (6, 7, 14 March 1990).

Government regulation of pollution has not adequately addressed the severe threats to the planet posed by global warming, acid rain, urban smog, and other forms of toxic contamination of our air, water, and food supplies. These environmental insults have taken place despite the array of regulations designed to control them. Economic growth, in both developing and industrialized countries, has received higher government priority than pollution control and has outpaced governmental efforts to require pollution control technology or a switch to less polluting industrial resources. For example, in the United States, even after implementation of the Clean Air Act revisions recently passed by Congress (with all their attendant political compromises), substantial environmental costs still will be imposed on society by residual, uncontrolled impacts, many of which would be economic to mitigate.

The marketplace could powerfully influence the environmental effects of industrial decision making. If industry were required to pay the costs of the pollution it imposes on society, economics could induce industry to choose more environmentally benign resources.

Environmental organizations, which historically have resisted pricing environmental impacts on grounds that so doing would constitute a license to pollute, have now embraced the idea. The Environmental Defense Fund's Daniel Dudek, a leading advocate of letting the market reflect the costs of pollution, helped draft the Administration's Clean Air Act proposal to create emissions trading rights.

Traditionally conservative electric utilities and state utility regulatory commissions have pioneered in applying marketplace principles to valuation of environmental externalities. In 26 jurisdictions, utility regulatory commissions have started to consider whether and how to incorporate these externalities into utility planning and resource selection procedures.

Until recently, most utilities, with the approval of state regulatory commissions, have selected supply- and demand-side resources on a least-cost basis, without considering the environmental costs that persist even after utilities have complied with all applicable environmental regulations. This practice effectively assigns a value of zero to residual environmental externality costs, which may be substantial.

As difficult as it is to compute the exact costs of environmental damages, "a crude approximation, made as exact as possible and changed over time to reflect new information, would be preferable to the manifestly unjust approximation caused by ignoring these costs" (Bland 1986).

Why Utilities and Commissions Should Consider Environmental Externality Costs

Incorporating the monetary costs of environmental externalities into utility resource selection procedures will encourage utilities to invest in more environmentally benign resources. Nevertheless, it can be argued that environmental externalities should be addressed solely by pollution controls or by legislated taxes or fees rather than by utility actions and regulatory commission orders affecting resource selection. Indeed, environmental costs of other industries, such as automobile and chemical manufacturing and smelting, are addressed only through legislated regulation. National pollution fees could internalize environmental externality costs for all polluting sources, and thereby send correct price signals to the marketplace. Likewise, regulations eliminating all economically unacceptable pollution would also internalize environmental externality costs.

Neither action is likely, however. Complete economic pollution controls are likely to be politically infeasible; national pollution taxes or fees are unlikely because the administration and Congress have been reluctant to impose new taxes. Furthermore, the command-and-control structure of existing environmental regulations ignores cost-effective pollution prevention opportunities. National policymakers committed

192 — Chapter Ten

to this form of regulation would, in the presence of political forces, select a suboptimal level of environmental quality.

Assuming that legislation does not compel utilities to internalize environmental externality costs, utilities will go on selecting resources that, after regulation, still impose significant environmental damages on society. This practice will continue unless environmental costs are included in these considerations, because cost is the principal criterion for selecting a resource.

There are two justifications for asking utilities to factor externalities into their resource selection processes: (1) utilities are franchised monopolies vested with a public interest that includes environmental protection;[1] (2) foreseeable international, federal, and state environmental laws and regulations are likely to impose more stringent environmental controls over the 30- to 40-year life span of electric power plants, making it imprudent for utilities to invest in resources that will have to be abandoned or retrofitted at great expense. The traditional role of utility regulatory commissions includes overseeing utility public interest obligations and preventing imprudent investments.

Note that when commissions require consideration of environmental externalities in utility planning and resource selection, they are not internalizing these costs but merely ensuring prudent selection of new resources. This policy is a good prelude to actual internalization.

Costs to Be Included

Commissions and utilities that decide to consider externalities face the daunting task of determining those costs. The first question they must answer is what kinds of costs to include. Most commissions that have addressed externalities have determined that only environmental damages should be included. However, the most thorough study of externality costs, completed recently for the European Economic Community, seeks to value not only environmental damage but also impacts on production, employment, and trade balance; depletion of nonrenewable resources; public subsidies and R&D expenditures; and "induced public expenditures" such as defense costs (Hohmeyer 1989).

The most severe environmental costs imposed on society by elec-

---

[1] The state of New York, for example, requires all state agencies to minimize or avoid environmental impacts "to the maximum extent practicable." See, for example, the New York Environmental Quality Review Act (SEQRA), New York Environmental Conservation Law, Section 8-0107, and its implementing regulation, 6 NYCRR, Part 617.9 (c) (3).

tric utility operations derive from the risk of damages to human health and the environment from air pollutants emitted by fossil-fuel-fired generation and from the radiation emitted by nuclear plant operations.

The principal culprits among air pollutant emissions are the greenhouse gases, most significantly carbon dioxide (CO_2); sulphur dioxide (SO_2) and nitrogen oxides (NO_x), the principal precursors of acid rain; tropospheric ozone resulting from chemical interactions of NO_x and volatile organic compounds in the presence of sunlight; and particulates, which provide the medium for ingestion and inspiration of toxic co-pollutants. Nationally, electric utilities accounted in 1988 for 33% of CO_2 emissions (Machado and Piltz 1988) and in 1985 for 68% of SO_2 emissions (16.2 out of 23.7 tons) and 33% of NO_x emissions (7.0 out of 21.1 tons) (NAPAP 1987).

Many of the above air pollutants react synergistically after release, forming new chemical combinations that react together to inflict increased environmental damage. Furthermore, emission damages from future power plants will be cumulative, not simply additive, to the damages from pollutants already emitted by existing plants: that is, they will do more damage than they would have done if no previous pollution had been present. These synergistic and cumulative effects, not just the additive damages from each new pollutant, should be considered.

Risks of nuclear-powered resources include radiation from uranium milling and mining, low-level radiation in and around operating plants and accidental contamination of plant personnel, catastrophic accidents such as occurred at Chernobyl and Three Mile Island, contamination from plant decommissioning, disposal of mill tailings and of high- and low-level nuclear waste, and impact on fish at nuclear facilities. These risks are largely unaccounted for under current regulation. A recent British report asserts that the nuclear cycle also emits carbon dioxide at levels potentially comparable to fossil-fuel plants (Hill 1990).

In addition to damage from air pollution and risks from radiation, societal costs imposed by many electric service supply resources may include water pollution, land deprivation, agricultural losses (from flooding by dams, for example), and contamination from solid waste disposal. Damage from electromagnetic fields has also been asserted. These costs are generally much less than those resulting from air pollution and radiation damage, but they are significant enough to merit consideration.

Air- and water-polluting emissions from generating plants often extend beyond the state or country in which the plant is located. Most

194 — Chapter Ten

state commissions that have addressed environmental impact costs have included all costs to society, not just those affecting their own states.[2]

Environmental externality costs should also reflect differences between pollution costs in urban and rural settings; where appropriate, emissions should be calculated per unit of population. Pollutants emitted near a heavily populated metropolis will produce vastly greater human health costs than those emitted near unpopulated areas. Similarly, agricultural damage costs will be higher where emissions are deposited on farming communities rather than on urban areas.

Costs from the entire fuel cycle should be considered (front-end, operational, and back-end costs). It is difficult to know how far to pursue front-end costs, of course. One could go back infinitely far, estimating the social costs of manufacturing all the equipment and machinery necessary to manufacture the equipment and machinery, and so on, for each stage of the fabrication process. At the least, the first-generation costs of plant construction and fuel transport and the production costs of demand-saving or renewable equipment and facilities should be considered.

Control versus Damage Costs

Having decided which costs to include, the next major problem facing a commission or utility is how to calculate them. The first major issue is whether to use the costs of actual damages or the costs of controlling pollutants before the damage occurs. There is considerable difference among experts on this subject.[3]

The advantage of using control costs is that readily available data make these costs easier to determine and thus more defensible. Costs derived from legislative standards, it can be argued, reflect the level of protection the relevant agency experts and the legislative body consider safe, though in fact standards tend to be determined politically rather than scientifically. The relative ease of determining control costs has caused all 11 state commissions that have ordered consideration of environmental externality costs to date to use them both as the basis for

---

[2] Vermont is the only exception, accounting for only that portion of out-of-state pollution from Vermont power plants that impacts Vermont residents. However, the Vermont commission seems to be backing off this position in its most recent proceedings: Vermont PSB, *Re Least-Cost Investments, Energy Efficiency, Conservation, and Management of Demand for Energy*, Docket No. 5330.

[3] For arguments supporting use of control costs, see Chernick and Caverhill 1989; for the case for using damage costs, see EPRI 1988.

Consideration of Environmental Externality Costs — 195

quantification of such costs and, where used, in the calculation of adders to the cost of polluting resources.[4]

The disadvantage of using control costs is that they may bear little or no relationship to the actual costs of damages imposed on society by the relevant pollutants, and they seldom cover all the risks involved because elected representatives tend, for political reasons, to enact controls well below marginal damage costs. Furthermore, control standards such as those established by the National Ambient Air Quality Standards of the Clean Air Act are adopted at a level to protect the public health and welfare with "an adequate margin of safety," often without regard to the costs to society of health and welfare damages. On the other hand, control costs might exceed damage costs, as in cases where standards are set to protect the most sensitive individuals in society. And there are many power plant pollutants, such as CO_2, for which no standards have been set.

The main advantage of using damage costs is that they are the relevant costs to be considered. It is the risk of damage to society, rather than the cost of controls, that the practice of incorporating environmental externality costs into utility resource selection seeks to address. Damage costs are useful as well for determining how much it is worth spending to institute additional controls.

The main disadvantage of using damage costs is the difficulty of calculating and defending them. Some experts feel that utility regulatory commissions (as opposed to agencies with environmental expertise like EPA) would have difficulty dealing with technical matters like valuation of human life and nonmonetarized costs like valuation of recreational facilities. On the other hand, there are some adequate scientific studies. Defense of a legal challenge to these values should be no more difficult than the generally successful defense of EPA health and safety standards that are based on similar kinds of scientific studies.

In view of the advantages and disadvantages described above, the most beneficial policy is to use damage costs when feasible, since these costs are most relevant to the impacts on society. Where adequate studies exist, damage costs should be used. Where studies on damage costs are inadequate, as in the case of global warming research, then

---

[4] The Oregon Public Service Commission, however, has ordered its utilities to seek to quantify damage costs in evaluating resource selections, *Re Least Cost Planning*, UM 180, OR PSC Order No. 89-507 (20 April 1989). The New York Public Service Commission has ordered a pooled study by its utilities of their environmental externality damage costs, NY PSC Case 28223, *Electric Utility Conservation Programs*, Opinion and Order 89-15 (23 May 1989).

196 — Chapter Ten

control costs should be used as the best available substitute, far superior to setting damage costs at zero by ignoring them.

Where impending legislative controls are reasonably ascertainable, as was the case with the Clean Air Act Amendments, the effects of the new controls on damage costs must be taken into account, since the pollutant costs covered will be internalized once the legislation is enacted. Of course, if controls like scrubbers or bag houses (a technology for removing particulates from flue gas) are required by statutory or regulatory mandate, the costs imposed on society by the pollutants controlled will no longer be external costs.

Major Issues in Damage Risk Valuation

General Considerations

In the valuing of environmental damages, the objective is to determine the risk of damage rather than to assess the damages themselves. This is a most important principle. We are seeking to define the cost of the *risk* to life, health, and the environment, and the costs that people are willing to pay to avoid such risks or assume them.

For example, it would be inappropriate to measure mortality damages by seeking to measure the value of a human life, say by adding up the reasonably expected lost lifetime earnings of the individual or individuals affected. The value would vary by earning power, with rich people valued more than poor ones and with housewives and the elderly considered to be of negligible value. Similarly, it is inappropriate to use only mitigation costs as a measure of externality values. Adding up the doctors' bills is inadequate, for example, in valuing human health damages. Who among us would be willing to incur the hardships associated with cancer or debilitating injury even if fully compensated for the cost of treatment? But for each individual or population of individuals, it would be appropriate to measure the value of the risk posed to their lives by determining what they would be willing to pay to avoid the risk or what they would be willing to be compensated to assume the risk.

Also, risks must be quantified for populations rather than for individual people, crops, or animals. An event that kills or harms a very small number of individuals may be very costly to them, but the social costs of this damage will be insignificant. This distinction does not have anything to do with the potential criminal or civil liability for taking an individual life or with the value society places on every human being; it has only to do with valuation of risks to life for purposes of influencing utility resource selections. The value of the risk of loss to

Consideration of Environmental Externality Costs — 197

a very few individuals simply is not large enough to affect the economics of choosing one kind of utility resource over another.

Likewise, damage awards by a judge or jury for particular environmental damages are generally not very useful because they value the harm to an individual of an actual event, not the risks to populations of possible events. Also, particularly in the case of jury awards, damage awards are not scientifically derived.

Measurement Methodologies

Assigning monetary values to risks poses the following issues:

- which measurement methodologies to use (how and when to use market prices, revealed preference, hedonic pricing, awards, or contingent valuation);

- how to allocate costs to joint projects;

- what discount rates and real value escalation to apply; and

- how to adjust for uncertainty.

(For a good discussion of all the externality costing methodologies and their applications, advantages, and disadvantages, see EPRI 1988; Freeman 1979; and, with respect to contingent valuation, Mitchell and Carson 1989.)

Market prices, where available, are useful in determining environmental damages. Knowing there is a 100% risk that a particular crop will be affected by a power plant and knowing the extent of the harm that will be imposed, one can multiply the crop loss by the market value to obtain the damages. Unfortunately, in most cases, risk and yield loss are seldom known with certainty, and, with large losses, the market price may be affected by the loss.

Revealed preference values are based on observed behavior. They are derived from the costs individuals, by their actions, have revealed they are willing to pay to avoid environmental damages or to accept as compensation for suffering such damages. Thus, in the case of loss of fishing opportunities in a lake because of acid deposition, travel costs fishermen are actually paying to reach alternative fishing areas might be used to value the damages. Travel costs, however, fail to take into account values not encompassed by the particular behavior measured. For example, travel costs would not accurately value the destruction of a unique historic resource even though there were other historic resources that could be visited. They may also fail to account for characteristics, such as an individual's age or income, that might prevent his or her traveling to alternative sites.

198 — Chapter Ten

Hedonic pricing is a form of revealed preference that uses market-based prices to infer prices of nonpriced goods and services. For example, selling prices of comparable homes with and without a scenic view can be compared to determine the value of the scenic view. Great care must be exercised to determine that the values compared are truly comparable.

Contingent valuation seeks to determine by surveys the value individuals would pay to avoid, or to accept as compensation for, an environmental hazard. Great care must be exercised to eliminate biases in the framing of questions, and even then respondents' answers may be colored by strategic motivations to influence a particular outcome. Nevertheless, for many nonmarket effects, contingent valuation is the only or best means of valuing risks and can be used to value multiple aspects of a complex risk without having to value each aspect separately.

Allocation of Costs to Joint Projects

In allocating costs to multipurpose projects, such as cogeneration or waste-to-energy facilities, the most important initial consideration is whether purposes other than generating electricity dominate the project. If the project would operate regardless of whether it generated electricity—for example, if producing heat were the project's dominant purpose—then none of the project's environmental costs should be allocated to production of electricity. If electricity production is one of the project's main purposes, however, environmental costs can be allocated according to

- the separable costs of fulfilling each purpose;
- the value of the product of each process;
- the relative importance of the purpose of each process to the plant;
- the added emissions from electricity production where calculable; or
- the heat rate of each process.

Since environmental costs are being valued, using emissions contributions, if ascertainable, is best. Emissions levels can be measured directly or derived from the amounts of fuel used in producing each product or from the proportional heat rates of each process.

Discount Rates and Real Value Escalation

There is much controversy among economists and other experts about what discount rate, if any, should be applied to environmental externalities. Discount rates are used to compare future economic benefits and costs to today's benefits and costs. Low discount rates weigh the future more heavily (and the present less heavily) than high discount rates do. Thus if a zero discount rate were used, damages to future generations would be counted the same as damages to today's populations. If a 6% discount rate were used, damage to future populations would be decreased 6% each year.

Some maintain that a zero discount rate should be used, particularly for risks to human life and health, because a life in the future is as valuable as a present life (Shuman and Cavanagh 1984). They maintain that sound stewardship of the environment mandates that the value we put on future lives and other environmental assets be considered as highly as present values. Furthermore, they maintain that discounting double-counts the possibility of avoiding future risks since the calculation of present risk already takes into account future events that may diminish the chances of the risk being realized. Lastly, they assert that it is unrealistic to discount long-lasting risks, such as those from high-level nuclear wastes, which pose risks for millennia.

Others reject a zero discount rate for risks to human life and health and argue that the value of damages decreases over time and that a zero discount rate places the present value of future environmental damages much too high. In an extreme case, for example, they assert that few would be willing to pay anything substantial for the risk of human fatality 10,000 years from now: long before that remote time, technology will likely resolve the environmental threat (or the world will be destroyed). In a less extreme example, if forced to choose between a risk of death today and the same risk 50 years from now, it seems likely that the delayed health risk would be preferred (although if the issue were protecting one's own life versus protecting the lives of one's children, an individual might well choose the latter). Use of discount rates higher than zero takes into account the lesser values that may be put on future risks. Placing lower values on future lives than on present ones, they assert, takes into account the lesser willingness of the public to pay for damages far into the future and the likelihood that the risks will be alleviated during long time periods.

Some experts maintain that as a matter of consistency the discount rate applied by utilities to their capital investments (approximately 12%) should be applied to environmental risks. They also argue that using lower discount rates will place the present value of environmental risks too low to have a meaningful influence on resource selection (Chernick and Caverhill 1989).

Many economists adopt a middle ground, using a "social rate of time preference" discount rate (social discount rate), usually in the

200 — Chapter Ten

neighborhood of 3%—lower than utility investment rates, but higher than a zero discount rate. The social rate of time preference is the rate at which society is willing to exchange consumption now for consumption in the future. It reflects the ability of society to remedy environmental hazards over time. The main reason asserted for using a social discount rate rather than the utility discount rate is that the value of environmental costs and benefits *to the public* is being evaluated (and discounted), not the investments of the utility. For the same reason, the social discount rate is preferable to the opportunity cost of public investment and the consumption rate of interest, since these rates measure the costs and benefits of risks to investors and individuals, not to society. Social discount rates should be calculated from the time environmental risk is created (BPA 1986).[5]

Real value escalation estimates the increases in price that will take place over time in environmental and energy resources due to inflation and increased scarcity of finite resources. Real value escalation must be used in valuing environmental externalities.

Uncertainty

Valuation of all environmental externality costs must deal with a considerable margin of uncertainty. Often wide ranges of costs are advanced in different studies. An example is the enormous range in estimates concerning the probability of nuclear accidents. These uncertainties should be dealt with by showing the full range of cost estimates and the bases for those estimates. Then sensitivity analysis should be applied to select a reasonable point within the ranges of the studies, and rationales for the selection should be made explicit.

Despite the uncertainties, the damage costs of most pollution produced by electric utilities can be estimated. The uncertainties involved usually are no greater than the uncertainty in selecting pollution standards to avoid "significant risk to human health and the environment" or "with a reasonable margin of safety," as prescribed by the environmental protection statutes and upheld by the courts.

The principal problem with using environmental damage costs to determine electric utility externality costs is that too few of these damages have been valued adequately. Major research is vitally needed in this area. Reliable damage figures are important not only for incorporating accurate externality costs in utility resource selections but also for imposing accurate pollution fees and pollution control standards.

---

[5] Pace 1990 used both a 3% social discount rate and a 12% utility discount rate applied to all damage studies it reviewed.

Consideration of Environmental Externality Costs — 201

State Incorporation of Environmental Externalities

Presently Used Methods

The methodologies presently used by states to incorporate environmental externality costs include quantitative, qualitative, rate-of-return, and avoided-cost consideration (see Table 10-1). Some states use more than one method; New York, for example, uses quantitative, rate-of-return, avoided-cost (by statute), collaborative, planning, and bidding consideration.

These methodologies have been implemented by the states in planning, bidding, and other resource selection determinations. Collaborative processes between utilities, state agencies, and intervenors for determination of environmental costs and their application have also been used, but with little success. In addition, several innovative methodologies have been proposed for incorporation of environmental externalities in utility planning and resource selection, but no state has yet adopted these methodologies. Table 10-2 shows methods used by the states that have begun to act on this issue and by Washington, D.C., Bonneville Power Administration (BPA), and Northwest Power Planning Council (NWPPC).

Orders for Consideration

Of the remarkable number of 29 state public service commissions (PSCs) or legislatures that have taken some action to incorporate environmental externality costs, 19 have issued orders or passed legislation requiring their utilities to take into account these costs in planning and/or bidding. Two states have such orders pending, meaning that a costing proceeding has been established and hearings are under way or that an Administrative Law Judge decision is pending or has been issued and is awaiting commission action. Eight states are actively considering orders for consideration of environmental externalities, meaning that a commission has explicitly stated that it intends to consider externalities.[6]

Quantitative Consideration

In quantitative consideration, a commission or utility under commission order establishes dollar values for environmental costs. The val-

---

[6] *Orders:* Calif., Colo., Idaho, Kan., Mass., Mich., Minn., Nev., N.J., N.Y., Ohio, Ore., Penn., Texas, Vt., Va., Wis. *Statutes:* Alaska, Nev. *Orders pending:* Conn., Iowa. *Orders under consideration:* Hawaii, Maine, Md., Mont., N.C., R.I., Utah, Wash., D.C.

Table 10-1. Status of Actions Incorporating Environmental Externality Costs

N = No action
O = Incorporation ordered
P = Incorporation order pending
U = Under consideration

	O	P	U	N		O	P	U	N
Ala.				X	Nev.[a]	X			
Alaska[a]	X				N.H.				X
Ariz.	X				N.J.	X			
Ark.				X	N.M.				X
Calif.[b]	X				N.Y.	X			
Colo.	X				N.C.			X	
Conn.		X			N.D.				X
Del.				X	Ohio	X			
Fla.				X	Okla.				X
Ga.				X	Ore.	X			
Hawaii			X		Penn.	X			
Idaho	X				R.I.			X	
Ill.				X	S.C.				X
Ind.				X	S.D.				X
Iowa		X			Tenn.				X
Kans.	X				Texas	X			
Ky.				X	Utah			X	
La.				X	Vt.	X			
Maine			X		Va.	X			
Md.			X		Wash.				X
Mass.[b]	X				Wash., D.C.			X	
Mich.[b]	X				W. Va.				X
Minn.[b]	X				Wis.	X			
Miss.				X	Wy.				X
Mo.[c]				X	BPA[a]	X			
Mont.			X		NWPPC[a]	X			
Neb.				X					

Notes:
[a] Established by legislation.
[b] Order issued to consider externalities; implementation pending.
[c] Commission has stated that it may consider externalities.

Source: Pace 1990

Table 10-2. Types of Actions to Incorporate Environmental Externality Costs

AC = Avoided-cost consideration	P = Incorporation order pending
B = Bidding consideration	PL = Planning consideration
C = Collaborative action	QL = Qualitative consideration
ED = Environmental dispatch	QN = Quantitative consideration
L = Legislative action	ROR = Rate-of-return consideration
O = Incorporation ordered	U = Under consideration

	QN	QL	ROR	AC	ED	C	PL	B	Comment
Alaska				O			L		
Ariz.		O	O					O	
Calif.	O				U	C*	O		*Collab. ended
Colo.	O*							O*	*By fuel type
Conn.	P		L*			C	P	P	*5% ROR adder
Hawaii							U		
Idaho		O		O					
Iowa				P		C			
Kan.	P		O						
Maine		U					U		
Md.	U					C		U	
Mass.	O*					C	P	P	*DSM evaluations
Mich.	U			O			U		
Minn.*		O					P		*Law caps SO_2
Mont.		U	L*				U		*2% adder for DSM
Nev.	O	L					O		
N.J.	O			O				O	
N.Y.	O		O	L		C	O	O	
Ohio		O					O		
Okla.			O*						*ROR, trash only
Ore.*	O	O					O		*Law caps CO_2
Penn.		O*					O*		*Not implemented
R.I.							U		
Texas		O					O		
Utah							U		
Vt.	O					C	O	P	
Va.				O*					*15% DSM adder to AC
Wash.			L*						*2% ROR law for DSM
Wash., D.C.	U							O	
Wis.	O	O			U		O		
BPA	L						L	U	
NWPPC	L						L		

Note: In the categories listed there is overlap; thus New York, for example, is listed as having quantitative, rate-of-return, avoided-cost (by statute), collaborative, planning, and bidding considerations.
Source: Pace 1990.

204 — Chapter Ten

ues calculated are then added to the cost of resources in the selection process or used in a resource rating system. Some commissions attach a proxy percentage adder to polluting resources, a percentage credit to nonpolluting resources, or both.

As Table 10-2 shows, 13 states and Washington, D.C. have adopted quantification or have quantitative orders pending or under active consideration. Nine states plus the BPA and NWPPC have acted to consider environmental externality costs quantitatively or to use a proxy adder to represent these costs. Two states have such quantification orders pending, and two states and Washington, D.C. have them under active consideration.[7]

The New York PSC has been the pioneer in incorporating quantified environmental externality costs, requiring its utilities to assign about 15% of total bid evaluation scoring points to environmental externality costs (about 24% of price scoring points), calculated at 1.405¢/kWh total environmental externality costs, based on a coal-fired plant meeting New Source Performance Standards (NSPS).[8] These same environmental costs must be used in valuing demand-side management (DSM) investments in integrated resource planning.[9] The PSC has also ordered all New York utilities to do a pooled study, with participation of outside experts and the public, to quantify the environmental externality costs of pollution from their utilities' operations.[10]

In Wisconsin, a noncombustion credit/adder has been adopted for screening of all utility resource acquisitions, so that a noncombustion source that costs 15% more than a combustion source will be considered on a par with the latter; this screening is followed by a qualitative

---

[7] *Quantified orders:* Calif., Colo., Mass., Nev., N.J., N.Y., Ore., Vt., Wis. *Limited order with more extensive order pending:* Calif. *Quantified orders pending:* Conn., Mass. *Under consideration:* Md., Mich., Wash., D.C.

[8] See N.Y. PSC Case 88-E-241, *Proceeding on Motion of the Commission (established in Opinion 88-15) as to the guidelines for bidding to meet future electric capacity needs of Orange & Rockland Utilities, Inc.*, Order Issuing a Final Environmental Impact Statement and Adopting Staff's Response to Agency Comments (24 March 1989).

[9] *Formats and Guidelines for July 23, 1990, DSM Plan Filing in Case 29223.* N.Y. Department of Public Service Staff (23 February 1990). The guidelines provide that
Environmental benefits are to be explicitly quantified in the total resource cost test. The Staff estimates of environmental costs, developed initially in the electric capacity bidding cases, should be used in the assessments for the 23 July, 1990 plan.
The environmental benefits to be used are
1.4¢/kWh for programs that promote energy efficiency
0.9¢/kWh for programs that are aimed at peak clipping
0.4¢/kWh for programs that are aimed at load shifting

[10] NY PSC Case 28223, *Electric Utility Conservation Programs*, Opinion and Order 89-15 (23 May 1989).

test requiring consideration of environmental impacts.[11] Also, in integrated planning, resource valuations must assume that carbon dioxide emissions will have to be reduced to 80% of their 1985 level by 2000 and to 50% of the 1985 level in the long run.[12] Wisconsin legislation concerning acid rain also requires utilities to cut 1980 levels of sulphur dioxide emissions by 50% by 1993 (Cohen and Eto 1989); while not directly used in valuation, this statute is relevant in that it internalizes some of the costs of acid rain.

The Oregon commission has not quantified environmental externality damage costs but has ordered the utilities to do so "to the fullest extent practicable" and to "indicate ranges of costs where definite damage costs cannot be determined." The utilities are required to consider these damage costs in resource selection.[13]

The Massachusetts Department of Public Utilities has just issued an order that values environmental externalities for an NSPS coal plant about 4¢/kWh.[14] The California Energy Commission has set a similar value on NO_x power plant pollution, and the California PUC is completing a proceeding requiring that these values be incorporated in utility resource selection.

Qualitative Consideration

Nine states presently have ordered, or are considering ordering, that their utilities take into account environmental externality costs—in planning or resource selection or both—without specifying how these costs are to be calculated or considered.[15]

Rate-of-Return Consideration

Rate-of-return consideration involves an award by a commission of an increased rate of return to utilities as an incentive to install nonpolluting resources. Increased rate of return is typically allowed either on particular nonpolluting resource investments—usually DSM, renewables, or resource recovery plants—or on total investments. While DSM incentives are adopted primarily to make DSM investments as profitable as supply investments, not necessarily to capture environmental externality costs, most commissions cite environmental bene-

---

[11] Wis. PUC *Re Advance Plans*, Docket 05-EP-5, 102, P.U.R. 4th 245 (6 April 1989).

[12] Wis. Stat. Ann. Sec. 144.385-389 (1989).

[13] Ore. PSC, *Re Least Cost Planning*, UM 180, Order No. 89-507 (20 April 1989).

[14] Mass. DPU Order 89-239 (31 August 1990).

[15] *Orders:* Ariz., Minn., Nev., Ohio, Ore., Penn., Texas. *Under consideration:* Maine, Mo. Wisconsin uses qualitative consideration after applying a 15% noncombustion credit/ adder (see footnote 11).

206 — Chapter Ten

fits as one of the reasons for offering such incentives. We have included in this compilation the states where this is so.

As Table 10-2 shows, nine states give rate-of-return consideration to environmental externality costs; Oklahoma does so for resource recovery plants only.[16]

Washington and Montana have statutes providing a 2% additional rate of return on energy conservation investments and Connecticut a 5% DSM rate-of-return adder, citing environmental externality costs as a justification. Kansas gives a 0.5%–2% increased rate of return on renewable and conservation resources. The Idaho commission has announced that in future rate cases it will take into account utility conservation efforts in determining the allowed rate of return on total investments. The Wisconsin commission experimentally has allowed Wisconsin Electric Power Co. to earn an additional 1% rate of return for each 125-MW reduction in peak load the utility achieves through efficiency investments. New York has adopted, as a temporary DSM incentive, the return of lost revenues from DSM investments plus a performance-based incentive.

Avoided-Cost Consideration

Seven state legislatures or commissions have required, or are considering requiring, adding a premium to utility-calculated avoided power plant capacity and energy costs, to help account for environmental externalities.[17] Thus, New Jersey has established avoided cost under the Public Utility Regulatory Policies Act of 1978 (PURPA)[18] at 10% over the regional power pool's energy billing rate to reflect the potential cost savings to society from the presumably more environmentally benign "qualifying facilities" (as defined in PURPA). The Virginia Commission requires addition of 15% to a utility's avoided-cost submission, also based on societal costs.

The New York State Legislature, citing environmental considerations, set a statutory 6¢/kWh PURPA avoided-cost rate, which is above most utility-calculated avoided costs; the Federal Energy Regulatory Commission (FERC) voided application of this rate, interpreting PURPA to prohibit reimbursement in excess of avoided cost. Many state commissions objected to this decision as an unwarranted usurpation of state rights. The FERC decision is being appealed, and one of the FERC commissioners has stated that application of the decision to other situations would be decided on a generic basis.

---

[16] *Orders:* Idaho, Kan., N.J., N.Y., Okla., Wis. *Statutes:* Conn., Mont., Wash.

[17] *Ordered:* Alaska, Idaho, Mich., N.J., N.Y., Va. *Pending:* Iowa.

[18] 16 U.S.C. 2601 et. seq.

Collaborative Consideration

Six states are involved in collaborative efforts among utilities, state regulators, and intervenors to determine how environmental externality costs will be calculated and incorporated.[19] None of these efforts has yet come to fruition, and a recent collaborative DSM effort in California resulted in inability of the parties to agree on externality values or incorporation methodologies (CEC 1990).

Planning Consideration

Twenty-two states, Washington, D.C., and the BPA and NWPPC require or are contemplating requiring consideration of environmental externality costs in least-cost planning; twelve states now have requirements, and eleven states have requirements pending or under consideration.[20]

Bid Evaluation Consideration

Eight states and BPA require, are considering requiring, or have orders pending to take into account environmental externality costs in evaluating bid scores.[21] Three state commissions—New York's, New Jersey's, and Colorado's—have provided for specific points to be assigned in bid evaluations to account for environmental externality costs, but only New York's effort is substantial.

Proposed Incorporation Methods

Environmental Dispatch

In environmental dispatch, a commission orders a utility or power pool either to dispatch environmentally benign resources ahead of more polluting resources—even though the latter may cost less—or to dispatch resources according to least cost, with environmental costs included in the determination of least cost. Environmental dispatch alleviates environmental damages and costs while displacing production from the most heavily polluting power plants and thus encouraging early closure of those plants.

No states currently are employing environmental dispatch as a

---

[19] Conn., Iowa, Md., Mass., N.Y., Vt.

[20] *Orders:* Ariz., Calif., Nev., N.Y., Ohio, Ore., Penn., Texas, Wash., D.C., Wis. *Statute:* Alaska, Vt. *Orders pending:* Conn., Mass., Minn. *Orders under consideration:* Hawaii, Maine, Mich., Mo., Mont., N.H., R.I., Utah (statute).

[21] *Orders:* Colo. (by fuel type), N.J., N.Y. *Pending:* Conn., Mass., Vt. *Under consideration:* Md.

208 — Chapter Ten

means of incorporating environmental externality costs. All the methods presently adopted by states to incorporate environmental externality costs address only resource selection to meet new capacity or new energy needs. In its mandated bidding regime, however, the New York commission requires utilities to include life extension of existing plants as a resource option and to determine the appropriateness of continuing to operate those plants by comparing their costs with the prices bid for new resources. Wisconsin's recently enacted acid rain statute includes environmental dispatch among the compliance options for meeting the SO_2 and NO_x standards.[22]

The Ohio Office of Consumers' Counsel did a recent study on cleaning up the state's very substantial contribution of acid rain precursors, finding that a combination of least-emissions dispatching and aggressive investment in energy efficiency could prevent increases in SO_2, reduce cleanup costs by more than 60%, and reduce cumulative costs for electric energy services by as much as $3 billion through 2005 (Centolella 1988). A model has been developed for analyzing the environmental costs and benefits of environmental dispatch (Heslin and Hobbs 1989). BPA's production models and resources planning models are capable of analyzing environmental dispatch but are usually run to determine the lowest social costs of meeting utility system loads. The Cornell Carnegie-Mellon model developed to model New York utility emission impacts could also accommodate environmental dispatch (Pace 1990).

Ranking

The Center for Global Change at the University of Maryland has worked out an innovative ranking and weighting methodology for evaluating environmental externality costs.[23] The problem with all ranking systems, however, is that their accuracy must inevitably be judged on the degree to which they approximate costs. To the extent that they depart from costs, they produce significant ranking and cost distortions. Using the best cost data available is easier to understand and can be varied more readily as new cost data become available.

---

[22] Wis. Stats. Ann., Secs. 144.385–144.387; see also Secs. 15.347, 16.02.

[23] Vermont Public Service Board, *Application of Twenty-four Electric Utilities . . . for a Certificate of Public Good Authorizing Execution and Performance of a Firm Power and Energy Contract with Hydro-Quebec and a Hydro-Quebec Participation Agreement*, Docket No. 5330, testimony of Susan Hedman, Alan S. Miller, and Irving Mintzer (January 11, 1990).

"Environmental LCUP"

"Environmental LCUP (Least Cost Utility Planning)" is an innovative concept proposed by Florentin Krause of Lawrence Berkeley Laboratory for incorporating environmental externalities. Under the proposal, emission reduction targets would be set for principal power plant pollutants like CO_2, SO_2, NO_x, and particulates, and the utilities would be required to meet these targets in a least-cost manner. As an enforcement mechanism, utilities could receive a positive rate-of-return incentive for meeting or exceeding the targets and a negative incentive for failing to meet them. Wisconsin, with its new acid rain law, comes as close as any state has to using this proposed method of setting emission standards that utilities would have to meet at least cost.

The advantage of environmental LCUP is that it avoids the necessity for calculating environmental externality costs and is designed to achieve specific emission reduction targets. It lends itself well to valuing hard-to-calculate regional and global externalities. The disadvantage is that it requires the setting of emission reduction targets, which may be as difficult as environmental externality costs for commissions to calculate. Their doing so also may be viewed as invading legislative prerogatives. However, environmental LCUP could be used for hard-to-value resources only, with more readily established values used for other resources. Environmental LCUP might be used for CO_2 costing—requiring, for example, a 20% reduction in a least-cost manner—and conventional methods used for valuing other pollutants.

Assessment of Environmental Costs Against Resources and Creation of a Pollution Mitigation Fund

An innovative proposal by former Maine Public Utilities Commissioner David Moskovitz would charge resource owners with the quantified environmental costs of each resource selected and deposit the proceeds in a pollution mitigation fund, thus internalizing the environmental costs. This proposal has the enormous advantage of making resource owners pay the costs of the environmental damages they impose on society instead of just using these costs in resource selection. It would also create a very substantial fund that could be used for environmental mitigation and for promotion of use of environmentally benign renewable resources and marginally cost-effective DSM programs. Instituting such a system may be beyond the statutory authority of many commissions, although the New York commission did require its utilities to devote 0.25% of gross revenues to establish a fund for

210 — Chapter Ten

DSM research and experimentation, including the assessment of environmental costs.[24]

Recommendations for Incorporation

There has not yet been sufficient experience with incorporating environmental externality costs under any of the statutes or state commission orders described above to be able to ascertain which methodology will work best. Considering all the pros and cons of the various proposals, we propose the following recommendations:

1. Environmental externality costs should be incorporated in all utility planning, bidding, and other resource selection.

2. Quantified environmental externality costs should be used, based on damage costs where adequate valuation studies are available, and otherwise based on control costs.

3. A major research effort is critically important to better determine environmental damage costs.

4. Rate-of-return incentives should be provided for acquisition of energy-efficient resources, such that a kWh saved will be as profitable as a kWh sold; this step may require decoupling profits from sales.

5. Environmental externalities should be included in setting avoided costs.

6. Environmental costs should be internalized by an assessment against resources selected and should be placed in a pollution mitigation fund.

7. Environmental dispatch and environmental LCUP should be tested to determine their environmental and ratepayer effects and applicability in cases of hard-to-quantify pollution costs.

Next Steps

Environmental externality valuation is still at an early stage of development. Much research is needed to get firm and defensible costing figures. The U.S. Department of Energy and the U.S. Environmental Protection Agency should perform a thorough study of quantifying

---

[24] N.Y. PSC, Case 28223, Opinion and Order 84-15, *Requiring the Development of Conservation Programs*, 21 May 1984.

environmental damage costs, on a scale comparable to the congressionally mandated National Acid Precipitation Assessment Program (NAPAP) study of acid rain impacts and damages. The research area requiring greatest attention is dose-response relationships.

While 29 state jurisdictions and Washington, D.C., the BPA, and NWPPC have started to consider environmental externalities, many of their efforts are tentative. A great deal of experimentation is needed on the various means of incorporation that have been attempted and proposed. A concerted effort should be made to exchange information among state commissions and utilities and to prompt other commissions to incorporate environmental externalities. It is heartening that the National Association of Regulatory Utility Commissioners has taken a major interest in this area, having held a national conference on the subject in October 1990 in Jackson Hole, Wyoming.

Information must also be exchanged among environmental costing experts, a process requiring a unique collaboration of economists, scientists, and utility experts. In fall 1990, the Pace University Center for Environmental Legal Studies and Fraunhofer Institut held an international costing conference sponsored by the German Marshall Fund of the United States and the Daimler-Stiftung Foundation. Academic, utility (Electric Power Research Institute, Gas Research Institute), and government research institutes should devote major efforts to both quantification and incorporation issues.

Acknowledgments

This paper is based on a report prepared by the Pace University Center for Environmental Legal Studies for the New York State Energy Research and Development Authority, supplemented by funding from the U.S. Department of Energy and the National Audubon Society (Pace 1990). The report was prepared by Pace Center staff; Pace University Law School students; and Shepard C. Buchanan of Bonneville Power Administration, Paul Chernick and Emily Caverhill of PLC, Inc., and Alan Krupnick of Resources for the Future.

References

Bland, P. 1986. "Problems of Price and Transportation: Two Proposals to Encourage Competition from Alternative Energy Resources." *Harvard Law Review* 10: 2.

Bonneville Power Administration 1986. "Cost-Effectiveness Methodology." In *Bonneville Power Administration Resource Strat-*

212 — Chapter Ten

egy. Report DOE/BP/751. Portland, Ore.: Bonneville Power Administration.

BPA. See Bonneville Power Administration.

Brick, S. 1989. "Strategies and Methods for Incorporating Environmental Concerns into Utility Planning and Utility Regulatory Decision-Making." Middleton, Wis.: MSB Energy Associates.

California Energy Commission *et al.* 1990. *An Energy Efficiency Blueprint for California: Report of the Statewide Collaborative Process.* Sacramento: California Energy Commission.

CEC. See California Energy Commission.

Centolella, P. 1988. "Clearing the Air: Using Energy Conservation to Reduce Acid Rain Compliance Costs in Ohio." Columbus: Ohio Office of the Consumers' Counsel.

Chernick, P., and E. Caverhill 1989. "The Valuation of Externalities from Energy Production, Delivery, and Use." Boston: PLC, Inc. Photocopy.

Cohen, S. D., and J. Eto 1989. "Preliminary Results of LBL/NARUC Externalities Survey." Berkeley: Lawrence Berkeley Laboratory.

Economist 1988. "Money from Greenery." *The Economist,* 21 October 1988.

Electric Power Research Institute 1988. *Benefits of Environmental Controls: Measures, Methods and Applications.* Palo Alto, Calif.: Electric Power Research Institute.

EPRI. See Electric Power Research Institute.

Freeman, A. 1979. *The Benefits of Environmental Improvement: Theory and Practice.* Baltimore: Johns Hopkins Press.

Heslin, J., and B. Hobbs 1989. "A Multiobjective Production Costing Model for Analyzing Emissions Dispatching and Fuel Switching." *IEEE Transactions on Power Systems* 4 (3).

Hill, R. 1990. "The Impact of Energy on Environment and Development." In *IVth Nobel Prize Winners Meeting, December 1989, Man, Environment, and Development—Towards a Global Approach.* Nova Spes Meeting Proceedings.

Hohmeyer, O. 1989. *Social Costs of Energy Consumption.* Berlin: Springer-Verlag. (Prepared under contract for the Commission of the European Communities, Directorate-General for Science, Research and Development, Document No. EUR 11519, by Fraunhofer-Institut für Systemtechnik und Innovationsforschung, Karlsruhe, Federal Republic of Germany.)

Machado, S., and R. Piltz 1988. "Reducing the Rate of Global Warming: The States' Role." Washington, D.C.: Renew America.

Mitchell, R., and R. Carson 1989. "Using Surveys to Value Public

Goods: The Contingent Valuation Method." Washington, D.C.: Resources for the Future.

NAPAP. See National Acid Precipitation Assessment Program.

National Acid Precipitation Assessment Program 1987. *Interim Assessment*. Vol. 1, U.S. Government Printing Office, Washington, D.C.

Pace. See Pace University Center for Environmental Legal Studies.

Pace University Center for Environmental Legal Studies 1990. *Environmental Costs of Electricity*. Report for the New York State Energy Research & Development Authority. Dobbs Ferry, N.Y.: Oceana Publications.

Shuman, M., and R. Cavanagh 1984. *A Model Electric Power and Conservation Plan for the Pacific Northwest, Environmental Costs*. Portland, Ore.: Bonneville Power Administration.

Richard L. Ottinger is director of The Energy Project at Pace University Center for Environmental Legal Studies; he is also Professor of Law at Pace Law School. As a member of Congress and Chairman of the House Energy Conservation and Power Subcommittee of the Energy and Commerce Committee (1964–1970 and 1974–1984), he was the principal author of the Public Utilities Regulatory Policy Act (PURPA).

Chapter 11

Valuation of Environmental Externalities in Energy Conservation Planning

Paul L. Chernick and Emily J. Caverhill, *Resource Insight, Inc.*[1]

Introduction

The construction and operation of energy supply facilities have negative effects on human health and the environment through intermediate effects on air, water, and land. These effects are not considered in traditional utility planning; therefore, they are considered external costs or "externalities."

An important benefit of clean resources such as energy conservation or demand-side management (DSM) is avoiding or reducing the environmental effects of traditional power supply options. In order to include these benefits in evaluating the cost-effectiveness of DSM measures, a value for these benefits must be estimated. We can estimate this value by establishing a list of externalities that will be avoided by the DSM measure, such as the air emissions from the avoided power plant, and valuing those externalities using one of the valuation techniques described in this paper.

This chapter describes the applications, strengths, and weaknesses of four methods for estimating the value of reducing externalities. The dollar values of externalities are estimated so that external costs can be considered in utility resource planning the same way traditional capital and operating costs are considered. The values of externalities are generally expressed in dollars per unit of externality; for example, dollars per pound of an air pollutant. Table 11-1 shows the monetary values adopted by four state utility commissions for use in utility resource

---

[1] Formerly PLC, Inc.

215

216 — Chapter Eleven

Table 11-1. Comparison of Monetary Values Assigned to Externalities in the United States (1990$/lb of pollutant emitted)

	New York PSC	Massachusetts DPU	California Energy Commission (in-state)	California Energy Commission (out-of-state)	Nevada PSC
Date Adopted	1989	Aug. 1990	Oct. 1990	Oct. 1990	Feb. 1991
Externality					
SO_2	0.43	0.78	6.48	0.56	0.78
NO_x	0.93	3.4	6.53	1.52	3.40
VOCs	NE	2.76	1.83	0.17	0.59
CO	NE	0.46	NE	NE	0.46
Particulates	0.17	2.09	4.39	0.45	2.09
CO_2	0.001	0.011	0.0036	0.0036	0.011
CH_4	NE	0.11	NE	NE	0.11
N_2O	NE	2.07	NE	NE	2.07
Water use (cents/kWh)	0.10	NE	NE	NE	NE
Land use (cents/kWh)	0.42	NE	NE	NE	NE

Notes:
1. NE = not estimated
2. Massachusetts values have been inflated from 1989$.

Source: Resource Insight, Inc.

planning. The methodologies discussed in this chapter are relevant to determining the social cost-effectiveness of all utility demand- and supply-side resources, including DSM.

Definition of Environmental Externalities

The term "externality" can refer to any cost or benefit that is not reflected in the price paid by a utility or its customers for energy-related goods or services. Only *environmental* externalities are considered in this paper, although other important economic and social externalities also exist.

An exhaustive list of environmental externalities from energy production and consumption would include disturbances to air, water, and land. Examples include the health and environmental damages caused by the emission of air pollutants and the impacts on aquatic ecosystems from the water consumption of a power plant cooling system.

Some of the environmental effects are well understood and predictable, such as the amount of land required by a facility. Other effects are strongly supported by empirical evidence, such as ambient pollution effects on human health. Some represent risks, which may be well understood, both as to probability and effect, or may be highly uncertain. For example, the designers of a dam may know that there is a one-in-a-million chance of the dam's failing and killing 2,000 people. Similarly, the number of people who will be killed in grade-crossing accidents involving coal trains is unknown, but the probability distribution may be highly predictable. Still other consequences are not fully understood in a technical sense, such as the effect of trace gas emissions on global warming and the effect of global warming on human and ecological systems. Finally, the net effects may be difficult to determine: the construction of a water reservoir may provide recreational benefits and habitat for waterfowl but destroy other recreational opportunities, flood wetlands, and disrupt the habitat of other ecosystems.

Scope of an Externalities Analysis

The complexity and results of an analysis of externalities are influenced by geographic scope. Some analyses of externalities are specifically designed to evaluate only those effects that occur in a specific service area or state. For example, the Massachusetts Department of Public Utilities has determined that externalities of electric power sources must be considered regardless of where the externalities occur. In contrast, the Vermont Public Service Board will include only costs that have some connection to state residents. This type of limitation can simplify the assessment process, by excluding some impacts, or complicate the process, by requiring the identification of the location of each externality.[2]

Geographic scope also affects the evaluation of specific externalities in other ways. For instance, some results of energy production, such as suspended particulate emissions, have varying effects and costs, depending on the area in which they are deposited.

The scope of an externality analysis may also be affected by decisions to exclude classes of effects or to limit the method of valuation. For example, an analysis that valued all effects at market prices (for example, an otter at the price of its pelt, a human life at the present

---

[2] For example, if oil spills were to be included in the analysis only if they occurred in New England waters, it would be necessary to determine the port through which marginal supplies of oil were imported.

218 — Chapter Eleven

value of lost wages, wilderness at the income it generates in tourism) would miss some important aspects of many effects but would be somewhat simpler to perform than a study that attempted to determine a total social value for each affected system.

The choice of externalities to be included in utility planning and the scope of the evaluation are largely independent of the method of estimating the externalities. The methods discussed in this chapter were developed to estimate externalities for decisions concerning utility and nonutility fossil-fired generation facilities, DSM, and off-system power purchases that are fossil-fuel-fired at the margin. However, these methods are also applicable to the evaluation of the externalities of hydroelectric, nuclear, or other energy supply or demand-reducing options.

Valuation Methods

In general, four basic approaches have been used in estimating values for environmental externalities:

1. estimating the relative physical, chemical, or toxicological potency of various pollutants;

2. polling of experts or other people;

3. direct estimation of the environmental effects and the costs of those effects; and

4. determination of the marginal cost of control for the pollutant to estimate the maximum cost society has committed (or appears about to commit) to avoid the pollutant.

Relative Potency

The first approach, relative potency, works quite well for estimating the relative importance, or value, of reducing emissions of the major energy-related greenhouse gases carbon dioxide (CO_2), methane (CH_4), nitrous oxide (N_2O) and carbon monoxide (CO). Lashof and Ahuja (1990) estimate the instantaneous global warming potentials (GWPs) of the greenhouse gases relative to CO_2. The relative value of reducing emissions of each greenhouse gas can be related to the value of CO_2 by correcting the instantaneous GWPs to reflect the shorter lives and lower molecular weights of the other greenhouse gases and properly discounting the effects of the gases over time. The first two corrections are important to reflect the total contribution to global warming of emissions that will have persistent effects depending upon the life and nature of the molecule. To reflect the value of delaying

Valuation of Environmental Externalities — 219

emissions of the greenhouse gases and slowing the *rate* of global warming, the value of the emissions is discounted over time.[3] Using Lashof and Ahuja's estimates of the GWPs and assuming a real discount rate of 6% give relative values for the greenhouse gases as 76 for CH_4, 320 for N_2O, and 6 for CO.

Relative potency might also be used for estimating the relative importance of other externalities. The value of reducing emissions of toxic air pollutants such as heavy metals can be estimated through the relative toxicities of these hazardous, and sometimes carcinogenic, pollutants. For example, in Massachusetts the acceptable ambient levels (AALs) are 0.003 and 1.36 (ug/m³) for cadmium and chromium, respectively. These AALs imply that cadmium is on the order of 450 times more toxic than chromium on a weight basis, and reductions of cadmium emissions should be correspondingly more valuable. Similarly, data in Hendrey (1986) imply that the acid rain damages caused by NO_x emissions are 83% those of SO_2, per pound of emissions (Chernick and Caverhill 1989). Hohmeyer (1988) cites work by Grupp (1986), which finds that the toxicities of SO_2, particulates, and VOCs are equal per pound of emissions and that NO_x is 25% more toxic per pound.[4]

While relative potency provides probably our best estimates of the relative value of reducing greenhouse-gas emissions, it may be inferior to the more direct estimation techniques, described below, for regulated air pollutants.

Polling

The second approach, polling, can be employed in valuation if the polling instrument is designed carefully and if the poll respondents are experts in the field in question and are informed of the intended use of the poll results. For example, it may be useful to poll epidemiologists on the coefficients of the dose-response functions relating human illness and death to pollution levels. However, the results of a polling effort can be useless or, worse, misleading. A good polling effort must (1) poll a representative sample of experts in the field of interest and publish the names and credentials of the respondents for public review;

---

[3] Emissions can be delayed by reducing the growth in demand for fossil-fuel-fired power through the adoption of aggressive DSM programs, for instance. Krause, Bach and Koomey (1989) suggest that the rate of global warming should be limited to about 0.1°C per decade, down significantly from our current commitment of 0.2–0.5°C per decade.

[4] Grupp's estimates appear to refer only to acute human toxicity, although Hohmeyer applies them for all purposes. Since these pollutants also have effects on materials, crops, visibility, and natural ecosystems, this assumption may be too rough to be useful for determining the relative value of reducing emissions of these pollutants.

220 — Chapter Eleven

(2) provide a clear statement of the question to be answered in the poll, including the *specific* basis for the requested response, and the intended use of the answer; and (3) make the polling instrument and the respondents' answers available for public review. Polling is unlikely to provide better information on valuation than the other methods of estimating externalities presented here. Certainly, its usefulness is critically dependent on the quality of the polling instrument and credentials of the respondents.

New England Electric System (1989) used polling to develop a worksheet for including externalities in its supply planning process. Although this initial exercise was probably carried out with good intentions, it had several problems, including the following:

1. The polling instrument asked for the relative importance of externalities without specifying whether the weights were to be stated in terms of damage for each pound of emissions, of the total damages from a typical new plant, of the total global damages from all current (or future) emissions, or of the regulatory difficulty of reducing emissions from new plants.

2. The qualifications of the respondents for this exercise were not clear.

3. The respondents replied to the poll on the condition of anonymity, and the results of the poll are not available for review.

4. A very small number of respondents were polled, and there was no clear derivation of NEES's results from the limited polling results reported.

Ultimately, the Massachusetts Department of Public Utilities rejected this approach to valuation in favor of the cost-of-control approach discussed below. Clearly, if it is to be useful, polling must be better focused than in this initial effort.

Direct Estimation of Costs

Direct estimation, the third approach, estimates the dollar costs of damages caused by power plant operation. For environmental externalities, such as air pollution and water pollution, the relevant effects might include impacts on:

• human life expectancy and health;

• human comfort and pleasure (impacts might include higher noise levels, altered visibility, and odors);

Valuation of Environmental Externalities — 221

- domesticated plants and animals, such as commercial forests, livestock, crops, and lawns;

- wild plants and animals in a variety of habitats; and

- nonliving materials, such as surfaces on buildings (especially stone and paint), at archeological and historical sites, on monuments, on vehicles, and in natural landforms.

In general, each of these effects must be quantified, and unit costs for each must be established. This procedure is generally treated as if it were a technical exercise, but the definition of relevant externalities, selection of important effects, choice of quantification measures and techniques, and determination of unit cost are often highly judgmental and subjective.

Determination of at least some of the social costs of some environmental externalities is feasible, although this process is complex. For example, the direct estimation of the costs related to an air pollutant requires detailed research and modeling of the five steps between the emission of the pollutant from a power plant stack and the resulting effects. Those steps are:

1. The movement of the pollutant through the atmosphere; modeling must take into account the effects of stack height and temperature, of wind speed and direction, and of ambient temperature and humidity.

2. The chemical changes that the pollutant—especially a precursor to ozone, sulfates, or nitrates—undergoes in the atmosphere.

3. The deposition of the pollutant on various surfaces, including lungs, soil, water, and material such as exposed metal and rubber.

4. The dose-response relationship between the pollutant and each system, including impacts on human health, visibility, crops, materials, and wildlife.

5. The unit values of the effects on each system—for instance, $/life lost or $/km-yr of reduced visibility.

In fact, the transport calculations are very complicated, the atmospheric chemistry and dose-response relationships may be highly uncertain, and the composition of the exposed population is highly site-specific.

Even when these difficult technical issues have been resolved, tough policy questions remain in the valuation of effects. Costing of market-valued goods, such as agricultural production and paint, can be

222 — Chapter Eleven

relatively straightforward; assigning values to human morbidity and mortality (which are themselves complicated by many subcategories including age, health, duration of illness, and the extent of suffering and dread) and to impacts on visibility, historic monuments and forests, and wildlife and ecosystems can be more controversial.

Direct estimation of environmental costs is a daunting and time-consuming task, given the multidisciplinary background needed by those responsible for assigning values to human life and health, visibility, and other nonmarket costs. Yet, if the assessors exercise caution in understanding data sources, correcting errors, and reflecting regional differences, direct estimation serves as a highly clear and workable method to value what seem to be intangible external costs. For a more detailed discussion of direct estimation of externalities see Ottinger *et al.* 1990, Hohmeyer 1988, or Chernick and Caverhill 1989.

Cost of Control

The fourth approach, cost of control, relies on the costs of existing or anticipated regulations or control measures to estimate the value society places on residual environmental impacts. For instance, the Clean Air Act Amendments of 1990 give us specific information about what society is willing to pay to reduce SO_2, NO_x, VOCs, and other toxic air emissions. In this method, externality values are expressed in such terms as \$/lb emitted or \$/unit of the externality. The combined value of all of the externalities of a power source is then:

$$\$/kWh\ generated = units\ of\ the\ externality/kWh \times \$/unit$$

This method is also referred to as shadow pricing, revealed preference, and marginal cost of abatement, and it has been used by several analysts to estimate the societal value of reducing residual emissions, including Chernick and Caverhill (1989, 1990a and 1990b); Shimshak *et al.* (1990); Shilberg *et al.* (1989); the New York State Energy Office (1989); and the New York Public Service Commission staff (Putta 1990). SCAQMD (1990) also suggests valuing NO_x emissions reductions from the costs of control of stationary sources.

Two Rationales. Deriving externality values from the cost of pollution control is sometimes described as if the resulting values were proxies for the direct costs of emissions. In fact, the cost of control technique uses the costs of control to provide direct information on the societal value of reducing emissions at the margin under either of two theoretical rationales.

The first rationale is that the cost of required controls serves as an estimate of the price that society is willing to pay to reduce the pollutant. This is the rationale behind the "implied valuation" or "revealed preference" approach to the use of control costs for valuing externalities. For instance, if society is willing to pay as much as \$2/lb for current or planned emissions control, then avoiding a pound of emissions should also be worth \$2/lb. For a more detailed discussion see Chernick and Caverhill 1989 and Schilberg et al. 1989.

The second, and perhaps the more compelling, rationale is that the costs of required controls may directly establish the social benefits of reducing emissions. For example, if emissions in a particular air basin are reduced through the aggressive use of DSM programs, then the most expensive control measures otherwise required for a particular pollutant can be avoided. Therefore, the benefit of the DSM program, in \$/kWh saved, is exactly equal to the emissions avoided (lbs/kWh) times the unit cost of the control equipment that would otherwise have been installed (\$/lb pollutant controlled). Similarly, if the objective of a particular regulation is to maintain a given level of ambient air quality, either construction of a less polluting plant or reduction in energy output by means of conservation will allow regulators to avoid the most expensive control measures that would otherwise have been required.

Marginal Cost. In the cost-of-control method, only the marginal cost of control is important. From the "implied preference" perspective, the fact that many required controls are inexpensive, or even that some inexpensive controls have not yet been required, is irrelevant to the determination of the price society is willing to pay to reduce emissions at the margin. From the avoided-control-cost perspective, the appropriate estimate is the cost of the highest-cost control measure, since this measure will be avoided by the use of clean resources such as DSM before cheaper measures are avoided.

Defining the Margin. Determining the marginal unit of externality control is difficult for at least three reasons. First, legislative and regulatory requirements for the control of externalities often seem mutually inconsistent, with some required measures having much higher costs than much less expensive measures that are not required. Second, the margin is often in flux. Third, the complexity of pollution control complicates the computation of the cost of controlling an environmental externality.

The first difficulty, apparent inconsistencies in regulatory behavior, will generally produce underestimates of externality costs. Controls are often not required on all sources of an externality, for any of several reasons. Legislation, regulation, or both may attempt to protect vulnerable (or powerful) economic interests or sectors and thus may exempt such groups as small businesses, marginal industries, and

224 — Chapter Eleven

households from controls imposed on other generators of the same externalities. Similar exemptions may be granted to sectors that are perceived to be contributing in other ways to the solution of the same or related problems. Exceptions are especially likely to be granted if those contributions are viewed as burdensome to the contributor.[5]

The cost of administration may also restrict the application of controls on externalities. Some producers of externalities, especially producers that are small and dispersed, may be very difficult and expensive to police. Regulators may also find their scarce resources are more efficiently used on new proposed sources (where the applicant must receive the regulator's active approval to proceed) than on existing sources (where the regulator may need to expend significant effort to force a change in the status quo).

One could argue that pollution control measures that are not societally cost-effective are sometimes required on small pollution sources for administrative simplicity or equity purposes. For instance, fuel efficiency standards apply equally to all vehicles in a class. For vehicles that are operated very little, the unit control costs can be very high, possibly higher than the regulator would otherwise require. In this case, the incremental cost of the control measure on the smallest (lowest-mileage) source would be somewhat higher than the marginal value used by the regulators in their decision, and higher than the implied societal value. To estimate the marginal societal value in such a case, one must probe further into the regulator's intentions. For pollution control measures that are not bundled in this way, including many requirements on utility plants and industrial sources, this is not generally a problem.

The second problem, that of a potentially changing margin, is particularly clear in the present case of acid rain. Acid rain has been a heated issue for at least a decade, and national legislation addressing acid rain has recently been adopted. This legislation redefines the margin by establishing a national SO_2 emissions cap and requiring large sources to hold SO_2 emissions allowances. The costs of these requirements suggest a new value of the associated SO_2 reductions. Once the allowance requirements are in place (in the year 2000), at least a portion of the SO_2 externality cost will be internalized.

---

[5] For example, the Clean Air Act Amendments of 1990 defer application of many of the air toxics provisions to electric utilities, even though arsenic emitted by a power plant is as dangerous as arsenic emitted by a smelter. The apparent rationale for the exemption is that the utilities will be responsible for most of the acid rain reduction and, in some cases, for many of the local air quality improvements mandated in other sections of the same bill. The amendments seem to assume that equity requires that utilities be temporarily exempted from the burdens of complying with the air toxics provisions.

Valuation of Environmental Externalities — 225

The third problem is the complexity of control measure costs. Some control measures, such as catalytic convertors on automobiles, reduce the emission levels of multiple pollutants, including CO, NO_x and VOCs. In such cases, it is sometimes difficult to determine which pollutants have motivated the imposition of the controls and to apportion the cost of the controls among pollutants.

Multiple Effects of Externalities. A single pollutant may have several ultimate effects, each imposing its own costs on society. As a result, the same pollutant may be regulated under several different rules. For example, nitrogen oxides are regulated as a respiratory pollutant and as a precursor to smog, and the pending acid rain legislation will regulate nitrogen oxides as a precursor to acid rain. In addition, recent evidence indicates that nitrogen oxides promote cancer; they may also contribute to the release of methane from soils, an effect that would make nitrogen oxides a candidate for regulation as a greenhouse gas (*Science News* 1989). These multiple effects may confuse the valuation of the externalities, due to the need to distinguish between additive and cumulative controls.

Imperfections in the Regulatory System. The decisions of the legislative and regulatory system provide us with useful information, including some sense as to how social and political structures (sometimes at local, state, federal, and international levels) have valued the externalities in which we are interested. However, we must recall that these structures are imperfect. We should not delude ourselves into believing that required levels of pollution controls, for example, represent revealed truth regarding the social valuation of pollution.

Environmental regulation, like any other kind of regulation, occurs in a political context, as the balancing process of many complex interests. For the most part, society does not collectively pay the cost of reducing externalities; individual persons and corporations pay different shares of the costs and receive different shares of the benefits. The distribution of the costs and benefits across interest groups and constituencies may have a greater effect on the eventual level of regulation than does the aggregate social value. Regulation may be too weak (in which case the true value of the externality may be much higher than the value implied by regulatory standards), too strong (in which case reductions of the externality through choice of energy technology may still save society the costs of the controls, and the cost of control may still be appropriate), or a mixture of the two.

Even if legislators and regulators strove solely to identify and serve the social good, no one is omniscient. Given the multitude of

226 — Chapter Eleven

interests and valuations, the choice made by the regulators may not match a reasonably defined social value.

In practice, neither the regulatory agencies nor the courts are normally given the responsibility of determining the public interest and pursuing it without restraint. Legislatures enact laws, and regulators and courts put them into effect. The regulatory and legal interpretation of the statute may not always be what the legislature expected or intended.[6]

Although rapidly changing regulations inevitably create some uncertainty, most of the above issues are quite tractable. With a proper conceptual framework and due diligence, the cost-of-control technique can produce highly relevant and useful estimates of the social value of reducing externalities. One of the most desirable features from the perspective of the analyst and decision-maker is that the toughest social choices, assigning dollar values to human life and health, to wildlife, to natural ecosystems, to historical monuments, and to visibility, are made by the legislators and by environmental regulators rather than by utilities and their rate-regulating agencies.

Conclusion

The environmental benefits of DSM measures are important benefits that should be included in DSM measure evaluation. A rational and consistent way of including these benefits is to estimate their dollar value. This chapter described four methods for estimating externality values that have different applications, strengths, and weaknesses.

Cost of control has proven to be the most tractable of the four methods for estimating externality values for many important regulated air emissions from energy production facilities such as SO_2, NO_x, and VOCs. It is also useful for estimating values of a variety of other externalities, including oil spills and water use. It is more difficult to apply to valuation of emissions of the greenhouse gases and other unregulated pollutants. For these pollutants, the relative potency method is useful if coupled with another valuation technique.

Direct costing may be useful once it is more fully developed, and it will be particularly useful for valuing site-specific externalities such as impacts on fish from water consumption and pollution. Direct costing will be very complex for valuing pollutants such as SO_2—pollut-

---

[6] Indeed, in some cases, such as the National Environmental Protection Act that established the requirement of environmental impact statements for major federal actions, it is clear that the legislature as a whole did not know how its language would be interpreted, and various legislators held very different views of just what the law would require.

ants with effects that are numerous and, in many cases, indistinguishable from those of several other pollutants or natural causes; that may be cumulative; and that are very difficult to measure or value (effects, for example, such as visibility and damages to ecosystems).

Despite the weaknesses inherent in these methods, several utility regulators have adopted externality values based on these techniques. The California Energy Commission, the New York Public Service Commission, the Nevada Public Service Commission and the Massachusetts Department of Public Utilities all have adopted specific $/lb values for several important regulated air pollutants. Each of these commissions relied on slight variations of the cost-of-control method.

For the greenhouse gas CO_2, Massachusetts estimated a value based on tree planting within the U.S. This is a variation, which is appropriate in the absence of regulations, on the cost-of-control approach. Basically, the Massachusetts commission determined that tree-planting is likely to be a cost-effective reduction method of Massachusetts' contribution to global warming and adopted a reasonable estimate of tree-planting costs for the value of reducing CO_2. From this, the values of the other greenhouse gases were estimated by relating them to the value for CO_2 using the relative potency approach. California based its CO_2 value on the energy saved through urban tree-planting, also a variation on the cost-of-control approach. New York simply adopted a place-holding value that was one-tenth of the CO_2 abatement costs developed by the New York State Energy Office. Several other utility commissions are considering the question of how to incorporate externalities into utility resource planning.

References

Chernick, P., and E. Caverhill 1989. *The Valuation of Externalities from Energy Production, Delivery and Use.* Boston: Resource Insight, Inc.

────── 1990a. *Report to the Boston Gas Company on Including Externalities in 89-239.* Boston: Resource Insight, Inc.

────── 1990b. *Comparison of Total Costs of Supply Options for 89-239.* Boston: Resource Insight, Inc.

Hendrey, G. 1986. "Acid Deposition: A National Problem." *Acid Deposition: Environmental, Economic and Policy Issues.* New York: Plenum Press.

Hohmeyer, O. 1988. *Social Costs of Energy Consumption: The External Effects of Electricity Generation in the Federal Republic of Germany.* Berlin: Springer-Verlag.

228 — Chapter Eleven

Krause, F., *et al.* 1989. *Energy Policy in the Greenhouse, Volume 1,* September.

Lashof, D., and D. Ahuja 1990. "Relative Global Warming Potentials of Greenhouse Gas Emissions." *Nature,* Vol. 344, Iss. 6266, pp. 529–31.

New England Electric System (NEES) 1991. *Comments by Massachusetts Electric Company* in Massachusetts Dept. of Public Utilities Order in Docket 89-239, Aug. 31.

New York State Energy Office (NYSEO) 1990. *Environmental Externality Issue Report.*

Ottinger, *et al.* 1990. *Environmental Costs of Electricity.* Dobbs Ferry, New York: Oceana Publications.

Putta, S. 1990. *Consideration of Environmental Externalities in New York State Utilities' Bidding Programs for Acquiring Future Electric Capacity.* Albany: New York Public Service Commission.

Schilberg, G., *et al.* 1989. *Valuing Reductions in Air Emissions and Incorporation into Electric Resource Planning: Theoretical and Quantitative Aspects.* Sacramento, Calif.: JBS Energy, Inc. for Independent Energy Producers.

Science News Sept. 30, 1989, p. 213.

Shimshak, R., *et al.* 1990. *Comments of the Division of Energy Resources on Environmental Externalities.* Commonwealth of Massachusetts Department of Public Utilities.

South Coast Air Quality Management District (SCAQMD) 1990. "Comments of the South Coast Air Quality Management District on Draft Electricity Report 90" (ER-90) (Draft). Sacramento, Calif.: California Energy Commission.

Paul L. Chernick is president of Resource Insight, Inc. He has worked on the design and valuation of energy conservation programs, utility costing and pricing and integrated demand and supply planning. He holds B.S. and M.S. degrees from the Massachusetts Institute of Technology.

Emily J. Caverhill is a research associate at Resource Insight, Inc. Her responsibilities include determining how external costs should be incorporated into utility planning. She has a B.Sc. in Engineering and an M.B.A. from Queen's University in Ontario, Canada.

Chapter **12**

Incorporating Environmental Externalities in Integrated Resource Planning: One Utility's Experience

Dean S. White and Timothy M. Stout, *New England Power Service Company*

Mary Sharpe Hayes, *Tennessee Valley Authority*

Introduction

In the late 1970s, many U.S. electric utilities and utility commissions began adopting least-cost, integrated resource planning (IRP) strategies. These strategies encouraged utilities to begin considering demand-side as well as traditional supply-side options in their resource plans. The goal of this shift away from strictly supply-side planning was to select resources that would result in the lowest cost of electricity services for consumers: many demand-side programs were significantly less expensive, per kilowatt-hour and per kilowatt, than many supply-side options. Consequently, demand-side options offered utilities a potentially enormous resource for meeting future demand.

Along with this new focus on least-cost options, the definition of "cost" was extended to include costs external to the finances of the utility. Known as external costs or externalities, these costs range from the costs of damage from the emissions of power plants to the effect on local property values from new generating facilities. While state and federal regulations require utilities to internalize some of these costs through, for example, emission controls such as electrostatic precipitators (which remove a certain amount of particulates from flue gas), regulations do not lead to total internalization. The objective of state

229

230 — Chapter Twelve

utility commissions, when promulgating orders that require some level of internalization of external costs, is to guide utilities in selecting resource options that minimize these costs.

This chapter provides an overview of how New England Electric (NEES),[1] a medium-sized electric utility in New England, has incorporated external costs, principally environmental externalities, in its IRP process. The chapter describes the development of a methodology for evaluating externalities, the application of this methodology to all resource options under consideration for the company's resource plan, and the impact of this application on resource selection.

Background

In November 1988, the Massachusetts Department of Public Utilities issued an order (Docket No. 86-36F) that required electric utilities to consider externalities in their analyses of resource options to meet future demand. Specifically, the order stated that the "evaluations of the cost-effectiveness of conservation and load management programs and other resource options should include, to the fullest extent practical and quantifiable, costs and benefits external to the transaction, most notably environmental externalities." The department recognized the difficulty of identifying and quantifying externalities but expected the electric utilities to work with "other interested persons to develop suitable measurement methodologies over time" (DPU 86-36-F, 22).

In response to this order, NEES, on behalf of its retail subsidiary, Massachusetts Electric, considered different methodologies for evaluating environmental externalities for the purpose of incorporating them in its long-range resource plan. In reviewing different methodologies, the company evaluated a wide spectrum of options. These options ranged from qualitative approaches—essentially lists of externalities—to quantitative approaches, in which monetary values are assigned to specific externalities. The company was specifically interested in a method that met the following objectives:

- consistency with the most current, precise, and detailed knowledge about externalities;

- ease of applicability to New England Electric's integrated resource planning process;

---

[1] The term "New England Electric System" is used loosely to encompass three retail electric companies (Massachusetts Electric, Narragansett Electric, and Granite State Electric), New England Power Company, and their affiliates.

Incorporating Environmental Externalities in Resource Planning — 231

- objectivity in its application to all supply- and demand-side resource options considered by the company; and

- flexibility to be refined or replaced as the company's experience with externalities broadens, as the current knowledge about environmental externalities improves, and as the company's resource planning process further evolves.

The three primary categories of methods reviewed by NEES are discussed below.

Qualitative Approaches

In a truly qualitative approach, resource options or portfolios of options are described in terms of their environmental attributes, such as emissions types and rates, effects on water and land use, and solid waste production. Decision makers review the list of environmental attributes and other decision criteria (prices and risk, for example) and develop a strategy for ranking resource options based on the qualitative assessment of their externalities. The benefits of qualitative approaches are flexibility and adaptability. The disadvantage is subjectivity, since decision makers must make judgments about costs and benefits of trade-offs among different resources.

Quantitative Approaches

Unlike the subjective, qualitative approach, a quantitative approach attempts to provide absolute objectivity by assigning monetary values to all environmental externalities. These values are commonly based on either cost of damage or cost of control. Cost of damage attempts to quantify and assign monetary value to the impacts of externalities associated with different resource options. Cost of control attempts to use the costs of controlling pollution as a proxy for the actual damages that may result. While further discussion of these paths is beyond the scope of this paper, a number of studies on externalities have documented the advantages and limitations of these different approaches (Bernow and Marron 1990; Hohmeyer 1988; ECO Northwest 1984; Krupnick 1989; Chernick and Caverhill 1989a, 1990. [Ed. note: See also in this book the chapters by Bernow and Marron, by Chernick and Caverhill, and by Ottinger.]).

The benefits of the quantitative approach, if it can be successfully implemented, are its objectivity and the ease of comparing the options' externality costs with their direct costs. The disadvantage of this approach—as suggested by a review of current literature on the subject and by discussions with acknowledged experts in the field—is that there are still insufficient data to enable damage costs to be estimated

232 — Chapter Twelve

at a level of accuracy appropriate for utility resource planning (NEES 1989).

Hybrid Approaches

The alternative to the above approaches is the use of hybrid approaches that attempt to capture the objectivity of the quantitative and the pragmatism of the qualitative approaches. The primary advantage of hybrid approaches is that they may be easily refined as data on externalities improve. These schemes generally involve weighting a range of environmental externalities based on the relative severity of their impacts. These ratings are then applied to each resource. The resulting externality score reflects the severity of each resource option's environmental impacts.

Among the hybrid approaches, rating and weighting schemes have received considerable attention. Specifically, beginning with Orange and Rockland (O&R), all New York electric utilities have developed rating and weighting approaches for including environmental externalities in their all-resource bidding schemes (NYSPSC 1988, 1989). Other states, including California and Connecticut, are currently considering similar methods.

The New England Electric Methodology

After reviewing the alternatives, the company chose to develop a hybrid approach, which the company decided would best meet the goals stated above and would allow immediate incorporation of environmental externalities in the company's resource plan, NEESPLAN 1990, which was under development. The company recognized that this approach initially would be geared toward providing usable short-run results but would be improved by increasing quantitative sophistication as additional data on actual social damages became available.

The company based its approach on the framework of the O&R rating and weighting scheme. This scheme was modified to allow (1) a focus on issues, (2) the use of several approaches to defining individual ratings and weightings,[2] and (3) the flexibility to use expert polling to arrive at weights for individual issues.

Focus on Issues

In deciding how to categorize externalities, NEES was persuaded to take an issue-focused approach by work conducted at the U.S. Envi-

---

[2] In the New York State Public Service Commission comments on O&R, a $.01405/ kWh externality adder is derived, relying principally on a cost of control methodology.

Incorporating Environmental Externalities in Resource Planning — 233

ronmental Protection Agency (EPA) and presented in a series of reports entitled "Unfinished Business" (EPA 1987). In these reports, EPA staff ranked numerous environmental problems posing four major types of health and environmental risks. In order to compare very different types of risks, the EPA study group elected not to define environmental problems by sources, pollutants, pathways, or receptors. Instead, the group chose to define the problems on the "general basis of how the laws are written and environmental programs are organized. Since the goal of the project was to put together a useful tool to compare the risks with which EPA was concerned, the project team decided to draw up a list of environmental problems that reflect how people think of the problems" (EPA 1987).

Multiple Approaches to Rating and Weighting

Rather than adopt a single approach to selecting ratings and weightings, NEES chose a more holistic approach. Instead of using only negotiation or trying to estimate damage costs using inadequate data, the company recognized that numerous elements could contribute to the determination of an issue's rating and weighting structure. For example, relative toxicities, where available, may be used to help establish the trade-off value of two pollutants. For other externalities like land use, relative toxicity is meaningless, so other approaches must be used. Several of the approaches considered in this methodology are discussed in the next section of this chapter.

Polling of Experts

NEES designed an experiment to assess whether polling of experts to assign weights to environmental externalities was feasible. As far as the company was aware, this technique had not yet been applied to the task of designing externality evaluation methodologies. Briefly, the policy instrument asked participants to assign relative weights to a list of environmental attributes by dividing 100 points among those attributes (see NEES 1989 for a detailed explanation of the methodology).

Overview of Matrix

The end product of the company's work on developing its externalities methodology is a relatively simple matrix providing the means for rating and weighting each option on a single page (Table 12-1). Column A of the matrix lists ten externality issues against which each option is evaluated. Each issue receives a weight (column B) reflecting the relative significance of the issue's environmental impacts. Column C lists the contributing factors (environmental agents or subissues) for each

Table 12-1. The NEES Matrix for Weighing and Rating Externalities

A	B	C	D	E	F	G	H	I	J	K
				LEAST		RATINGS		---> MOST		
Issue	Weight	Contributing Factors	Weight	0	1	2	3	4	Impact	Issue Score
Global Warming	9%	Carbon Dioxide (lb/kWh)	100%	0	0.63	1.25	1.88	2.5		
Acid Rain	14%	Sulphur Dioxide (lb/MMBtu)[a]	50%	0	0.3	0.6	0.9	1.2		
		Nitrogen Oxides (lb/MMBtu)	50%	0	0.15	0.3	0.45	0.6		
Land Use	7%	Amount (acres)		0	2000	4000	*multiply "amount" score by "character" score to calculate impact			
		Character		industrial	res/com/farm	pristine				
Solid Waste	14%	Overall Assessment (lbs/kWh)	100%	0	0.05	0.1	.015	0.2		
		Adjustment								
Water Use/Quality	16%	Amount (gallons/MWh)	20%	0	150	300	450	600		
		Attributes of Input Water	40%	no water used	minor loss of recreational area	minor loss of habitat	major loss of recreational area	major loss of habitat		
		Thermal Attributes of Receiving Water	30%	no change		receiving water with minor thermal change		receiving water with major thermal change		
		Chemical Attributes of Receiving Water	10%	no change		receiving water with minor chemical change		receiving water with major chemical change		

			no fuel		natural gas		coal or oil
Emissions to Air 16%	Air Toxics (fuel type)	25%	no fuel		natural gas		coal or oil
	Particulates (lb/MMBtu)	25%	0	0.0075	0.015	0.0225	0.03
	Sulphur Dioxide (lb/MMBtu)	25%	0	0.3	0.6	0.9	1.2
	Nitrogen Oxides (lb/MMBtu)	25%	0	0.15	0.3	0.45	0.6
Aesthetics 2%	Visual Effects	40%	not visible or obtrusive	seasonally visible	visible from major roads	visible from recreation areas	visible from sensitive areas
	Noise	40%	no noise	below nighttime ambient	below daytime above nighttime	above daytime ambient	meets property line limit
	Signal Interference (households affected)	20%	0	10	20	30	40
Indoor Air Quality 2%	Creation	10%	reduces production		no effect		increases production
	Retention (air changes)	90%	increases number		no effect		reduces number
Fuel Issues 4%	Overall Assessment	100%	no fuel		natural gas		coal/oil
Ozone 16%	VOCs (lb/MMBtu)	45%	0	0.0015	0.003	0.0045	0.006
	Nitrogen Oxides (lb/MMBtu)	45%	0	0.15	0.3	0.45	0.6
	CFCs	10%	no release				CFCs released

Note:
a MMBtu = million Btu.
Source: NEES 1990.

236 — Chapter Twelve

issue; each factor is assigned a weight (column D) reflecting the factor's relative importance within the issue to which it is assigned. Columns E through I provide the ratings from zero to four for each contributing factor.

After the contributing factors are rated for each resource option, weights are applied to the ratings, giving a weighted impact score (column J). For some of the environmental issues, several "contributing factors" are separately rated, weighted, and then summed to form the issue score. These scores are then added to yield a composite score for each resource option (column K). A brief description of the process by which issues, factors, weights, and rates were determined is provided below. For example, a source emitting 0.9 lb/MMBtu of SO_2 and 0.3 lb/MMBtu of NO_x would have an impact score (column J) of 0.45 for SO_2 (50% × 0.9) and 0.15 for NO_x (50% × 0.3). The issue score (column K) would be 0.6 for acid rain.

Selection of Issues

The first step in developing the matrix was the compilation of a list of major environmental externalities for which each demand-side and supply-side resource option would be evaluated. The externalities were then grouped into issues such as global warming and acid rain (Table 12-1, column A). Issues were grouped and described so as to match the way the public thinks about environmental impacts. For example, while lay persons may not have an opinion on the importance of mitigating sulfur dioxide or carbon dioxide, they may have opinions on the importance of controlling acid rain or global warming.

Weighting of Issues

The issue weights in the matrix (Table 12-1, column B) were determined by blending two different types of information, including polling of experts and weights assigned in previously developed (mostly by Orange and Rockland) rating and weighting schemes. In addition, a number of studies and reports on individual issues and combinations of issues contributed to the determination of the final weights used in the matrix (NEES 1989). A strictly quantitative approach to defining weights, though preferred, was impossible due to the current lack of reliable quantitative studies.

Selection of Contributing Factors

The key utility-caused factors contributing to each issue are selected and weighted according to their relative importance. For example, sulfur dioxide and nitrogen oxides are the contributing factors used to

Incorporating Environmental Externalities in Resource Planning — 237

assess a resource option's effects on acid rain. These two factors were chosen from the list of acidic precursors from man-made and natural resources because these factors are most closely linked to utility sources. Table 12-1 shows the contributing factors for each issue (Table 12-1, column C).

In some cases, proxies for the true environmental agents have been chosen as contributing factors. For example, under the "Emissions to Air" issue, fuel type is used as a proxy for air toxics. Rather than listing the many specific air toxics that can be emitted in minute quantities, the framework takes advantage of their correlation to fuel type.

Weighting of Contributing Factors

Weights for each contributing factor represent the factor's contribution to an issue (Table 12-1, column D). Within an issue, contributing factor weights must sum to 100%. Several approaches may contribute to the estimation of factor weights:

1. Multiplicative. This approach is useful when contributing factors interact, as in the cases of land use and water use/quality. For example, land use depends both on the nature and on the amount of land affected.

2. Regulatory. The regulatory weighting methodology relies on current or pending regulations to provide issue weights.

3. Polling of Experts. The use of polling for some issues is supported by the current lack of specific data for evaluating trade-offs between some agents. The approach is holistic, calling on the professional experience of the environmental policy expert, and is consistent with the complex and diverse nature of certain externalities.

Rating of Contributing Factors

The scale used to rate the impact of each contributing factor in concrete terms must be able to measure the range of impact of all options that might reasonably be proposed by independent power producers. Therefore, the upper limit of the scale must be high enough to measure the impact of the resource option having the maximum impact for that contributing factor. While the range of options that may be proposed by independent power producers cannot be foreseen, the upper limit of the measurement scale must be set to accommodate all probable options. For example, the "Emissions to Air" scale's upper limit might

238 — Chapter Twelve

be based on NSPS[3] limits for a pulverized coal station. The lower limit of each scale may be set at zero or lower where positive effects are achievable. The approaches used for rating factors are discussed below.

1. Regulatory. Regulatory ratings and weights are derived from existing or potential legislation. Assuming legislation reflects society's values, issues and contributing factors can be compared at the margin on the basis of society's preferences. For example, SO_2 emissions have been rated on the basis of the 1.2 lb/MMBtu limit cited in the NSPS and the Clean Air Act.

2. Range of Options. This approach normalizes the ratings to cover the typical range of electric utility options. For example, the carbon dioxide upper limit could be the carbon dioxide output of a pulverized coal plant.

3. Root Cause. This approach simplifies the rating of certain contributing factors by reducing complicated aftereffects to their root causes. For example, fuel type itself, rather than the effects of burning the fuel, was used to rate air toxics emissions.

4. Grading. The central notion of grading approaches is that several features of a resource can lead it to earn a particular rating. For example, it is a particularly useful concept for rating complex, site-specific externalities such as land use and water use/quality. The advantage of the approach is that it can accommodate a large number of dissimilar elements simultaneously. It must be recognized that to generate an issue score, it may be necessary to combine ratings on very different contributing factors. New England Electric made initial choices for weighting factors, recognizing that these would be refined as experience is gained.

5. State of the Art. This approach recognizes that the rating scheme should not attempt to assign a precise rating to an issue that is either not well understood or not well linked to utility options. Indoor air quality is a good example. Air quality can be linked qualitatively to the number of air changes, but quantitative links have not yet been established.

6. Reference to Other Work. This approach takes advantage of prior thought by incorporating the rating approaches developed in other

---

[3] The New Source Performance Standard is an air emissions standard or limit applied to a category of sources. A category can be small industrial boilers, large utility boilers, combustion turbines, and so forth.

Incorporating Environmental Externalities in Resource Planning — 239

jurisdictions. For example, the rating design for aesthetics filed by O&R has been adopted in the NEES methodology.

Application of Methodology to Integrated Resource Plan

To develop NEESPLAN 1990, NEES applied the weighting and rating approach to resource options from both New England and Canada. Project-specific environmental scores can be found in Table 12-2. As one would expect, coal-fired resources received the highest environmental scores, thus indicating a higher level of potential environmental degradation. Conservation programs and distribution system upgrades received the lowest scores, reflecting little, if any, potential for environmental harm.

Several methods for factoring environmental scores into cost-effectiveness ratios were considered. The first was to simply let the scores determine the ranking of resources within a particular range of cost-effectiveness. For instance, all projects with cost-benefit ratios of 0.7 to 0.8 could be ranked by lowest to highest environmental score: an option with a cost-benefit ratio of 0.8 and with a low environmental score would be selected before an option with a 0.7 cost-benefit ratio and a higher environmental score. This option was not adopted because it could give the environmental scores very little weight in the resource selection process and lead to inconsistent results. Using the example above, a project with a 0.69 cost-benefit ratio and a high environmental score would be chosen before a project with a 0.70 cost-benefit ratio and a very low environmental score. Clearly, this approach could defeat the purpose of including environmental externalities in the planning process.

The second option examined a point scheme similar to that used in new resource bidding mechanisms. Points would be assigned to all attributes (price, dispatchabiity, environmental score, and so on), and the resource would be selected on the basis of its overall point score. This option, however, although potentially useful in a bidding framework, did not fit the company's planning techniques and models, which use cost estimates, not scores, to examine resources.

The third, and chosen, option was to attempt to place a monetary value on the environmental score and add that expense to the cost of the resource. Translating the scores into cost adders made the scores compatible with the planning process and planning models. This method allows resource planners to simply include one additional cost in the costs tracked by the planning model.

However, converting the scores to costs is obviously not an easy

240 — Chapter Twelve

Table 12-2. Comparison of NEES Resource Options with and without Externalities

Demand-Side Options	Environmental Scores	% Cost Adder	Cost-Benefit Ratio Excl. Externalities	Cost-Benefit Ratio Incl. Externalities
Commercial and Industrial Programs	(0 = no impact; 4 = maximum impact)			
—Retrofit Conservation	0.04	0.2%	0.35	0.35
—Retrofit Load Management	0.07	0.4%	0.32	0.32
—New Construction Conservation	0.04	0.2%	0.25	0.25
—New Construction Load Management	0.07	0.4%	0.32	0.32
—Interruptible Rates	0.68	3.8%	0.57	0.59
—Standby Generation	1.68	9.5%	0.57	0.63
Residential Programs				
—Low-Income	0.04	0.2%	0.69	0.70
—New Construction	0.09	0.5%	0.93	0.93
—Electric Space Heat Retrofit	0.08	0.5%	0.56	0.57
Supply-Side Options				
Power Purchase from an Unscrubbed Coal Unit	2.66	15.0%	0.83	1.01
Atmospheric Fluidized Bed Coal	2.57	14.5%	0.94	1.11
Natural Gas-Fired Combined Cycle	1.07	6.1%	0.75	0.79
15 MW #2 Oil-Fired Peaking Unit	1.44	8.1%	0.87	0.92
Purchase from a Biomass Unit	2.05	11.6%	0.78	0.89
Canadian Hydro	0.66	3.7%	1.04	1.09

Incorporating Environmental Externalities in Resource Planning — 241

task. The chosen method assigned the project with the highest environmental score—and hence, the highest level of potential environmental degradation—a cost penalty of 15% (for reasons explained in the next section). The present value of all direct costs associated with the project (capital, O&M, fuel, and other associated costs) was increased by 15% to reflect environmental externalities. Other projects received an adder based on the ratio of their score to the highest score. For instance, one option considered in the development of the company's long-range plan was a 20-year power purchase from a yet-to-be constructed coal-fired unit. Under the company's rating and weighting methodology, this option received the highest environmental score at 2.66 and, therefore, was assigned an adder of 15%. If another project received a score of 1.33, the present values of its direct costs were increased by 7.5% ($1.33/2.66 \times 15\%$). Hence, the environmental cost adder received by a project was directly related both to its environmental score and to its direct costs. If two projects received similar scores but one was more costly than the other, the project with higher costs would also receive a higher environmental cost penalty. A list of the externality percent adders are shown in Table 12-2.

One potential problem with this approach is that by linking the externality cost estimate to the direct costs of a project, illogical results can occur if two similar technologies are considered but one has significantly more environmental controls and hence is more expensive. The externality cost adder would be higher for the project with the better environmental controls. This was not a problem in the company's use of the method because the company had enough knowledge about the projects being considered to recognize that no similar technologies were significantly different in their level of environmental controls.

Choice of Maximum 15 Percent Adder

As mentioned earlier, establishing accurate environmental damage costs is the preferable way to quantify externalities. To do so, all externalities have to be identified and the costs, or benefits, of each determined. However, current data do not allow such an approach. Although much work has been done to define the direct costs of specific externalities, the tremendous variance in the resulting estimates does not encourage confidence in their usefulness. For instance, a state study concluded that the value of fish loss in Lake Michigan due to acid rain was $6.2 million per year. Another evaluation indicated the loss was $50,000 per year, less than one percent of the state estimate. Moreover, much of the direct cost research is not only uncertain but also area-specific and not directly transferable from one region to another.

242 — Chapter Twelve

NEES chose a 15% penalty for several reasons. First, it is large enough to lead to changes in the resource plan because of the extent to which it improves conservation program cost-effectiveness and limits the cost-effectiveness of some supply-side options. At the same time, the 15% penalty is not so large that erroneous assumptions in the methodology would significantly distort resource plans in the short term. Second, 15% is merely an initial estimate that can be revised as the company gains more experience with environmental externalities, as more research becomes available, and as feedback from regulators, customers, and interested parties is received. Finally, 15% is within the range of adders adopted in other jurisdictions, including Wisconsin, Vermont, and New York (Wisconsin PSC 1989; Vermont PSB 1988; Putta 1989).

Cost-Effectiveness Determination

The traditional aim of determining resource cost-effectiveness is to minimize the cumulative present value of direct costs to a utility's customers (that is, the utility's revenue requirements). To include societal costs, such as environmental externalities, requires an expanded, or societal, definition of cost-effectiveness. In Order 86-36-F, the Massachusetts Department of Public Utilities specified a cost-effectiveness test that can be called a "modified total-resource cost test." Under this order, the costs of any resource include the utility's revenue requirements, environmental externalities, and the customer costs associated with demand-side programs. The benefits include the value of avoided capacity and energy costs, and any customer savings from demand-side programs, not including bill savings. (Bill savings are excluded because they are already included in avoided capacity and energy costs.)

Effect on Resource Cost-Effectiveness and Long-Range Plan

Using the modified total-resource cost test tended to improve demand-side and worsen supply-side cost-effectiveness. In particular, the cost-effectiveness of new coal-fired resources was jeopardized by the environmental cost penalty. Before the environmental penalty was applied to a purchase from a third-party coal-fired unit, the cost-benefit ratio for the resource was 0.83. After the penalty, the resource's cost-benefit climbed to 1.01, indicating that the resource was not cost-effective. The results for an atmospheric fluidized bed unit were similar: 0.94 before and 1.11 after. All other supply-side options retained their cost-effectiveness after the penalty was applied. Natural gas–fired units, in particular, remained cost-effective.

All demand-side programs examined during the planning process

Incorporating Environmental Externalities in Resource Planning — 243

remained cost-effective and were included in the long-range resource plan. In the present resource plan, one-third of the company's incremental capacity needs are to be met by demand-side programs. Because the inclusion of environmental externalities tended to improve demand-side cost-effectiveness, there are now more cost-effective demand-side opportunities than previously. Opportunities that were marginal or slightly non-cost-effective from a revenue-requirements-only perspective became cost-effective when externalities were considered. NEES continues to identify these resources so that future plans can reflect their improved attractiveness.

For the long-range plan, a small amount (<10% of incremental needs) of new coal-fired generation from third-party suppliers has been included, even though the cost-effectiveness of this source appears jeopardized by the consideration of environmental externalities. One of the reasons for this decision is fuel diversity. Natural gas is expected to climb from essentially 0% of the company's fuel mix at present to 25% by 1999 and to 36% by 2009. Excluding new coal-fired resources from the long-range plan would mean replacing them with additional gas-fired resources, in which case, fuel diversity could be threatened. NEES expects coal prices to be more stable than other fuel prices. Price stability and protection against rate volatility are worthy goals, which, at this time, could be jeopardized by excluding coal entirely from the resource plan.

Another reason coal-fired generation has been retained in the resource plan is that societal and economic as well as environmental externalities must be considered. Moving towards a societal perspective requires addressing many difficult issues. Societal and economic impacts, such as economic development effects, deserve further consideration before any resource is excluded from the company's resource plan.

Limitations and Planned Revisions to the Methodology

In using the rating and weighting approach during the company's planning process, NEES planners discovered limitations in both the methodology and its application.[4] Many changes will be undertaken prior to the development of the company's next resource plan. A revised version of the matrix will remove the upper bound on the rating's scale. Impact ratings that cannot exceed a score of four will be removed

---

[4] Critical reviews have also been completed by others (Shimshak *et al.* 1990).

244 — Chapter Twelve

because some resources were found to have effects that exceeded the ratings' upper limits. Removing the limits will allow greater variances among projects to be considered. The weightings for both the issues and contributing factors will also be reexamined to determine if they should be revised to reflect new information, scientific opinion, and public perception. In addition, company planners will determine if additional externalities should be considered (such as methane) and if certain externalities (such as indoor air quality) should be removed. It also became quite clear that the approach should have the ability to factor in environmental improvements (offsets) brought about from any resource. For instance, repowering an existing unit may reduce emissions while adding capacity. Similarly, a power project may add a waste treatment facility to a town, thereby improving the quality of wastewater released to the environment. A matrix approach that examines only increases in externalities misses these important changes. Finally, the relationship of the methodology to the planning environment is also being reviewed.

Conclusion

In developing this methodology, NEES has taken the first step at incorporating environmental externalities in its resource planning process. Although the approach is not without shortcomings, it does provide a mechanism for addressing a difficult and complex issue. By accounting for a broad range of environmental effects in a manner that is simple, explainable, and reasonable, given present information, the approach provides a framework for discussion that various parties can understand and agree upon. While a quantitative approach based on accurate costs of externalities may be the ultimate goal, the hybrid approach described in this paper offers an effective short-term alternative. As public utility commissions, utilities, national laboratories, universities, and other organizations delve further into damage costs, the resultant data can be used to continually refine the accuracy of this approach.

A Final Note

The NEES method received considerable encouragement from the Massachusetts regulatory commission. In Docket No. 86-36G the commission stated that "we are of the opinion that a weighting and rating system is an appropriate method for addressing the question of how to treat environmental externalities" but also went on to say that "monetarization of externalities may fit into such a framework for calculating a direct adder to price." Overall, the commission appeared quite receptive to the rating and weighting approach and proposed three possible ways to organize such an approach. However, in late August of 1990, the commission issued an order in Docket No. 89-239 that ordered utilities to use as initial estimates of the cost of environmental externalities, dollar values for eight pollutants using a cost-of-control method. These values imply an environmental cost adder of $0.048 per kilowatt-hour for a new coal plant meeting the New Source Performance Standards. The order effectively excluded the type of method described in this chapter. As a result, although a revised matrix was developed, it was never used. The company is now reconsidering its approach to addressing environmental externalities in its resource planning process.

Acknowledgments

The authors would like to acknowledge the assistance of the following individuals who assisted in the preparation of this chapter: Alan Destribats, Elizabeth Hicks, and Jon Lowell of New England Power Service Company; and Eileen Brusger of Temple, Barker, and Sloane. We would also like to thank Richard Ottinger of the Pace University Center for Environmental Legal Studies for cooperation and assistance in the early development of the New England Electric methodology.

References

Bernow, S., and D. Marron 1990. *Valuation of Environmental Externalities for Energy Planning and Operations*. Boston: Tellus Institute.

Chernick, P., and E. Caverhill 1989a. *The Valuation of Externalities from Energy Production, Delivery, and Use, Fall 1989 Update*. Report to the Boston Gas Company. Boston: PLC, Inc.

———— 1989b. Memos to New England Power Service Company on the New England Electric Externalities Methodology (January 5, February 14).

———— 1990. *Report to the Boston Gas Company on Including Environmental Externalities in 89-239*. Boston: PLC, Inc.

ECO Northwest; Shapiro and Associates; and Seton, Johnson, and Odell 1984. *Economic Analysis of the Environmental Effects of a Combustion-Turbine Generating Station at Frederickson Industrial Park, Pierce County, Washington*. Final report submitted to the Bonneville Power Administration. Portland: Bonneville Power Administration.

246 — Chapter Twelve

———— 1986. *Estimating Environmental Costs and Benefits for Five Generating Resources*. Final report and technical appendices submitted to the Bonneville Power Administration. Portland: Bonneville Power Administration.

Environmental Protection Agency 1987. *Unfinished Business: A Comparative Assessment of Environmental Problems Overview Report*. Washington, D.C.: Environmental Protection Agency.

EPA. See Environmental Protection Agency.

Hohmeyer, O. 1988. *Social Costs of Energy Consumption*. New York: Springer-Verlag.

Krupnick, A. 1989. *The Environmental Costs of Energy Supply: A Framework for Estimation*. Washington, D.C.: Resources for the Future.

Massachusetts Department of Public Utilities 1988. Docket No. 86-36-F. Boston.

Meridian Corp. 1989. *Energy System Emissions and Material Requirements*. Report prepared for the Deputy Assistant Secretary for Renewable Energy. Washington, D.C.: U.S. Department of Energy.

NEES. See New England Electric System.

New England Electric System 1989. *Integrating Environmental Externalities into Resource Planning at New England Electric*. Study prepared by Temple, Barker and Sloane. Lexington, Mass.: New England Electric System.

New York State Public Service Commission 1988, 1989. Cases 88-E-240 through 88-E-245. Albany.

NYSPSC. See New York State Public Service Commission.

Putta, S. 1989. "Consideration of Environmental Externalities in New York State Utilities' Bidding Programs for Acquiring Future Electricity Capacity." Albany: New York Public Service Commission.

———— 1990. "Need and a Method for Valuing Environmental Costs in Deregulated Power Generation." Paper presented at the Eastern Regional Business and Economics Utilities Conference, 9–12 April, New York.

Shimshak, R., B. Biewald, H. Salgo, D. Marron, and S. Bernow 1990. "Comments of the Division of Energy Resources on Environmental Externalities." Testimony before the Massachusetts Department of Public Utilities, Case Nos. DPU 86-36-G and DPU 89-239, March 2.

State of Vermont Public Sector Board 1988. *Investigation into Least Cost Investments, Energy Efficiency, Conservation and Load Man-*

agement of Demand for Energy. Hearing Officer's Report and Proposal for Decision. Docket No. 5270. Montpelier.

State of Wisconsin Public Service Commission 1989. *Advance Plan 5
Order.* Docket No. 05-EP 5. Madison.

Vermont PSB. See State of Vermont Public Sector Board.

Wisconsin PSC. See State of Wisconsin Public Service Commission.

*Dean S. White is a senior analyst in the Least Cost Planning Group at
New England Power Service Company. His responsibilities include
development of New England Electric System's long range resource
plan, economic analysis of new resource options and strategic planning. He received a B.A. from the University of California and an
M.P.P. from the John F. Kennedy School of Government at Harvard
University.*

*Timothy M. Stout is the coordinator of Planning in the Demand Plan-
ning Department at New England Power Service Company. Prior to his
current position, he worked at the Conservation Law Foundation of
New England. He holds a B.A. from Middlebury College and an M.A.
from Boston University.*

*Mary Sharpe Hayes is president of the Customer Group at the Tennes-
see Valley Authority. She oversees the transmission system, rate design
and energy service programs. She has an M.B.A. from Dartmouth College, where she was an Amos Tuck Scholar, and an M.A. in Microbi-
ology from the University of California, Davis where she was a
Regents Fellow.*

Chapter 13

The Inclusion of Environmental Goals in Electric Resource Evaluation: A Case Study in Vermont

Stephen Bernow, *Tellus Institute*

Donald Marron, *Massachusetts Institute of Technology*

Introduction

The scope of utility resource planning has broadened significantly, as the old focus on supply planning has been superseded by an integrated approach that includes demand-side measures (such as improved efficiency, load shifting, and fuel switching) and nontraditional supply options (such as cogeneration, solar, and other renewables). One important extension of this integrated resource planning (IRP) process would be to take account of environmental costs in comparing these resource options. The inclusion of environmental costs may result in significant changes in the relative ranking of energy resources.

Most recent thinking on how to include environmental costs in electric resource planning has focused on general methodologies for assessing and assigning monetary values to environmental impacts. These methodologies have thus far been rarely used in actual electric planning analyses. This chapter describes an IRP evaluation that explicitly incorporated environmental impacts. However, the methodology was unique: instead of attaching monetary values to emissions and other impacts, this study posited quantitative environmental goals and used the cost of meeting these goals as the basis for estimating environmental costs.

250 — Chapter Thirteen

Description of the Overall Study

On behalf of the Vermont Department of Public Service, a comprehensive evaluation of the Vermont electric system was prepared in order to determine whether a proposed power purchase from Hydro-Quebec was consistent with IRP for Vermont (Docket No. 5330). The power would be provided by a new hydro facility, including a new dam, proposed for development at James Bay in northern Quebec.

This planning study included load forecasts, demand-side management (DSM) program design, assessment of new power supply options, evaluation of nonutility resources, analysis of transmission needs, and a review of power markets (Tellus Institute 1990). It also included a detailed analysis of certain environmental effects associated with alternative resource plans for Vermont, both with and without the proposed Hydro-Quebec project. These environmental effects were included in the economic evaluation of the proposed power purchase and of alternative resources.

In this paper, we focus on the IRP study's environmental analysis, drawing upon other elements of the study as needed to provide context. We also discuss alternative approaches to treating environmental issues in IRP, and we introduce an approach—the energy/environment target IRP method—that permits straightforward incorporation of environmental goals in energy planning. In this method, all resource plans are evaluated in terms of the same energy and environmental objectives, and the overall costs—combining direct resource costs and the additional costs of meeting the environmental goals—of the plans are compared. Including these additional costs to meet environmental targets may reveal that certain options that would otherwise not be considered in a least-cost plan are in fact desirable.

Of particular interest here are the treatment and environmental performance of DSM, fuel switching for selected end uses, and cogeneration. Expanding the boundaries of electric system planning beyond power supply facilities and beyond the electric sector itself for certain end uses can produce both economic and environmental benefits.

Selection of Electric Resources

The existing electric system in Vermont, comprising 24 retail utilities and the Vermont Department of Public Service, serves a winter peak demand of 960 MW and annual energy requirements of 5,271 GWh. Resources are primarily a mix of nuclear, hydro purchases, coal, and residual oil. Load is projected to grow at an annual rate of about 1.7%

for peak demand and 2.3% for energy use over the next 20 years. Peak demand would reach 1,170 MW and energy use 6,850 GWh in the year 2000 (Tellus Institute 1990).

The overall IRP analysis considered a variety of new supply and demand-side options, including both utility and nonutility resources, to meet the power needs of Vermont over the next 20 years. These options included:

- the proposed Hydro-Quebec project;[1]

- new utility-owned generating facilities, primarily natural gas combined-cycle and distillate oil combustion turbines, both outfitted with selective catalytic reduction (SCR);[2]

- new nonutility generating facilities including cogenerators; and

- demand-side management programs, including fuel switching from electricity to natural gas.[3]

While the focus of the analysis was on the Vermont system itself, we also needed to model external resources in order to capture two effects. First, we needed to account for the fact that the set of new resources in competing resource plans (some plans with and some without the proposed Hydro-Quebec project) might differ, within a given year, in the amount of electricity generated or capacity provided. For this reason, it was necessary to identify the generating resources that would provide the marginal power. We assumed that this power would come from the marginal resources in the New England Power Pool (NEPOOL), since the Vermont utilities are part of that centrally dispatched system.[4] Thus, environmental impacts from NEPOOL marginal resources, as well as from new Vermont resources, were included in the analysis.

Second, we needed to consider what would happen to the power

---

[1] The proposed contract included provisions for up to 340 MW of committed purchases (the "minimum take") and up to 110 MW of additional cancelable capacity. The analysis focused on the committed purchases.

[2] Note, however, that use of SCR on oil-fired peakers is unproven.

[3] We considered a number of DSM programs in our study. In this chapter, we refer to a "strong" DSM program that would provide up to 300 MW of savings (about 21%) off a projected 2010 peak load of 1,400 MW. Energy savings were projected to be about 1,083 GWh in 2010, or about 12.5% of demand in that year.

[4] NEPOOL includes virtually all utilities in New England. For the economic analysis, the Vermont system could be modeled in isolation owing to the power-pricing protocols of NEPOOL. The environmental analysis, however, required that the actual pool resources and their emissions be estimated.

252 — Chapter Thirteen

proposed for purchase and to its environmental impacts if Vermont ultimately rejected the contract. We modeled three rejection cases in order to capture the range of possible effects and will discuss two of those cases here. In one case, we assumed that the development at James Bay would be unaffected by Vermont's rejection of the contract. In this case, the hydropower would be used to displace fossil genera tion in Canada and New England. For this case, we assumed that 2/3 of the power would flow back into New England and that 1/3 would flow to Ontario Hydro or another utility with similarly polluting coal-fired plants. In a second case, we assumed that development at James Bay would be deferred if the contract were rejected and thus that no hydro power would be used to displace fossil generation.[5]

Selection of Environmental Loadings to Be Evaluated

Within an integrated planning analysis, it would be appropriate, in principle, to include all the environmental impacts associated with electric resources—air emissions, land use, water emissions, thermal pollution, solid waste generation, noise, traffic, aesthetic degradation, and so forth. For purposes of the Vermont analysis, however, we needed to limit our focus to a subset of the environmental effects and chose to focus on the following air emissions:

- acid gases: sulfur oxides (SO_x) and nitrogen oxides (NO_x);

- greenhouse gases: carbon dioxide (CO_2), methane (CH_4), carbon monoxide (CO), and nitrous oxide (N_2O); and

- other emissions: total suspended particulates (TSP) and total hydrocarbons (THC).

In particular, we focused on those emissions that occur during energy conversion—that is, in the production of electricity or the serving of a direct end use. We did not include emissions that occur in the rest of the fuel cycle, either upstream (in the extraction, processing, and transport of fuels and in construction of power plants) or downstream (in the disposal of wastes and in decommissioning).

The study encompassed geographic boundaries beyond those of Vermont's electric system. First, the focus on acid and greenhouse

---

[5] After we completed the study, a Hydro-Quebec witness testified that the second case was more probable.

The Inclusion of Environmental Goals in Electric Resource — 253

gases necessitated attention to regional and global environmental impacts. Second, owing to Vermont's exchanges of power with utilities in Canada, New England, and New York, we included out-of-state sources of these pollutants if those sources were affected by Vermont's planning and operation.

Finally, since hydroelectric development in the James Bay region of Canada would serve most, if not all, of the proposed contract, we also modeled the land-use impacts associated with potential resource options. Analysis of the ecological and socioeconomic impacts of hydroelectric development in that region, while relevant to an overall assessment of Hydro-Quebec options, was beyond the scope of our study. However, as we discuss in a later section, our results can be used to estimate the cost to Vermont of avoiding or accepting these impacts.

Development of Environmental Loadings Coefficients

Environmental loadings coefficients for land use, expressed in acres per MW, were developed from U.S. Department of Energy (DOE 1983) and Michigan Electricity Options Study (MEOS 1986) figures. The land-use impacts of resource plans were expressed in acre-years, the cumulative land area allocated to electricity production times the number of years of such allocation. Some land-use impacts may involve large-scale and irreversible changes to ecosystems, habitats, animal and plant populations, and cultures. This has been a significant concern with regard to James Bay development as a whole, and the Vermont sale portion as well.[6]

Emissions factors for most new electric resources, expressed in lbs per million (MM) Btu of fuel input, were derived from a data base that summarizes U.S. Environmental Protection Agency (EPA), DOE, and other estimates of environmental impacts (Tellus Institute 1989). NEPOOL heat rate estimates were then used to convert these factors to lbs/MWh emissions coefficients (Table 13-1). Emissions from the Hydro-Quebec purchase itself were assumed to be zero, since that

---

[6] As noted earlier, our study incorporated resource costs and the additional costs of meeting the system's air emissions targets but did not directly ascribe costs to land-use or ensuing ecological and socioeconomic impacts. All costs, then, were direct monetary costs and were discounted accordingly. The plans were compared on the bases of their costs and land-use impacts (as both energy services provided and net emissions impacts were the same across all plans). The different land-use impacts—measured in acre-years—can be taken to represent the full range of potential impacts in James Bay associated with the construction and operation of the project.

254 — Chapter Thirteen

power is to be provided primarily by hydroelectricity.[7] Emissions coefficients for the Ontario Hydro system (to which some power would flow in one rejection scenario) were based on a 50–50 mix of scrubbed and unscrubbed coal plants. (See Tellus Institute 1990 for further detail on how emissions factors were computed.)

Emissions coefficients of resources for which there is not such a direct ratio between fuel consumption and electricity output—resources including cogeneration, DSM, fuel-switching options, and the NEPOOL margin—were derived as described below.

Cogeneration Facilities

Cogeneration facilities produce usable thermal energy as well as electricity. Therefore, in the development of their emissions factors, it would not be correct to attribute all their air emissions to the production of electricity. We therefore adjusted their emissions coefficients to account for the commercial and industrial boiler emissions avoided because of the steam output of cogeneration facilities. To calculate these emissions offsets, we developed emissions coefficients for medium-sized industrial boilers fueled by natural gas, distillate (#2) oil, and residual (#6) oil. Since the fuel type of future Qualifying Facility (QF) and independent power producer (IPP) resources is unknown, we modeled two generic QF-IPP scenarios for this particular analysis:

Scenario 1: All facilities are assumed to be natural gas combined-cycle with cogeneration. All avoided boilers are assumed to be fueled by residual oil.

Scenario 2: Facilities are assumed to be 50% natural gas combined-cycle, half with cogeneration and half without; 25% coal atmospheric fluidized bed with cogeneration; and 25% wood-fired generation. The cogeneration is assumed to avoid a mix of boilers characteristic of the region (EIA 1989): 41% natural gas, 33% distillate oil, and 25% residual oil.

---

[7] In testimony filed after our original analysis was completed, a number of parties argued that some fossil power might be necessary to serve the contract in its early years. Some parties also argued that flooding for hydro development would result in releases of carbon dioxide and methane over the next several decades from loss of standing biomass and of future CO_2 uptake. This impact is relevant to the analysis of Vermont resource plans only for the case in which rejection of the contract affects development at James Bay. We found that the magnitude of these two effects is relatively small: the annualized carbon emissions from biomass loss in James Bay is about 200 lb/MWh, about one-tenth that of a new combined-cycle gas plant. The impact of these carbon emissions on the overall results is given in Rosen (1990)

The Inclusion of Environmental Goals in Electric Resource — 255

Table 13-1. Environmental Loadings									
	Emissions in Pounds per MWH								**Land Use**
Resource	**SO_x**	**NO_x**	**CO_2**	**CH_4**	**TSP**	**CO**	**THC**	**N_2O**	**(acres/MW)**
Hydro Quebec	0.00	0.00	0	0.00	0.00	0.00	0.00	0.00	190.00
Natural Gas Comb. Cycle (CC)	0.00	0.64	953	0.11	0.01	0.51	0.01	0.06	1.00
Natural Gas CC, Cogen 1	−0.68	0.31	719	0.10	−0.03	0.45	0.01	0.03	1.00
Natural Gas CC, Cogen 2	−2.24	0.02	682	0.09	−0.12	0.45	0.01	0.01	1.00
Distillate Comb. Turbine (CT)	2.36	1.14	1828	0.02	0.33	1.29	0.40	0.24	0.10
Coal Fluidized Bed	5.31	5.37	2078	0.01	0.29	0.29	0.03	0.31	1.00
Coal Fluidized Bed, Cogen 1	4.43	4.95	1778	0.01	0.23	0.22	0.02	0.27	1.00
Wood Steam	0.13	2.27	3010	0.46	0.43	4.26	1.42	0.46	0.67
NEPOOL Margin	12.14	5.11	1813	0.02	0.84	0.54	0.11	0.34	0.10
Ontario Hydro Margin	12.10	2.97	2320	0.01	1.21	0.24	0.03	0.33	0.00
Non-Fuel Switching DSM	0.00	0.00	0	0.00	0.00	0.00	0.00	0.00	0.00
Fuel Switching (R/C WH)	0.00	0.47	153	0.00	0.02	0.02	0.01	0.00	0.00
Fuel Switching (CSH)	0.75	0.52	625	0.04	0.07	0.14	0.07	0.10	0.00
Fuel Switching (RSH)	0.30	0.33	655	0.02	0.08	0.09	0.05	0.03	0.00

Notes: Natural Gas CC, Cogen 1 is a cogenerating CC that displaces a mix of residual, distillate, and natural gas fuels in industrial boilers. Natural Gas CC, Cogen 2 displaces only residual oil. Coal Fluidized Bed, Cogen 1 displaces the same mix of fuels as the Natural Gas CC, Cogen 1.

R/C WH = Residential Small Commercial Water-Heating
CSH = Commercial Space Heating
RSH = Residential Space Heating

The wood steam coefficients assume unsustainable wood consumption. Sustainable burning would have substantially lower CO_2 emissions.

These two scenarios are used to illustrate a range of the environmental benefits and costs that may result from the QF-IPP development assumed in this IRP study.

In each scenario, determining the actual level of emissions avoided by the cogenerator required that we also know the overall steam efficiencies for the cogenerator type and for the avoided boiler. For the cogenerator, we assumed that 25% of the heat remaining after electricity production would be turned into usable steam. Thus, for a natural gas combined-cycle cogenerator with a heat rate of 8,214 Btu/

256 — Chapter Thirteen

kWh, the overall thermal efficiency (the fraction of energy input used for thermal end uses) would be about 15% ($= .25 \, [1 - 3,412/8,214]$; 1 kWh = 3,412 Btu). For industrial boilers, we assumed an overall efficiency of 75%. Given the gross emissions from a cogenerator (E_g, measured in lbs/MMBtu), its overall thermal efficiency (S_c), the gross emissions from the average avoided boiler (E_b, in lbs/MMBtu), and the boiler's overall thermal efficiency (S_b), we can calculate the cogenerator's "net" emissions (E_n, in lbs/MMBtu) as:

$$E_n = E_g - E_b \times (S_c/S_b)$$

where, in general, $S_c = (1 - 3,412/\text{Heat Rate} \times F$, and F is the fraction of thermal energy not used for electricity production that is captured for thermal end uses.

By multiplying emissions in lbs/Btu by the cogenerator's electric heat rate, we can then determine the facility's emissions coefficient in lbs/MWh.

NEPOOL Margin

Since different resource plans for Vermont imply different levels of generation from the rest of the NEPOOL system, it was necessary to consider how emissions from NEPOOL as a whole would vary under different resource plans. Based on a dispatch analysis of the NEPOOL system, we estimated the fraction of the NEPOOL margin that would be made up of various resources:[8] residual oil steam (78%), distillate oil steam (less than 1%), natural gas steam (3%), distillate combustion turbines (17%), and natural gas combustion turbines (2%). Based on emissions coefficients for each individual resource type, we used these fractions to develop a weighted average emissions rate to represent the NEPOOL margin.[9]

For the land-use impact associated with capacity differences, we

---

[8] We used a production costing model that "dispatches" the system power plants to meet load over the 8,760 hours of the year; such a model allocates plants starting with the lowest variable (fuel and O&M) cost of energy and moving towards higher-cost energy until the load is met over all hours. The system "margin" is the mix of resources that meets the highest increment of load for any hour (or group of hours). In this study, the system margin was estimated by comparing two dispatch runs, one with the expected loads and one with a modest load decrement. The difference is the system margin—that is, energy production from all those plants whose output decreases to satisfy the decrement.

[9] The emission rates for the NEPOOL system margin can be thought of as short-term *avoided emissions*, directly analogous to short-term avoided costs. Long-term avoided emissions are associated with different long-term resource plans. If costs are ascribed to these avoided emissions, adding the resultant *avoided emissions costs* to the direct avoided costs gives total (energy plus emissions) avoided costs.

assumed that the NEPOOL margin was peaking capacity, with a land-use factor of 0.10 acres per MW.

Demand-Side Management and Fuel Switching

In this study, the environmental benefits of DSM are the avoided emissions and avoided end-use impacts that would otherwise result from electric generation and the addition of new capacity. DSM resources that embodied efficiency improvements were assumed to have emissions coefficients of exactly zero (an assumption consistent with our decision to consider only emissions from the conversion stage of electricity production).

Fuel-switching programs (for example, switching from electricity to natural gas for residential water-heating) cause increased emissions at the end use; these emissions were counted as an environmental cost of DSM. The environmental benefits would be the avoided emissions from reduced electricity generation. In general, these avoided emissions would exceed the end-use emissions added by fuel switching owing to the lower thermodynamic efficiency and somewhat dirtier fuel mix of electricity production compared to direct use of fuels.

We considered three types of fuel switching programs:

- residential space heating: from electricity to natural gas (13%), propane (52%), and distillate oil (35%);

- residential/small commercial water-heating: from electricity to natural gas (25%) and propane (75%); and

- commercial space heating: from electricity to natural gas (13%) and distillate oil (87%).

Emissions factors, expressed in lbs/MMBtu of fuel input, for each type of heating unit were based on standard furnaces and water heaters. Aggregate emissions coefficients for each program were then developed as the weighted average of the individual emissions coefficients. Fuel switching "heat rates" were then developed based on (1) the efficiency of the end-use conversion device and (2) electric losses in transmission and distribution. For example, given an average space heater efficiency of 80% and a line loss factor of 7.9%, we calculated a space-heating fuel switching "heat rate" of 3,953 Btu/kWh, calculated as:

$$3{,}953 \text{ Btu/kWh} = \frac{3{,}412 \text{ Btu/kWh}}{1.079\,(.80)}$$

By multiplying these heat rates by the lbs/MMBtu emissions factors, emissions coefficients in lbs/MWh were calculated for the fuel switching programs.

258 — Chapter Thirteen

Valuation of Environmental Loadings: Evaluation Methodologies

The art and science of attributing costs to environmental impacts are still in early stages of development. Most studies have taken one of two general approaches, emphasizing either the costs of environmental damage or the costs of emissions controls. In this study, we have developed a third approach based on setting environmental targets.

The Damage Cost Approach

While significant efforts have been made using the damage costing approach (ECO Northwest *et al.* 1984, 1986, 1987; Hall *et al.* 1989; Hohmeyer 1988), we believe that these approaches do not provide a suitable basis for establishing policies in New England at this time. In our review of some such estimates, it soon became clear that environmental impacts and their costs are inherently complex and uncertain, are qualitatively and quantitatively site-specific, and depend not only on scientific and economic analysis but also on public perception and values. The direct ascription of costs to certain impacts—such as loss of species, degradation of habitats, loss of human life, disturbance of cultures—is itself controversial and arguably inappropriate. In our opinion, neither the science nor the economics of environmental damage assessment is sufficiently developed at this time—nor is public policy discussion sufficiently advanced—to assign acceptable damage costs to the air pollutants modeled in our Vermont analysis.

The Control Cost Approach

While abatement costs are generally well understood compared with damage costs, they do not necessarily bear any simple relation to the damage costs for which they are used as proxies. They may be inappropriate for representing both the overall and relative magnitudes of damages associated with different pollutants because the relative cost of controlling a pollutant does not necessarily reflect the pollutant's relative threat to the environment; a very damaging pollutant may be cheaper to control than a less damaging one. Thus, their use as a surrogate for the actual health, socioeconomic, and ecological damages associated with environmental impacts could result in inappropriate ranking of alternative resources and resource plans.

Because of this problem with simple cost-of-control valuation, a more nuanced control cost approach has been proposed, based on the notion of regulators' "revealed preferences" (Schilberg *et al.* 1989) or "shadow pricing" (Chernick and Caverhill 1989). In this approach, existing and proposed environmental regulations are analyzed in order to estimate the value that society implicitly places on specific environmental impacts. For example, acid rain legislation may be analyzed in order to estimate the costs society is willing to impose on itself to reduce emissions of SO_x. In analyzing the regulations, we can identify the highest (or marginal) cost reduction strategy required by the regulations. If we assume that regulators are "rational," this can then be taken as an estimate of the value that they (and society) have placed on air emissions. At the very least, it can be argued that this value represents the "revealed preferences" of regulators, and that, to be consistent, it ought to be applied when decisions affecting these environmental impacts are made.

While the "revealed preferences" method does have a number of difficulties,[10] we believe that in some instances it is a useful way to estimate the values that society places on air emissions. Indeed, in some of our more recent work (Shimshak *et al.* 1990), we explicitly recommended that Massachusetts adopt the "revealed preferences" approach in order to include environmental costs in an all-resources bidding system.[11] In a recent Order (89-259) affecting the least-cost planning concept, the Massachusetts Department of Public Utilities adopted a set of monetary values proposed by Tellus Institute for SO_2, NO_x, TSP, CO, CO_2, VOCs, and CH_4. Upon the recommendation of Tellus Institute (Tellus 1990b) and with support from other analysis, the Nevada PSC adopted essentially the same set of values in an Order (89-752) expanding the language of Nevada's IRP rule to include environmental externalities.

In the current instance, however, we decided, as discussed in the text, that the environmental standards approach was appropriate for the Vermont study.

---

[10] It assumes that regulators have made a rational assessment of society's "costs" and "benefits." These costs can be either higher or lower than actual damage costs, either through insufficient information or because a risk-averse or risk-accepting margin is adopted. Moreover, society's revealed preferences can change over time as information, analysis, and values change. Thus, a limitation of this approach is that past or current revealed preferences may bear little relation to actual impacts and their current value to society or to the value that further attention, scrutiny, and public debate might reveal. Both the acid rain policy debate of the last decade and a half and the nascent greenhouse gas discussion are examples of such evolution.

[11] It should be noted, however, that, ideally, one would not simply accept a single-point value based on current "revealed preferences," but would instead explore a range of such values and the direct cost implications of moving along that range.

260 — Chapter Thirteen

The Environmental Goals Approach

In order to incorporate environmental goals directly in our overall economic analysis, we decided to use emissions targets as an alternative to directly developing monetary costs for air emissions.[12] To do so, we estimated the abatement costs that would have to be incurred by Vermont utilities if they attempted to follow a policy of no net new emissions from electric resources during the 20-year planning period. The resulting abatement costs are a measure of the costs associated with the avoided air emissions.

While recognizing that this approach has drawbacks, as do other valuation methodologies, we believe that it provides one useful basis for accounting for environmental impacts within the IRP process. In this approach, it is recognized that the valuation of environmental impacts is so fraught with difficulty that it may be appropriate, at this time, for public policy on the environment to precede and motivate technical analysis. To that end, system emissions targets (including reductions) can be treated as goals of the resource evaluation process rather than as quantified outputs of that process. The resource evaluation process can then be seen as attempting to meet two goals: satisfying Vermont's electricity requirements and keeping emissions below a specified emissions target. The least-cost resource plan, including supply technologies, demand-side options, nonutility resources, fuel switching, and pollution control techniques (or offsets), would have to satisfy these criteria. Naturally, different plans that meet the criteria would have different costs and uncertainties. However, by using this approach, planners, regulators, and the public are informed about the overall costs that would be required to meet the two criteria and about the additional costs that would be incurred as a consequence of setting alternative environmental targets.

The environmental goals or standards approach puts the environmental targets forward as a matter of environmental policy and places the burden of proof on those who would argue that the costs of meeting the targets are too high or the benefits too low. Such an approach could take local, regional, and global ecological sustainability criteria as constraints on development choices generally and on energy resource decisions in particular. It would insist upon criteria that would prevent the undermining of environments and the well-being of future genera-

---

[12] We did not attempt to value land-use impacts. Assigning monetary values to land-use impacts is problematic because some environmental costs are already internalized in land costs (after all, land is a market good) and because of interregional difficulties, among other reasons.

The Inclusion of Environmental Goals in Electric Resource — 261

tions as a result of today's consumption, production, transport, and so forth.

It could be argued that the goals of intergenerational equity and/or placing an intrinsic value on preserving natural environments (that is, bequeathing to the future an environment that has not been seriously degraded) cannot be guaranteed by monetizing and discounting the current generation's environmental impacts. The environmental standards approach can take sustainability targets as inputs to resource decisions, whence only direct costs are represented and discounted. Sustainability targets could entail rollbacks from currently unacceptable environmental impacts.

Clearly, the contexts in which such targets are set (along with the instruments to implement them, whether taxes, regulations, tradable permits, and so forth) will vary, cutting across political and administrative jurisdictions. An important question arises here regarding the appropriate institutional basis for setting environmental targets for utility planning. Whether public utility commissions can assume this role, traditionally held by environmental regulatory and siting agencies, is not clear. Perhaps this will be tested in the near future. Perhaps, also, institutional cooperation should be pursued.

Derivation of Emission Avoidance Costs Using the Environmental Standards Approach

The environmental target established for all scenarios in the Vermont IRP study was no net new emissions for the Vermont system (including the impacts of unpurchased Hydro-Quebec power and the NEPOOL margin). Thus, we did not value emissions from the resources considered, since we assumed that no net emissions would occur; instead of estimating costs associated with air emissions externalities, we estimated the costs of avoiding the externalities. A range of estimates of pollution abatement costs was compiled from the literature to reflect the potential mix of abatement options that could be used to achieve this target. The average of high and low abatement costs across a range of technologies and techniques for emissions reduction were used here instead of New England–specific abatement costs. In the case of acid gases, this approach is consistent with emerging national strategies for achieving reduction targets at the lowest feasible costs through emissions trading. However, it would not be feasible for the region to always find the lowest-cost abatement technique elsewhere, since other utility systems would also be seeking to employ these lowest-cost solutions. Similar abatement cost increases could occur for afforestation—increasing net biomass for taking up carbon—as increasing amounts

262 — Chapter Thirteen

Table 13-2. Environmental Abatement Costs in 1989$/ton

Emission	Low	Medium	High
NO_x	200	2,600	5,000
SO_x	200	850	1,500
CO_2	3	7	11
TSP	220	360	500
THC	n/a	n/a	n/a
CO	7	15	24
CH_4	30	70	110
N_2O	540	1,260	1,980

Source: Tellus Institute 1990.

of land are brought under production. In both cases, the demand for the least-cost emissions reduction measures (in the case of afforestation, the land least expensive to use) would exceed the availability of such measures. Thus, we have taken the low and high abatement cost estimates to reflect a realistic range of options and their costs.[13]

Table 13-2 summarizes the results of our review of the literature. The low-cost estimates for SO_x abatement are based on the use of low-sulfur fuels, while the high-cost estimates are based on scrubber retrofits. Similarly, low-cost estimates for NO_x abatement are based on low-NO_x burner retrofits and the high-cost estimates on selective catalytic reduction. The costs of all of the greenhouse gases are based on uptake of carbon in afforestation,[14] with global warming potentials based on estimated long-term contributions.[15] Use of these figures captures only the greenhouse effect; other environmental effects, such as health effects associated with CO, are not included. Note that, because

---

[13] Note that our usage of average abatement costs differs substantially from the usages in New York (New York State Energy Office *et al.* 1989; Putta 1989) and in California (Therkelsen 1989; CEC 1990). In our case, we require that emissions be abated, while in the New York and California applications, emissions are still assumed to occur. Average abatement costs are not the correct measure to use for valuing unabated emissions. (See Bernow and Marron 1990.)

[14] The CO_2 cost figures are based on estimates of reforestation costs; there are little or no data on the actual cost of such an approach to offsetting CO_2 emissions. In later work (Shimshak *et al.* 1990; Bernow and Marron 1990), we have concluded that the CO_2 numbers presented here are somewhat low, although the $7 figure is the same as that recently adopted in California (CEC 1990).

[15] We used the following global warming potentials (by weight): $CO_2 = 1.0$, $CO = 2.2$, $CH_4 = 10$, $N_2 = 180$; these are based on Lashof and Ahuja 1990.

The Inclusion of Environmental Goals in Electric Resource — 263

of a lack of information, no cost figures were developed for hydrocarbon emissions; as a result, hydrocarbons were implicitly valued at $0. This omission clearly understates the environmental costs associated with electric resources. (For more information on the derivation of these cost figures, see Tellus Institute 1990.)

Consistent with our assumption that the emissions abatement costs will actually be incurred as abatement technologies are implemented, we chose to use our estimate of the Vermont utilities combined weighted cost of capital (about 10.5%) as the discount rate in this analysis. Use of this discount rate was appropriate for abatement costs that will actually be incurred, just as construction and fuel costs are incurred.[16]

Results

Results for the Hydro-Quebec Contract

The principal results of the Vermont analysis are presented in Table 13-3 for two cases: one in which rejection of the contract does not cause the Hydro-Quebec development to be deferred and one in which rejection causes a deferral. Costs are expressed in 1989 present-value dollars; they reflect the total costs over the 20-year study period. Note that the deferral assumption only affects the emissions costs in the cases in which the Hydro-Quebec project is assumed to be rejected. Each case includes one analysis for low fuel prices and one for high fuel prices.[17] Note that the variation in fuel prices has a significant impact on direct ratepayer revenue requirements (roughly $500 million in the Hydro-Quebec Out case), but only a small impact on emissions costs (the difference ranges from $4 million to $7 million, depending on the scenario). Emissions costs in the low-fuel-price cases are slightly lower than in the high-fuel-price cases because of slightly greater use of newer, cleaner facilities and a corresponding reduction in use of existing facilities.

In both fuel-price scenarios, revenue requirements are lower for resource plans including the contract. In the nondeferral case, emissions costs with the proposed contract are actually higher than they are

---

[16] By setting energy and environmental targets, all costs incurred in meeting these targets are conventional economic costs. Thus, conventional discounting would be appropriate. If the method adopted had ascribed values to the environment, a societal discount rate would be more appropriate. It is noteworthy, however, that even conventional economic analysis, without consideration of environmental impacts, may be undertaken using a societal discount rate instead of a private discount rate.

[17] In the study, many other scenarios were also modeled. (See Tellus Institute 1990.)

264 — Chapter Thirteen

Table 13-3. Results of the Analysis

Scenario	Revenue requirement	Emissions costs	Total costs	Land use (thousand acre-yrs)
Hydro-Quebec Accepted				
Low Fuel	7871	395	8266	1413.8
High Fuel	8184	399	8583	1417.4
Hydro-Quebec Rejected				
a. HQ Development Not Deferred				
Low Fuel	7914	331	8245	1416.3
High Fuel	8409	336	8745	1418.8
b. HQ Development Deferred				
Low Fuel	7914	579	8493	18.8
High Fuel	8409	586	8995	19.2

Note: All costs are in millions of 1989 present value dollars.

when the contract is rejected. This slightly counterintuitive result (why would purchasing hydro power increase emissions costs?) is explained by the fact that, in this case, the power would reduce emissions even more if it were used to back down existing generation in New England and Ontario Hydro rather than mostly in new facilities in Vermont. In the low-fuel-price scenario, this emissions cost increase from purchasing Hydro-Quebec power more than offsets the revenue requirements savings from doing so, and thus the contract appears uneconomic in this instance. In the high-fuel-price scenario, however, the direct economic benefits of the contract outweigh the increase in emissions costs. In the deferral cases, finally, emissions costs greatly favor the contract, and thus the net results also favor the contract.

Of course, the net economic results exclude many other impacts, most notably the land use and related environmental and sociocultural impacts that arise for the deferral case. The economic results can be used, however, to provide a framework for evaluating these concerns. For the low-fuel-price scenario, for example, we can pose the following question: Do we believe that the environmental impacts associated with 1.4 million acre-years of land use (roughly 47,000 acres per year for 30 years) are worth more than $43 million in direct economic and $184 million in emissions costs savings? If so, we should reject the contract despite these savings. Similarly, we can ask whether Vermont is willing to increase its incremental electric costs by about 2.7% (227/8,266) in order to avoid these other impacts.

Table 13-4. Emissions Costs for Electric Resources in 1989$/MWH

| | Natural Gas CC | | | Coal AFB | | Dist. | Fuel Switching DSM | | | |
	Non-Cogen	Cogen 1	Cogen 2	Non-Cogen	Cogen 1	CT	Water Heat	Commer SH	Resid SH	Other DSM
NO$_x$	$0.83	$0.40	$0.03	$ 6.98	$ 6.44	$1.48	$0.61	$0.68	$0.43	$0.00
SO$_x$	0.00	−0.29	−0.95	2.26	1.88	1.00	0.00	0.13	0.13	0.00
CO$_2$	3.34	2.52	2.39	7.27	6.22	6.40	2.93	2.19	2.29	0.00
TSP	0.00	−0.01	−0.02	0.05	0.04	0.06	0.02	0.01	0.01	0.00
CO	0.00	0.00	0.00	0.00	0.00	0.01	0.00	0.00	0.00	0.00
CH$_4$	0.00	0.00	0.00	0.00	0.00	0.00	0.00	0.00	0.00	0.00
N$_2$O	0.04	0.02	0.01	0.20	0.17	0.15	0.01	0.06	0.02	0.00
Total:	$4.21	$2.65	$1.45	$16.76	$14.75	$9.10	$3.57	$3.26	$2.88	$0.00

Note: As noted in Table 13-1, the Cogen 1 facilities displace a mix of residual, distillate, and natural gas in industrial boilers, while Cogen 2 displaces only residual oil.

CC = Combined Cycle, AFB = Atmospheric Fluidized Bed, CT = Combustion Turbine

Results for Efficient Resources

Based on the emissions coefficients and the average abatement costs, it is possible to estimate costs associated with specific electric resources. Table 13-4 presents disaggregated emissions costs for a variety of resources. These figures illustrate the significant environmental benefits associated with cogeneration and demand-side management. Note, for example, that the fuel-switching programs, which increase emissions at the end use, still have lower emissions costs (per MWh) than a noncogenerating natural gas combined-cycle facility. Note also how the emissions costs of cogenerating facilities vary significantly, depending on the fuel used by steam boilers that are displaced.

The results in Table 13-4 can be used to make a direct gross emissions comparison of DSM options and supply resources. They do not, however, give an exact picture of DSM's environmental benefits, since DSM will back down a mix of facilities (for example, a mix of new gas combined-cycles, new distillate combustion turbines, and old residual oil steam boilers). In order to capture this effect, we analyzed the overall dispatch of the Vermont system with different amounts of DSM. All DSM programs were found to provide net emissions benefits. Table 13-5 presents these benefits in levelized cents per kWh and as percentages of utility avoided costs (which were calculated based on changes in the entire Vermont system). Non-fuel-switching programs were found to

266 — Chapter Thirteen

Table 13-5. Avoided Costs (Direct Economic and Environmental) for DSM				
Program	Avoided Utility Costs	Avoided Emissions Costs	Total Avoided Costs	Emissions Costs as a % of Utility Costs
Water Heat Fuel-Switching	8.19	1.24	9.43	15%
Resid. ESH Fuel-Switching	11.37	1.07	12.44	9%
Commer. ESH Fuel-Switching	11.33	0.99	12.32	9%
Non-Fuel-Switching	7.73 to 11.72	1.40 to 2.13	9.13 to 13.85	18%
Source: Nichols 1990.				

save from 1.40 to 2.13 cents per kWh of emissions costs, roughly 18% of utility avoided costs. Not surprisingly, fuel-switching programs produced lower net savings, both in absolute terms (0.99 to 1.24 cents per kWh) and as a percentage of utility avoided costs (9% to 15%).

Conclusion

In this study we have adopted the energy/environment target IRP approach, simultaneously meeting end-use energy requirements and emissions targets. The environmental credits found for DSM indicate that additional DSM investments, beyond those considered in this study, may be effective for achieving even lower-cost resource plans that meet both energy needs and the emissions constraints.

We should reiterate that these cost adders are based on an estimate of average abatement costs. As we discussed above, this is appropriate under the premise that a "no net emissions" policy will be applied to the resource plans considered in this analysis. If such a policy of internalizing emissions costs through the establishment of targets were not carried out, then the monetary value of avoided air emissions could be significantly higher, insofar as the average cost of abatement is likely to be significantly less than actual damage costs. As a result, the environmental benefits of DSM and cogeneration would appear even larger: using the high-cost figures, for example, the environmental costs and benefits in Tables 13-4 and 13-5 would appear about 1.5 to 2 times larger.

The Inclusion of Environmental Goals in Electric Resource — 267

Acknowledgments

As noted in the text, this environmental analysis was only one part of a much larger planning exercise. For that reason, we would like to acknowledge the assistance of our colleagues at Tellus who worked on the study. We would also like to thank Bill Steinhurst and Doug Smith of the Vermont Department of Public Services for their effort in this case.

References

Bernow, S., and D. Marron 1990. *Valuation of Environmental Externalities for Energy Planning and Operations: May 1990 Update.* Boston: Tellus Institute.

California Energy Commission 1990. "Committee Order for Final Policy Analysis." Docket No. 88-ER-8 (the 1990 *Electricity Report*), March 27. Sacramento: California Energy Commission.

CEC. See California Energy Commission.

Chernick, P., and E. Caverhill 1989. *The Valuation of Externalities from Energy Production, Delivery, and Use, Fall 1989 Update.* Boston: Boston Gas Company.

DOE. See United States Department of Energy.

ECO Northwest; Shapiro and Associates; and Seton, Johnson, and Odell 1984. *Economic Analysis of the Environmental Effects of a Combustion-Turbine Generating Station at Frederickson Industrial Park, Pierce County, Washington.* Final report. Portland, Ore.: Bonneville Power Administration.

——— 1986. *Estimating Environmental Costs and Benefits for Five Generating Resources, Final Report and Technical Appendices.* Portland, Ore.: Bonneville Power Administration.

——— 1987. *Generic Coal Study: Quantification and Valuation of Environmental Impacts,* Portland, Ore.: Bonneville Power Administration.

EIA. See Energy Information Administration.

Energy Information Administration 1989. *State Energy Data Report: Consumption Estimates for 1960–1987.* Washington, D.C.: Energy Information Administration.

Hall, J., *et al.* 1989. *Economic Assessment of the Health Benefits in the South Coast Air Basin, from Improvements in Air Quality.* Final Report to the South Coast Air Quality Management District. Los Angeles, June.

Hohmeyer, O. 1988. *Social Costs of Energy Consumption: External*

268 — Chapter Thirteen

Effects of Electricity Generation in the Federal Republic of Germany. New York: Springer-Verlag.

Lashof, D., and D. Ahuja 1990. "Relative Contributions of Greenhouse Gas Emissions to Global Warming." *Nature* 344 (April): 529–31.

MEOS. See Michigan Electricity Options Study.

Michigan Electricity Options Study 1986. *Contract 4A: Assessment of New Utility Power Plants, Final Report*. Lansing: Mich. Dept. of Commerce.

New York State Energy Office, New York Department of Public Service, and New York Department of Environmental Conservation 1989. *New York State Energy Plan, Staff Report Volume IV: Electricity Supply Assessment*. Draft report. Albany: New York State Energy Office.

Nichols, D. 1990. Further Testimony Before the Vermont Public Service Board. Docket No. 5330, February 6.

Putta, S. 1989. "Competition in Electric Generation—Environmental Externalities." Paper Presented to the Power Planning Committee of the New England Governors' Conference in October in Boston.

Rosen, R. 1990. Surrebutter Before the Vermont Public Service Board. Docket No. 5330, February 16.

Schilberg, G., J. Nahigan, and W. Marcus 1989. "Review of CEC Staff's Revised 'Valuing Emissions Reductions for Electricity Report 90,' Staff Issue Report #3$." Prepared for the Independent Energy Producers Association, Sacramento.

Shimshak, R., B. Biewald, H. Salgo, D. Marron, and S. Bernow 1990. "Comments of the Division of Energy Resources on Environmental Externalities." Presented Before the Massachusetts Department of Public Utilities. Case No. DPU 86-36-G and DPU 89-239, March 2.

Tellus Institute 1989. *The Environmental Database* (EDB). Boston: Tellus Institute.

———— 1990a. *The Role of Hydro-Quebec Power in a Least-Cost Resource Plan for Vermont*. Report to the Vermont Department of Public Service. Boston: Tellus Institute.

———— 1990b. *Incorporating Environmental and Economic Goals into Nevada's Energy Planning Process*. S. Bernow, *et al.*, Report 89-209, July 30, 1990, Submitted in PSC Docket #89-752. Boston: Tellus Institute.

Therkelsen, R. 1989. "Valuing Emissions Reductions for Electricity Report 90." California Energy Commission Staff Paper #3R, November 21. Sacramento: California Energy Commission.

United States Department of Energy 1983. *Energy Technology Char-*

acterizations Handbook: Environmental Pollution and Control Factors. Washington, D.C.: U.S. Department of Energy.

Stephen Bernow is a founder and vice president of Tellus Institute for Resource and Environmental Strategies. He is also a senior scientist with the Stockholm Environment Institute-Boston Center at Tellus. He focuses on the inclusion of environmental goals and externalities in integrated resource planning and policies. He received a Ph.D. in Physics from Columbia University.

Donald Marron is currently working towards his Ph.D. in Economics at the Massachusetts Institute of Technology. Previously, he worked as a research associate at the Tellus Institute. He has a B.A. in Mathematics from Harvard College.

Chapter 14

Air Pollution Projection Methodologies: Integrating Emission Projections with Energy Forecasts

Michael R. Jaske, *California Energy Commission*

Introduction

Demand-side management (DSM) is increasingly being offered as a partial solution to ambient air quality problems in portions of the nation. Dispersed sources of emissions—from residential and commercial buildings and equipment, from vehicles, and from other small sources not regulated in the past—must be controlled if ambient standards are to be achieved in these regions. DSM is difficult to link directly with likely emission reductions because current emission projections are not based on energy consumption. Incorporating the effects of DSM into the air quality planning process requires improvement in emission projection methodologies and data. This chapter describes initial progress in developing a new methodology for projecting air pollution emissions that uses energy demand forecasts as the bases for fuel combustion emission projections.

Energy demand forecasting models are an intriguing starting point for emission projections, since the former are designed to project fuel consumption into the future as a function of economic and demographic growth, of shifts among energy forms, and of changes in energy use per unit of activity as a result of price-induced behavior and DSM programs. This chapter demonstrates that extensions of demand forecasting models to produce emission projections and energy demand forecasts simultaneously can be readily accomplished and that

272 — Chapter Fourteen

such models allow coordinated and consistent energy and air pollutant planning. The emission projection model based on energy demand has been used successfully to assess three emission reduction strategies for the utilities of the South Coast Air Basin in the Los Angeles region. This work directly links energy demand with emission projections in a modeling framework that allows "what if" scenarios to be evaluated. California Energy Commission (CEC) efforts have been coordinated with those of the South Coast Air Quality Management District (SCAQMD) via an interagency working group assessing energy/emission linkages in preparation for the 1991 revision of SCAQMD's air quality management plan (AQMP). A summary of the results is presented.

Current Practice in Emissions Projections

Air pollution emissions in California are projected by the Air Resources Board (ARB) using data and information prepared jointly by ARB and individual air quality management districts (ARB 1988). This method is used for individual air basin planning and in photochemical air shed modeling to determine the extent of ozone formation. California has not emphasized the geographically broad assessments of the type used in the National Acid Precipitation Assessment Project because of the absence of coal-burning power plants or other extremely large point sources. The ARB/SCAQMD method is simple and includes no direct role for energy consumption as the intermediary between economic activity and the emissions resulting from fuel combustion. Emissions are projected for a large number of categories of controlled sources for small geographic areas. Such categories might include boilers, styrofoam containers, and auto repainting spray booths. Because air shed models are used to compute ambient concentrations in small geographic areas, emission projections have emphasized a fine level of disaggregation. For example, the South Coast Air Base (SCAB), administered by SCAQMD, prepares emissions data for about 150 control categories in each of 600 grid areas. This approach is needed because determining the formation, concentration, and distribution of ozone requires elaborate air shed modeling of the photochemical reactions of NO_x (nitrogen oxides) and ROG (reactive organic gases).

SCAQMD Emission Projections

The ARB/SCAQMD method relies upon a base-year emission inventory[1] and projects future emissions from that base. Essentially, the method can be represented as:

$$EP(p,c,t) = EI(p,c,base) \times GF(c,t) \times CF(p,c,t)$$

where:

$EP(p,c,t)$ = emission projection for pollutant p in control category c in year t,
$EI(p,c,base)$ = emission inventory in the base year,
$GF(c,t)$ = growth factor for control category c for year t, and
$CF(p,c,t)$ = control factor for category c for pollutant for year t.

Total emissions for any pollutant in any year is simply the sum of emissions over the entire set of source control categories.

Base-Year Emission Inventory

As shown above, the current ARb/SCAQMD emission projection methodology requires a firm estimate of the base-year emission inventory of each pollutant. These data are developed jointly between ARB and each district both from point sources that are regulated and for which reasonably reliable information is obtained routinely and from area sources whose emissions are estimated through secondary information such as statewide fuel consumption, emission factors, and population distribution. Overall, emission inventories are known less precisely than energy consumption, although research studies are attempting to improve knowledge of emissions and ambient concentrations of regulated pollutants (Lawson 1990).

The emissions in the SCAB inventory, which was obtained from ARB with SCAQ 8 CAQMD's cooperation, have been organized to match classifications of sources in the CEC demand forecasting models. Further, these emissions have been categorized into those from fuel combustion and those from other major sources: evaporation of organic chemicals, natural decay and decomposition processes, and dust. Table 14-1 (Part A) shows emissions from fuel combustion and other sources for the two most important pollutants—NO_x and ROG, the precursors to ozone. Clearly, fuel combustion is the dominant source of NO_x and an important source for ROG in some customer sectors. The stationary sectors commonly addressed in utility-delivered

---

[1] The emission inventory is an estimate of annual emissions released within a given airshed from each source category.

274 — Chapter Fourteen

fuel-planning models are important for NO_x but secondary for ROG. Table 14-1 (Part B) summarizes these emissions for each sector for the three fuel groups. Natural gas is the dominant stationary fuel, while petroleum is the dominant transportation fuel.

Integrated Energy/Emission Projections

In an effort to link energy demand forecasts with emission projections, CEC is developing a new methodology, which treats fuel combustion emissions as by-products of energy demand. Energy demand is forecast first, and emissions projections are computed from the demand forecast using emissions factors associated with the nature of the fuel combustion process. The modeling approach used in this analysis will be described in both energy and emissions terms.

Energy Demand Forecasts

Energy demand forecasts serve a variety of roles in utility and energy agency planning. Over the past two decades, as use of computers has become ubiquitous, forecasting techniques have evolved from simple linear trending to complex simulation models.

Utility Forecasting. Utility energy demand forecasting is gradually standardizing the types of models used for various purposes, although diverse models remain in use. Long-run forecasting for electricity and natural gas is now emphasizing end-use simulation models for the residential, commercial, and even industrial sectors defined by economic activity. The Electric Power Research Institute has contributed to the development and commercialization of such models for the electric utility industry. Econometric modeling continues to dominate short-run forecasting for rate setting purposes, where the opportunity for structural change is limited. Spreadsheet-based simulation models are becoming important for assessing the possible consequences of demand-side management programs.

California Energy Commission Forecasting. CEC staff routinely prepare electricity and natural gas demand forecasts for use in electricity and natural gas resource planning. These forecasts are prepared using very complex simulation models for each customer sector (CEC 1989). Customer sectors aggregate customers based on the economic activity taking place on the customer's premise. For example, pulp and paper production plants are an industry distinct from the extruded plastic parts industry. In all cases, these models project energy demand for individual subsectors as a function of economic and demographic pro-

Air Pollution Projection Methodologies — 275

Table 14-1. 1987 Emission Inventory by Sector and Source

Part A: Combustion Share of Emissions
(tons/day)

	Fuels		Non-Fuels		Total		Fuel Share (%)	
	NO_x	ROG	NO_x	ROG	NO_x	ROG	NO_x	ROG
Residential	34.7	7.4	0.0	74.2	34.7	81.6	100.0%	9.1%
Comm. bldg	24.1	5.7	3.9	80.4	28.0	86.1	86.2%	6.6%
TCU[a]	42.3	15.0	0.5	32.5	42.8	47.5	98.9%	31.5%
Process ind	68.4	5.1	15.4	35.6	83.8	40.6	81.6%	12.5%
Assemb ind	33.3	2.6	4.2	183.2	37.5	185.8	88.8%	1.4%
Other ind	29.5	3.0	0.7	43.2	30.2	46.2	97.7%	6.5%
Ag & water	0.3	0.5	0.2	39.7	0.5	40.2	58.5%	1.1%
Other	68.4	45.3	0.2	82.2	68.6	127.5	99.8%	35.5%
Transportation	596.2	475.9	0.0	0.0	596.2	475.9	100.0%	100.0%
Elec generation	42.1	2.7	0.0	0.1	42.1	2.8	100.0%	95.9%
Total	939.4	563.2	25.1	571.0	964.4	1134.2	97.4%	49.7%

Part B: Combustion Emissions by Fuel Type
(tons/day)

	Natural Gas		Petroleum		Other Fuels		All Fuels	
	NO_x	ROG	NO_x	ROG	NO_x	ROG	NO_x	ROG
Residential	32.0	0.7	0.2	0.0	2.5	6.8	34.7	7.4
Comm. bldg	11.1	0.5	11.0	0.5	2.0	4.7	24.1	5.7
TCU[a]	12.1	1.0	28.0	0.9	2.2	13.0	42.3	15.0
Process ind	18.8	1.1	8.4	0.6	41.1	3.4	68.4	5.1
Assemb ind	22.8	1.4	7.6	0.4	2.9	0.8	33.3	2.6
Other ind	19.9	1.7	1.7	0.1	7.9	1.2	29.5	3.0
Ag & water	0.1	0.0	0.0	0.0	0.2	0.4	0.3	0.5
Other	0.0	0.0	42.7	4.6	25.7	40.7	68.4	45.3
Transportation	0.0	0.0	172.2	29.9	424.0	446.0	596.2	475.9
Elec generation	36.4	1.6	5.4	0.3	0.3	0.8	42.1	2.7
Total	153.3	8.0	277.2	37.4	508.8	517.8	939.4	563.2

Note:

[a] TCU = Transportation, Communication, and Utilities, a set of commercial industries including railroads, pipelines, TV and radio, sewage treatment, and so forth.

Source: ARB/SCAQMD 1987 Emission Inventory Data Base analyzed by commission staff, April 1990.

276 — Chapter Fourteen

Table 14-2. Available Disaggregation in Summary Models

Customer Sector	Level of Disaggregation
Residential	3 housing types with 20 end uses
Commercial building	11 building types with 10 end uses
Commercial industries	18 industries
Process industries	13 industries
Assembly industries	20 industries
Other industries	7 industries
Agriculture	2 industries with 10 crops
Water supply	2 industries

jections of the growth of subsector activity within a given geographic area. California utilities use similar modeling techniques.

Table 14-2 enumerates the eight distinct consumer-sector energy models used to project electricity, natural gas, coal, and stationary petroleum fuel energy requirements. The CEC produces a forecast of energy *consumption* (covering all uses of each fuel type, including self-generation and private as well as utility sales) rather than a forecast of *utility sales*; thus, the job of preparing emission projections is made easier because fuel use from all sources is already included. As Table 14-2 suggests, the model breaks the eight sectors into a large number of constituent elements. Residential, commercial-building, and assembly-industry models use end uses in developing overall demand. This approach contrasts with the aggregate econometric models still used by some utilities across the country. Such aggregate models would not be as useful a starting point for emission projections.

Commission Emission Projection History

Prior to the emergence of SCAQMD's draft 1988 AQMP, CEC assessed emissions only in certification proceedings for individual power plants. Emission studies for each power plant in question concentrated upon compliance with air quality regulations governing the site of the proposed facility. Once the scope of SCAQMD's proposed 1988 AQMP became clear, CEC realized that a more comprehensive emission projections capability was necessary in order to include emissions of pollutants in planning and policy decisions.

Emission Projection Model Design Considerations

In developing an emission projection capability, the commission had to consider the purpose for which the model would be employed.

Pollutants to Be Included. Initially, model building focused on local air pollution problems and the fuel-sourced pollutants contributing to global warming. This focus led to an initial selection of pollutants to include within the model: NO_x, ROG, PM_{10} (particulate matter less than 10 microns), SO_x (oxides of sulfur), CO, and CO_2. Further studies could include other greenhouse gases, such as N_2O, methane, and CFCs.

Scope of Emissions Included Within the Model. Design of the model critically depends upon determining the scope of emissions that the model is intended to address. If the model is intended to account for all stationary emissions (including, for example, those resulting from evaporation), then a design oriented around energy consumption is inadequate. If, however, the results need to show emissions from fuel combustion only, then a design based on energy use should suffice.

The commission is not an air quality management agency. Its charter is to balance several social considerations while determining the need for, and preferred mix of, new energy resource additions. Therefore, it does not appear proper for the commission to second guess SCAQMD or any other air pollution control district in its control measures for sources of emissions not attributable to energy consumption. On the other hand, the commission needs to be cognizant of two factors. First, SCAQMD emissions trade-offs among the source categories might be made differently by the commission. Second, even if CEC accepts SCAQMD's control measures, possible differences of economic and demographic projections could affect projections of emissions from sources other than energy consumption. On balance, the commission has decided, for now, to focus on fuel combustion emissions.

Level of Disaggregation. Whether or not to develop the model at the level of end uses is an important question. Emission factors across the residential and commercial building end uses should not vary greatly. Therefore, there is little justification for using end-use emission factors for these two sectors. For the industrial impacts of air quality control measures, having end-use levels appears useful, since emission factors are likely to vary considerably. The first-generation model should be developed at an intermediate level of end-use detail: for the residential and commercial sectors, each end use is not differ-

278 — Chapter Fourteen

entiated by building type; for the industrial and agricultural sectors, industries (see Table 14-2) are not differentiated by end use. The design of the computer code implementing the model should recognize that the second generation of the model probably will need to have full industry and end-use emission detail.

Emission Projection Algorithms

The emissions computation algorithm is extremely simple and straightforward. There are only a few basic steps, some of which have been added to existing subroutines and some of which take place in a new stand-alone emissions integration computer code. Figure 14-1 illustrates how the existing CEC demand forecasting models have been adapted and augmented to the project emissions. In this figure, dotted box outlines indicate new modules that have been added to prepare emission projections. Most of the needed modeling apparatus already exists. The greatest efforts are required for transportation fuels, since the transportation models are not as mature as those for the stationary energy-consuming sectors.

Raw Emission Calculations. For natural gas–sourced emissions, the following equation defines the most disaggregated level of emission projection.

$$EMISSG(t,p,i,e) = EMFACG(t,p,i,e) \times GAS(t,i,e)$$

where:

$EMISSG(t,p,i,e)$ = natural gas–sourced emissions in year t for pollutant p for industry i for end-use e,
$EMFACG(t,p,i,e)$ = emissions factor for natural gas consumption (note that the time dimension allows for change over time), and
$GAS(t,i,e)$ = natural gas consumption.

For petroleum-sourced emissions, the following equation defines the most disaggregate level of emission projection.

$$EMISSO(t,p,i,e) = EMFACO(t,p,i,e) \times OIL(t,i,e)$$

where:

$EMISSO(t,p,i,e)$ = petroleum-sourced emissions in year t for pollutant p for industry i for end-use e
$EMFACO(t,p,i,e)$ = emission factor for petroleum combustion (note the time dimension allows for change over time)
$OIL(t,i,e)$ = petroleum consumption.

Aggregate emissions for each fuel type for each pollutant are prepared by summing all industries and end uses.

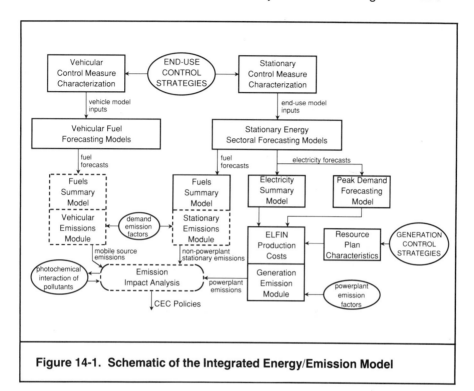

Figure 14-1. Schematic of the Integrated Energy/Emission Model

Calibrating Emissions to Known Inventories. Within the fuels summary model (integration, calibration, and report writing code), a series of calculations is required. Total emission of a given pollutant is simply the sum over each fell type.

$$AGEMS(t,p) = AGEMSG(t,p) + AGEMSO(t,p)$$

where:

$AGEMS(t,p)$ = total raw stationary emissions in year t for pollutant p.

These raw-emission projections for each pollutant must be calibrated to the level representing emissions inventory as defined by ARB for the district in question. Given the softness of the emissions factors, our reliance upon energy consumption as the precursor for emissions, and a host of other factors, it is obvious that some level of calibration will be needed; the principal question is how much. Since reliable emission inventories for multiple years do not exist, the calibration procedure simply scales pre-1987 emissions to the actual 1987 estimates shown in Table 14-1 (Part B).

280 — Chapter Fourteen

Known Limitations of This Model. The initial generation of an integrated energy demand/emission model has two limitations that require additional work in the future. First, the level of disaggregation is not sufficient to capture the impacts of many existing air quality control measures. For example, each industry is represented by a single emission factor multiplied by energy demand for each fuel type. SCAQMD's proposed Rule 1102.1 requires replacement of internal combustion engines by electric motors; this rule cannot be easily assessed within the model, because the emissions from the natural gas used to fire such engines are not projected separately from those of the much larger use of gas for process heat. Second, emission inventories have been developed in a manner that relies heavily on annual permit fee data for regulated point sources and merely estimates emissions for nonregulated sources from aggregate energy demand data. Part of the explanation for this data problem lies in California's air pollution regulatory structure, which maintains 41 distinct air pollution control districts, each of which requires emission inventory data from ARB. The emission estimates for small sources not subject to the permit process—gas water heaters, for example—are suspect but of considerable importance because of the large amount of fuel burned by these sources. The first of these limitations can be corrected in another generation of the model, but fundamental data problems of area sources will take much longer to resolve.

Air Quality Scenarios Project

The CEC's *1990 Electricity Report* (ER-90) proceeding motivated examination of air quality scenarios for SCAQMD sooner than was possible using the emissions projection model described above. A simplified, more aggregated model based on energy demand was prepared for this purpose. Results of this interim model are available for three scenarios reflecting control strategies presented in SCAQMD's adopted plan (SCAQMD 1989). This exploratory analysis is intended to illustrate consequences of the control strategies and to motivate additional analyses of greater depth and rigor where emission reduction strategies are integrated with energy planning.

Three scenarios were developed and analyzed:

- high levels of demand-side management program savings;
- very extensive industrial electrification; and
- high penetration of electric vehicles (EV).

For each scenario for each of the three Southern California utility planning areas, a demand forecast and associated resource plan were prepared to determine the energy demand, generation resource requirements, and emission consequences of the scenario in comparison to a baseline forecast and resource plan.

The purpose of this analysis is to explore the magnitudes of changes in key descriptors of the energy system and air quality of the South Coast region as a result of scenarios that are significant departures from the baseline. These baseline forecasts and resource plans incorporate CEC analyses of the energy impacts of adopted SCAQMD rules, but not speculative policy goals. The CEC electricity planning process results in commitments to fund additional DSM programs or to construct new generating facilities. The CEC considers the 1989 AQMP control strategies to be speculative because they have not received sufficient political and regulatory support from the entities that would have to cooperate in implementing them. The analyses of these scenarios contributes to better understanding of the full energy and emission consequences of SCAQMD strategies and provides part of the factual basis needed for endorsement of such proposals.

Stationary Source Emission Projections

End-User. End-user and power plant emission projections were developed separately, each using a methodology devised by CEC staff. CEC developed a simplified model for projecting emissions from fuel consumption at stationary sources. This model utilizes the 1987 base-year emission inventory described above, along with fuel demand forecasts (natural gas, petroleum, and other). For each fuel and pollutant, projected emissions can be expressed as

$$EP(f,p,s,t) = EI(f,p,s,1987) \times SCF(f,p,s,t) \times FUEL(f,s,t)/FUEL(f,s,1987)]$$

where:

$EP(f,p,s,t)$ = emission projections for fuel f for pollutant p in year t for sector s,

$EI(f,p,s,1987)$ = base year emission inventory,

$SCF(f,p,s,t)$ = sectoral emission control factor representing the influence of all *current control measures,* and

$FUEL(f,s,t)$ = amount of energy consumed in sector s for fuel f in year t.

The total fuel combustion emissions for each pollutant is simply the sum over the emissions for each of the three fuel types. Recall from Table 14-1 (Part B) that combustion of natural gas provides most of the energy used and produces most of the emissions from the station-

282 — Chapter Fourteen

ary sector. Petroleum use and emissions are much smaller in this sector. Other fuel energy use and emissions are generally smaller still, although some unusual categories of emissions appear in this later grouping of fuels (use of catalytic coke in refineries, for example).

Power Plant. CEC developed a method for projecting electric power plant emissions linked to a resource plan for each utility. McAuliffe (1990) reports details of the corresponding changes in emissions from electric power generation for the three scenarios. For each scenario a modified resource plan was developed to supply electricity at the level required. The baseline and each scenario were evaluated using the production cost model ELFIN to determine fuel usage and emissions.

Scenario Definitions

The three scenarios are closely related but not identical to components of the control strategy embedded in the SCAQMD 1989 AQMP. In the AQMP, all three elements—DSM, industrial electrification, and electric vehicles—are pursued simultaneously. Each of these CEC scenarios pursues different emission source targets, and the combined case represents the influence of each of the three individual elements.

High-DSM Scenario. In this scenario, by the year 2009 the residential, commercial building, and commercial industry (TCU) sectors each reduce electricity and natural gas usage by 30% compared to the baseline demand forecast. This reduction would be two to three times greater than the savings that will have been accomplished by 2009 by California's building and appliance standards and all other programs to which CEC is committed, and that are included within the baseline demand forecasts. CEC staff are working with SCAQMD to identify additional feasible, cost-effective DSM actions, given the DSM efforts already under way in California.

Industrial Electrification Scenario. This scenario assumes both major fuel switching from natural gas to electricity and substantial combustion efficiency improvements. By the year 2009, industrial electricity usage is 50% higher than in the baseline, while natural gas consumption is 35% lower.

High-EV Scenario. This scenario assumes that electric automobiles, light trucks, and medium-duty trucks achieve 20% penetration by 2000 and 70% penetration by 2010. The remaining vehicle fleet is assumed to be fueled by gasoline and diesel fuels.

Air Pollution Projection Methodologies — 283

Results

Table 14-3 provides summary results for the baseline forecast for several key energy and emission variables and the impact of each of the three scenarios (and the combination of them) on this baseline. Results are shown for end users (residential, commercial, industrial) and for power plants to provide the electricity consumed by end users.

End-User Results. For the high-DSM scenario, substantial savings of natural gas energy result in small reductions of emissions for regulated pollutants and CO_2. This result—major changes in fuel usage with minimal change in emissions—may not seem plausible, but as Table 14-1 shows, very small proportions of emissions come from fuels sold by utilities; most emissions come from vehicle fuels.

For the industrial electrification scenario, the natural gas reductions are only half of those in the high-DSM case, but the emission reductions for the criteria pollutants are nearly as large. Substantial increases in electricity demand occur as a result of substituting electricity for natural gas in production processes and powering of emission control equipment. This scenario suggests that for an equivalent level of natural gas reduction, industrial electrification reduces emissions of criteria pollutants more than does high DSM. Industrial combustion processes are currently "dirtier" per unit of natural gas combusted.

The high-EV scenario increases use of electric energy somewhat more than does industrial electrification, but has essentially no peak demand impacts and causes far greater reductions in fuel combustion emissions. The small peak demand impacts relative to energy increases in the high-EV scenario result from the highly optimistic assumption that utility load control will ensure that the great majority of the battery recharging load will take place at night. For the air basin under study, the high-EV case is the only scenario that significantly reduces criteria pollutants and CO_2 for each unit of reduction in end-use consumption. Demand forecasting for transportation fuel merits greater attention due to the large share of emissions from vehicle fuel usage.

Electric Power Plant Impacts. Electric power plant emissions of the regulated pollutants are a small share of the base-year emission inventory for SCAQMD, and the absolute value of these emissions is expected to decline for some years as a result of control measures imposed on utilities by SCAQMD rules. For CO_2, however, power plants are a very large component of the total emissions, as is suggested by their large consumption of natural gas in the baseline case.

284 — Chapter Fourteen

Table 14-3. Summary of End-User and Power Plant Results for the SCAQMD Region

	End-User		Power Plant	
	2001	**2009**	**2001**	**2009**
Electric Energy (Gwh)				
Baseline forecast	128,260	147,962	0	0
High-DSM impact	−15,661	−31,824	0	0
Indust elec impact	+8,580	+16,277	0	0
High-EV impact	+18,812	+29,438	0	0
Combined impact	+11,731	+13,891	0	0
Peak Demand (MW)				
Baseline forecast	29,521	34,682	0	0
High-DSM impact	−4,099	−8,435	0	0
Indust elec impact	+1,443	+2,766	0	0
High-EV impact	+243	+380	0	0
Combined impact	−2,413	−5,288	0	0
Natural Gas (mill. therm)				
Baseline forecast	8,543	8,888	4,641	5,261
High-DSM impact	−726	−1,317	−926	−1,860
Indust elec impact	−386	−722	+452	+764
High-EV impact	0	0	+1,261	+1,862
Combined impact	−1,112	−2,039	+35	−931
Transport Fuels (mill. gallons)				
Baseline forecast	6,135	5,422	0	0
High-DSM impact	0	0	0	0
Indust elec impact	0	0	0	0
High-EV impact	−1,226	−2,645	0	0
Combined impact	−1,226	−2,645	0	0
NO_x Emissions (tons/day)				
Baseline forecast	721.1	694.8	330.5	339.7
High-DSM impact	−8.1	−14.8	−15.9	−23.1
Indust elec impact	−6.8	−12.7	+7.5	+10.3
High-EV impact	−66.3	−142.6	+20.5	+15.1
Combined impact	−81.2	−170.1	+20.1	+12.3
ROG Emissions (tons/day)				
Baseline forecast	337.0	305.8	12.2	15.4
High-DSM impact	−0.5	−0.8	−0.4	−2.2
Indust elec impact	−0.6	−1.2	+1.0	+1.8
High-EV impact	−51.1	−107.9	+0.6	+2.3
Combined impact	−52.2	−109.9	+0.4	+1.5
CO_2 Emissions (tons/day)				
Baseline forecast	320,990	308,275	183,717	200,351
High-DSM impact	−11,779	−21,360	−17,633	−35,764
Indust elec impact	−6,272	−11,711	+15,101	+26,126
High-EV impact	−31,970	−69,208	+23,776	+33,442
Combined impact	−50,023	−102,278	+21,963	+21,357

Note: Several different regions are used in this analysis. For the end user, electricity and natural gas are forecast at the level of planning areas for major utilities. NO_x and ROG are projected for the SCAB, while CO_2 is projected from natural gas demand at the planning area level. For power plants, fuel consumption and emissions are computed for all power plants serving load within the combined utility planning areas.

Table 14-4. Implied Power Plant Capacity Additions			
	Actual	**1989 AQMP**	
Generating Capacity (MW)	**1988**	**2001**	**2009**
Baseline Resource Plan	29,700	36,519	40,507
Scenario Impacts			
High DSM		−800	−4,091
Industrial electrification		+1,720	+3,220
High electric vehicles		+540	+2,888
Combined		+1,460	+1,778

For the high-DSM scenario, power plant natural gas demand decreases as electricity conservation reduces total generating requirements. This decrease contributes to decreased emissions of all pollutants, and Table 14-4 indicates that some generating facilities that would have been added can be foregone.

For the industrial electrification scenario, the power plant impacts work to offset the end-user reductions in natural gas demand and end-user emissions. The electricity substituted for natural gas requires sufficiently large amounts of power plant fuel, in the form of natural gas, that total natural gas demand increases slightly, as do CO_2 emissions. About 3,200 MW of additional power plant capacity is also needed.

For the high-EV scenario, power plant natural gas required to supply the additional electricity is a major increase over base case power plant requirements. No offsetting end-user reductions exist to lessen this increase. The emission consequences of this scenario are quite beneficial, because the vehicle fuels burn so poorly that very large net reductions in regulated pollutants and CO_2 can be expected.

The combined scenario results indicate that on a composite basis the SCAQMD AQMP control strategies reduce natural gas and transportation fuel demand and decrease emissions of both the regulated pollutants and CO_2. The emission consequences of the DSM activities of the residential and commercial sectors are minimal, but they do provide a means of allowing fuel substitution in the industrial and transportation sectors.

Further Development

The results of this exploratory analysis indicate that energy demand forecasts can be successfully used as a basis for emission projections

286 — Chapter Fourteen

and that emission control measures now under discussion in Los Angeles would have a major effect on energy demand. Not surprisingly, changes in energy demand can have significant impacts on end-user emissions. The scenarios assessed here provided some preliminary indications of the consequences of SCAQMD's control strategies. DSM control strategies in the residential and commercial sectors appear to offer limited benefits because their share of the emission inventory is small. More in-depth analysis is needed to develop policy recommendations for these control strategies.

Follow-up Recommendations

As a result of this analysis, several additional steps need to be taken:

1. completion of the initial version of the integrated energy/emission projection model described above;

2. further communication between the CEC, SCAQMD, and ARB regarding base-year emission inventories and development of improvements to the current SCAQMD/ARB method of projecting emissions for fuel combustion sources;

3. improved understanding of stationary fuels usage within the SCAQMD region, especially for petroleum and other liquid fuels;

4. improvements in the SCAQMD/ARB emission inventory to resolve ambiguities regarding the correlation of fuel combustion and emissions; and

5. additional studies to assess the energy demand, emission, and cost implications of specific air quality control strategies.

Action is proceeding on several of these items as CEC assists SCAQMD in preparing energy impact analyses for its 1991 AQMP.

Acknowledgments

The project described here has been undertaken by the commission staff for its own needs and in cooperation with SCAQMD. Patrick K. McAuliffe developed the electric resource plans consistent with these demand forecast scenarios. Hebert Diaz-Flores performed the work to develop 1987 emission inventories in customer sectors consistent with commission demand forecasting models. Glen Sharp is developing the formal emission projection model. Elaine Chang (SCAQMD) provided useful review and comment for this scenario project. Despite this cooperation, this chapter is the responsibility of the author and not the

California Energy Commission or the South Coast Air Quality Management District.

References

ARB. See California Air Resources Board.

California Air Resources Board 1988. *Air Pollution Emission Inventory Program.* Sacramento: Air Resources Board.

California Energy Commission 1989. *California Energy Demand: 1989 to 2009, Electricity Demand Forecasting Methods.* CEC Publication No. P300-89-004. Sacramento: California Energy Commission.

CEC. See California Energy Commission.

Jaske, M. 1990. *Air Quality Scenario Project, Volume 2: Technical Documentation, Part A: Non-UEG Demand Forecasts and Emissions.* Sacramento: California Energy Commission.

Lawson, D. 1990. "The Southern California Air Quality Study." *Journal of the Air Waste Management Association* 40(2):156–65.

McAuliffe, P. 1990. *Air Quality Scenario Project, Volume II: Technical Documentation, Part B: Electric Generation and Emissions.* Sacramento: California Energy Commission.

SCAQMD. See South Coast Air Quality Management District.

South Coast Air Quality Management District 1989. *1989 Air Quality Management Plan.* El Monte, Calif.: SCAQMD.

Michael R. Jaske is the chief energy forecaster for the California Energy Commission. He has provided technical direction to several air quality and energy impact studies, and contributes to inter-agency energy and air quality working groups. He has a B.S. from Oregon State University and a Ph.D. from Michigan State University.

Chapter 15

Conserving Energy to Reduce SO₂ Emissions in Ohio: An Evaluation Using a Multiobjective Electric Power Production Costing Model

Benjamin F. Hobbs, *Department of Systems Engineering and Center for Regional Economic Issues, Case Western Reserve University, Cleveland, Ohio*

James S. Heslin, *Energy Management Associates, Inc., Atlanta, Georgia*

Introduction

The purpose of this chapter is to examine the cost-effectiveness of energy conservation, together with emissions dispatching (ED), as parts of a comprehensive SO₂ emissions reductions strategy in Ohio, the state having the highest SO₂ emissions levels in the United States. Conservation both reduces emissions and lowers costs for fuel and capacity expansion by decreasing the output of the power generation system. ED also decreases emissions but increases generation costs by operating dirtier plants less and cleaner (and more expensive) plants more (Delson 1974; Bernow *et al.* 1990; Depenbrock *et al.* 1990). Various degrees of ED are possible, ranging from slight deviations from traditional least-cost dispatching (by shifting the output of, say, just two plants) to least-emissions dispatching (ranking plants in order of increasing emissions per megawatt-hour (MWh) and using the cleanest plants first).

Most analyses of the cost of compliance with acid rain legislation

290 — Chapter Fifteen

have not explicitly considered conservation or ED. However, interest in those approaches has grown for two reasons. First, their potential benefits have been highlighted by recent studies. For example, analyses using linear programming or systems dynamics models have concluded that ED and energy conservation could cost-effectively lower SO_2 emissions by 50% or more in Ohio and elsewhere in the midwestern United States (Centolella *et al.* 1988; Nixon and Neme 1989). Conservation and ED can be used together with other measures, such as fuel switching, new power plants, retrofit of emissions controls, and coal cleaning, to reach emissions reduction targets in a cost-effective manner (Heslin and Hobbs 1991).

The second reason for heightened interest in conservation and ED is that, unlike the 1978 Clean Air Act Amendments, the 1990 Amendments give credit for emissions reductions achieved by those strategies. In particular, the acid rain control title of the 1990 Amendments has two innovative features that drastically alter the nature of air quality compliance planning for electric utilities. These are

- A volume cap on SO_2 emissions for each of the generation units named in the law. This cap is implemented by allocating "emissions allowances," measured in tons/year, to each unit by formulae specified in the act. Formerly, units were instead subjected to an emissions rate cap, measured in lb/MMBtu, and to mandates for particular control technologies.

- The creation of a market in emissions allowances, which allows units whose allowances exceed their SO_2 emissions to trade their excess allowances to other units which have too few allowances.

Because it is now a system's total tonnage of emissions that matters, and not emissions rates at particular units, three new options for compliance have opened up: purchases of allowances from other utilities, energy conservation, and emissions dispatching. In addition, the new law gives additional incentive to conservation by granting "bonus" allowances to successful conservation programs initiated after the act's passage.

This chapter builds upon previous studies of conservation and ED in two ways. First, these emission reduction methods are analyzed together with other approaches using recent estimates of the costs of emerging control technologies on specific generation units (PEI Associates, Inc. 1989). Second, we use a state-of-the-art method, probabilistic production costing, for estimating power plant output and emissions.

Unlike most models that are used for analyzing emissions control strategies, probabilistic production costing can accurately model, for

Conserving Energy to Reduce SO$_2$ Emissions in Ohio — 291

example, random outages of generating units and unit operating constraints. An example of an operating constraint is the requirement that a unit run at a minimum level for mechanical or system reliability reasons. Our model, called MODES (Multiple Objective Dispatch Evaluation System), generates curves showing the cost of lowering emissions by ED and fuel switching for a given set of power plants, fuel prices, control technologies, and conservation programs. Since the model interfaces with widely used microcomputer database and spreadsheet software, users can quickly modify the inputs to reflect different technology choices, demand growth rates, or fuel prices.

In the next two sections, we review our methodology and assumptions. Later sections present the results of the Ohio analyses.

Methodology for Estimating the Output and Emissions of Generating Units

MODES's purpose is to estimate the emissions and costs of ED and fuel switching for a coordinated power system. The software is intended for use in detailed state- or utility-level policy analysis and planning. Previous models for analysis of ED were designed for broad-brush regional or national assessments or real-time commitment of power plants (e.g., Yokoyama *et al.* 1988). MODES yields more accurate estimates of production costs and emissions than other policy models, while achieving the quick execution times and the long-term perspective required for policy analysis.

Probabilistic production costing, the methodological basis of MODES, was developed by the utility industry in order to obtain more accurate estimates of expected generation costs for a system that must meet a varying level of power demand using a set of generation units that are subject to random forced outages (Baleriaux *et al.* 1967; Stremel *et al.* 1980). Such outages are common occurrences and are caused by mechanical failures, safety-motivated shutdowns, and fuel shortages. Probabilistic production costing is a reasonable compromise between the need for more realistic dispatch models and the desire to avoid the computational difficulties inherent in detailed unit commitment models (Talukdar and Wu 1981).

MODES extends probabilistic production costing to achieve several objectives in addition to generating costs. For example, an analysis of the short-term potential of ED and fuel switching in Ohio modeled generation costs, SO$_2$ emissions, and employment in the Ohio coal fields (Heslin and Hobbs 1989).

MODES generates a number of different dispatch orders for a given set of generation units. Each order represents a different weight-

292 — Chapter Fifteen

ing of the objectives being considered.[1] In this case, this composite objective is a weighted sum of the variable cost of MWh and the SO_2 emissions per MWh. A weight of 1 for cost and 0 for emissions results in the traditional "least-cost" dispatch order; placing higher weights on emissions yields increasing degrees of ED. To determine weights, the programmer experiments with different weights until trade-off curves representing the desired range of emissions levels are generated. After the expected output of the units under each dispatch order is calculated, the values of objectives can be calculated and the trade-offs displayed.

Each solution is efficient in that, for a specific set of loads and generation units, no other solution could improve on one of the values without sacrificing another value. For instance, if, for a given set of unit outputs, the emission level was 1.1 million tons and the incremental cost was $2.7 billion, we could not lower the tonnage without increasing the cost, and vice versa. A set of such solutions is called an efficient set; if only two objectives (cost and emissions, for example) are considered, then this set is a two-dimensional trade-off curve. A distinct trade-off curve can be obtained for each set of assumptions concerning conservation programs, generation plant additions and retirements, and emissions control technologies. By plotting several such curves together, as in Figure 15-1, strategies that achieve the desired emissions reductions at least cost can be identified. (See White 1981 for a similarly framed analysis of trade-offs between SO_2 and electricity costs.) Alternatively, given a market price for emission allowances, such curves can be used to determine which strategy minimizes the utility's cost, net of any purchases or sales of allowances (see Hobbs 1990).

Probabilistic production costing models such as MODES can accommodate minimum output ("must run") limitations on unit output (hydropower units whose output is limited not only by their capacity but also by the amount of water that is available), maintenance outages, nondispatchable units such as wind or cogeneration, and pumped storage (Yamayee 1985). An innovative feature of MODES is its inclusion of the possibility of fuel switching (e.g., from high- to low-sulfur coal). If fuel switching allows a generating unit to perform better on the composite objective, the fuel is changed.

Despite these complications, trade-off curves can be generated quickly on a microcomputer even for large power systems. Details on

---

[1] This approach is basically the same as Delson's (1974) "emissions tax strategy" formulation of the dispatching problem.

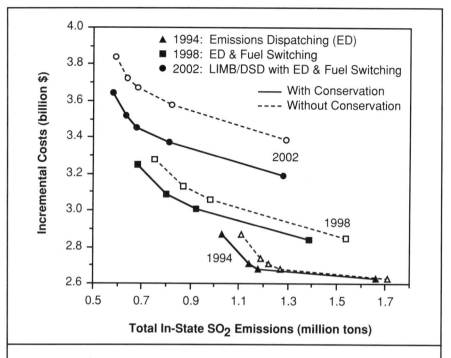

Figure 15-1. Incremental Cost of Reducing SO_2 Emissions for Ohio: 1994, 1988, 2002

MODES's computational procedures are available in Heslin and Hobbs 1989.

Assumptions

In this section, we summarize our assumptions concerning electric loads, conservation programs, generation units, fuel switching, and control technologies. Assumptions concerning interest rates and taxes are consistent with EPRI 1986. All costs are expressed in real 1987 dollars.

Compliance Targets

Here, strategies are developed to lower utility emissions in Ohio to compliance targets for each of three years (1994, 1998, and 2002). These targets are representative of SO_2 emissions targets in several previously proposed acid rain bills, such as U.S. Senate Bill 5562, and

294 — Chapter Fifteen

are comparable to the ceilings assumed in the analysis by Centolella *et al.* 1988. The compliance targets are 1,300,000 tons/yr of SO_2 in 1994, 1,050,000 tons/yr in 1998, and 525,000 tons/yr in 2002. As a comparison, Ohio's 1988 SO_2 emissions were 2,100,000 tons. We also show strategies that yield emissions reductions above and below those targets so that the costs of other targets can be estimated.

Base Case Load Projections

Projections for Ohio's load growth are taken from Centolella *et al.* 1988. In the absence of additional conservation programs, a growth rate of 1.5%/yr for both energy and the system peak was forecast. Future loads are projected from a base year of 1987 (133,470 GWh/yr with a peak of 21,553 MW).

Out-of-state sales of power are disregarded in this analysis. Since at least two Ohio utilities (Columbus and Southern Company and Ohio Power Company) export considerable amounts of power to other states, the generation and, thus, emissions of the Ohio system are underestimated in our analysis. Such sales are accounted for in our analysis of the American Electric Power System (Hobbs 1990).

Energy Conservation Programs

Our focus is on so-called strategic conservation programs whose aim is to reduce total energy use in all hours of the day. Other types of demand-side management programs—such as peak clipping, load management, and valley-filling (Gellings and Chamberlin 1986)—are disregarded, even though their impacts on emissions might be significant. For instance, load management might lower the output of clean peaking units, while increasing the generation from dirtier base-loaded units.

We use findings from Centolella *et al.* 1988 concerning the cost and effectiveness of a "moderate effort" conservation program. The program is based on the assumption that electric utilities will pay for 50% of the cost of any conservation measure whose levelized expense (based on a 6% real discount rate) is less than $0.06/kWh. We presume that the utility's costs are recovered from ratepayers at the time that the savings take place; i.e., program costs are included in the rate base and recovered over several years. Most of the savings come from industrial motors and lights; residential refrigerators, water- and space-heating, and lights; and commercial cooling and lights. Penetration rates are calculated assuming adoption of any conservation investment whose net perceived cost to the consumer (evaluated at the higher "implicit" discount rates of 35%-75%) is less than the price of electricity.

Centolella *et al.* (1988) conclude that such a program can lower the load growth from 1.5% down to 1.0%/yr at an average levelized cost of \$20.70/MWh saved. The MODES run shows that this cost is close to the short-run marginal cost of plant operation in Ohio and considerably less than the long-run expense of plant construction.

Included in Centolella et al.'s (1988) estimates of energy savings are those that result from additional conservation investments motivated by conservation-induced electric rate increases. However, no price-induced fuel substitution and energy service reductions beyond those reflected in the base-case load projections were considered in that study. Later in this paper, we examine the possible implications of additional price-induced load decreases.

We also undertook sensitivity analyses in which we looked at (1) the effect of a less successful program that lowers the growth rate to only 1.2%/yr at the same total cost and (2) an "aggressive" program that drops the growth rate to 0.7% at an assumed average cost of \$26.5/MWh (Centolella *et al.* 1988). Under the latter program, utilities are assumed to pay 90% of the cost of conservation measures (rather than 50%), and information programs lower implicit consumer discount rates to 15%–25%. The cost of the increment of savings over the "moderate" program is approximately \$36/MWh, which exceeds the short-run cost of power production.

Generation Units

The generation units considered are those that are owned and operated by Ohio utilities and that are expected to be on-line in the early 1990s. The total capacity is about 33,000 MW. Of this amount, 80% is coal-fired, with nuclear plants and combustion turbines making up most of the remainder. Power system and fuel use data are obtained from industry and federal sources (Ohio EPA 1986; Stone and Webster 1986; DOE 1986). "Must-run" constraints on each unit's output are included in the model. The amount of power provided by units owned by Ohio utilities but located out of state is assumed to be fixed, so that the analysis can focus on the costs and emissions of Ohio facilities. The emissions and costs of non-Ohio facilities are excluded from the results presented below. Future analyses should consider multistate systems.

Unit retirement data are taken from industry data (Stone and Webster 1986). Additional retirements assumed for 2002 include all small, old units with high heat rates. New coal-fired units with scrubbers, plus combustion turbines for meeting system peaks, are added to replace the retired capacity and to maintain a reserve margin of at least

296 — Chapter Fifteen

20%. Consistent with Centolella *et al.* 1988, the construction cost of new coal-fired units is assumed to be $1,216/kW in 1987, escalating at 2.1%/yr in real terms.

Emissions Control Retrofit Technologies

Energy conservation alone is unable to achieve the assumed compliance targets in 1998 and 2002, even in tandem with ED. Other measures will be needed in addition, including fuel switching, retrofits of emissions control technologies, or both. Elsewhere (Heslin and Hobbs 1991), we used a data base for a specific generating unit (PEI Associates, Inc. 1989) to investigate a wide range of emissions control technologies that can be retrofitted. In this paper, we combine conservation with the technologies that appeared most cost-effective in the three target years. In 1994, ED alone appeared to be the least expensive means of achieving the compliance target. By 1998, ED together with fuel switching seemed cheapest (but only by a slight margin compared to a mix of retrofitted flue gas desulfurization and ED). But by the year 2002, fuel switching and ED alone were found to be insufficient to achieve proposed SO_2 reductions. Control technology retrofits then became necessary for some generation units. The most attractive technologies are limestone injection multistage burners (LIMB) and duct spray-drying (DSD). Their SO_2 removal rates are only about 50%, but for some Ohio units, their estimated capital costs are much lower than for traditional scrubber technologies (PEI Associates, Inc. 1989). Heslin *et al.* (1989) list the generation units at which retrofits are assumed to take place.

Fuel Switching

Coal-fired units identified as feasible candidates for fuel switching (Ohio EPA 1986) are assumed to be able to switch to a generic imported low-sulfur coal. We assume that any unit that fuel switches would burn low-sulfur coal costing $58/ton. This price was derived by examining the prevailing prices of low-sulfur coal currently burned in Ohio and then adjusting for low-sulfur premiums projected to occur if an acid rain law is passed (Energy Ventures Analysis 1987). To this expense was added the levelized cost of modifying electrostatic precipitators (Ohio EPA 1986)

Results

Figure 15-1 summarizes the results of the analyses in which it is assumed that changes in electric rates affect power demands only to the extent estimated by Centolella *et al.* 1988.[2] Six curves are shown, one pair for each of the target years (1994, 1998, 2002). Within each pair, one curve represents the impact of increasing degrees of ED (and, in 1998 and 2002, fuel switching), assuming no additional conservation effort; the other curve indicates the effect of ED and, in the later years, fuel switching, together with the "moderate effort" conservation program.

The lower right-hand point on each curve is the least-cost dispatching point if generating unit merit orders are based only on cost. Due to recent additions of new, cleaner capacity, emissions in 1994 are lower than historical levels, even in the absence of any additional measures to reduce emissions. ED can lower those emissions by an additional 500,000 or so tons at relatively little cost (averaging $200/ton of SO_2 removed). No other combination of measures, either excluding or including conservation, is significantly less expensive for achieving the compliance target.

Figure 15-1 shows that the moderate-effort conservation program does not appreciably lower the costs of achieving the compliance target in 1994, primarily because (1) marginal supply costs are low (averaging $.02/kWh) and (2) programs have trimmed demand by only 2% by that point. The programs do not appear attractive until 1998, when desired emissions levels fall to 1.1 million tons/yr. The primary effect of conservation in that year is to lower emissions by about 60,000–80,000 tons per year (that is, conservation shifts the cost-emissions curve to the left by that amount). However, because conservation investments will yield savings for many years, efforts made before 1994 may still more than pay for themselves later on.

ED is not quite so cheap or effective after 1994 because the cushion of extra generating capacity is eaten up by demand growth. However, it is still a cost-effective part of any emissions control strategy, yielding an inexpensive 300,000 to 400,000 tons/yr of reductions. In 1998, some fuel switching is required, but no emissions control retrofits. By the year 2002, retrofits of LIMB and DSD prove attractive, together with ED and fuel switching.

Figure 15-1 reveals that in 1998 and 2002, unlike 1994, a strategy including conservation is less costly and leads to fewer emissions than otherwise identical strategies that exclude it. For a given level of emissions in 1998, a conservation program is about $70 million per year less costly (roughly $.50/MWh). (However, if the DSM programs perform disappointingly and achieve only a 0.3%/yr reduction in load

---

[2] Hobbs and Heslin 1990 consider cases in which price elasticities are higher than those implicit in Centolella *et al.* 1988.

298 — Chapter Fifteen

growth, rather than 0.5%/yr, this cost advantage disappears.) By the year 2002, conservation programs definitely appear worthwhile, even if they do not achieve the full 0.5%/yr decrease in load growth. In particular, a moderate but successful conservation effort could lower the cost of reaching a 600,000 ton/yr target by about $230 million/yr. A large part of the cost savings in the year 2002 case results from 1,940 fewer MW of capacity being needed in the conservation case.

But the year 2002 curves confirm a preliminary conclusion of the National Acid Precipitation Assessment Project (South 1990) that, in the long run, conservation merely replaces the relatively low emissions from new generating plants. This is shown by conservation's effect on the cost-emissions curve: by 2002, the right-hand point of each curve (the least-cost dispatch point) shifts downward, not leftward; costs are lowered, but not emissions. Only if the amount of new generating capacity is held constant does conservation also lower emissions. Thus, conservation is desirable in the year 2002 not because it decreases emissions, but because it saves capacity expansion costs. This conclusion will change if conservation is used to retire older, dirtier plants more quickly, rather than to cancel new facilities.

Under our assumptions, the aggressive conservation program (not shown in the figure), which achieves more savings at a higher price, is not more attractive than the moderate program in any year. It is considerably inferior in 1994 and 1998, because the incremental cost of the additional savings (compared to the moderate program) significantly exceeds the short-run cost of generation. However, such a program will become attractive after the year 2002, because it will defer costly capacity additions.

In Heslin and Hobbs 1991, we estimated the employment impacts of the conservation strategies using an economic input-output model of Ohio. Both backward linkages (stemming from inputs bought by Ohio utilities, including DSM programs, plant construction, and coal purchases) and forward linkages (resulting from the effects of electric rates upon industrial competitiveness and disposable income) were considered. In 1994 and 1998, strategies including conservation resulted in fewer net job losses in the state than strategies which excluded it. The difference in 1994 was about 3,000 jobs, while in 1998 it was more significant (20,000 jobs, out of a total of several million for the state). In the year 2002, however, conservation resulted in slightly fewer jobs, mainly because of decreases in plant construction. However, we judge that difference to be too small to be significant.

We note, however, that from a national perspective, such job losses might be compensated for by gains in employment in other states. The inclusion of such secondary benefits in benefit-cost analyses is controversial.

There are other considerations that apply in evaluating conservation programs. One is that conservation may lower load growth uncertainty, which is valuable to an industry that must make capacity decisions a decade ahead of time (Hirst and Schweitzer 1989). Another is that conservation also would increase electric rates in the short run in Ohio (Centolella *et al.* 1988), more so than other strategies, although it would lower rates in the long run. This is because, in the short run, a utility's fixed costs must be spread over fewer sales, if average-cost-based rates are greater than the marginal cost of providing power. These rate increases may motivate further decreases in electricity demand. This rate feedback is addressed by Hobbs and Heslin (1990), who use the method of Hobbs (1991) to estimate those decreases and assign an economic value to them. They find that those decreases can alter the attractiveness of conservation.

Conclusions

Our use of a probabilistic production costing model (MODES), together with a detailed data base of Ohio's generation plants and emissions control options, has made it possible to rigorously analyze the cost and SO_2 emissions benefits of conservation programs in Ohio. On an annualized cost basis, the moderate program appears justified in 1998 and 2002, but not in 1994. However, from a present worth standpoint, it is likely that a moderate program started in the early 1990s can be justified, because its later benefits would compensate for costs incurred earlier. A new version of MODES, called CERM (Comprehensive Emissions Reduction Model), has the capability of automatically evaluating the trade-offs involved in multiyear programs (Hobbs *et al.* 1991). It has been used to assess the net benefits of conservation for the American Electric Power System (Hobbs 1991).

An important conclusion is that the greatest benefit of conservation is its ability to defer the cost of new generation and to save on fuel costs. The SO_2 emissions reduction benefit of conservation is roughly an order of magnitude smaller (see also the analyses by Nixon and Neme 1989 and Hobbs 1990). The "bonus" allowances granted to new conservation programs under the 1990 Clean Air Act Amendments do not substantially change this conclusion. However, the environmental benefit of conservation may be much larger than merely the reductions in SO_2 emissions it yields, because conservation can also reduce the

300 — Chapter Fifteen

amount of CO_2, waste heat, ash, strip-mined land, scrubber sludge, and other externalities associated with power production.

Future work should examine the other environmental benefits of conservation and emissions dispatch using probabilistic production costing (Bernow et al. 1990). Research is also needed to determine the efficacy of these strategies in other regions of the country and to quantify the value of the flexibility they give to utilities that face gross uncertainties in future fuel prices, demands, and government regulations (Hirst and Schweitzer 1989). Finally, since electric rate changes caused by conservation programs can induce changes in loads that are of the same order of magnitude as the direct energy savings of the programs, research should also be directed at quantifying those load changes and assigning an economic value to them (Hobbs and Heslin 1990; Hobbs 1991).

Acknowledgments

Funding was provided by the Ameritech Corporation, through the Case Western Reserve University Center for Regional Economic Issues, the Ohio Air Quality Development Authority, and the National Science Foundation, grant ECE-8552524. Comments by reviewers, especially Paul Centolella, led to significant improvements in the paper.

References

Baleriaux, H., E. Jamoulle, and Fr. Linard de Guertechin 1967. "Establishment of a Mathematical Model Simulating Operation of Thermal Electricity-Generating Units Combined with Pumped Storage Plants." *Revue E* (Belgium) 5(7):225–45.

Bernow, S., B. Biewald, and D. Marron 1990. "Full Cost Environmental Dispatch: Recognizing Environmental Externalities in Electric Utility System Operation." Paper presented at National Association of Regulatory Utility Commissioners Conference on Environmental Externalities, Jackson Hole, Wyo., Oct. 1–3.

Centolella, P., et al. 1988. *Clearing the Air: Using Energy Conservation to Reduce Acid Rain Compliance Costs in Ohio*. Columbus: Ohio Office of the Consumers' Counsel.

Delson, J. 1974. "Controlled Emissions Dispatch." *IEEE Transactions on Power Application Systems* 93: 1359–66.

Depenbrock, F., J. Sinha, and A. Calafiore 1990. "Acid Rain Legislation and Utility Compliance Strategies." In *Proceedings: Conference on Innovations in Pricing and Planning.* EPRI CU-7013. Palo Alto, Calif.: Electric Power Research Institute.

DOE. See U.S. Department of Energy.

Electric Power Research Institute 1986. *TAG—Technical Assessment Guide.* Vol. 1: *Electricity Supply.* EPRI P-4463-SR. Palo Alto, Calif.: Electric Power Research Institute.

Energy Ventures Analysis, Inc. 1987. *Coal Markets and Utilities' Compliance Decisions.* EPRI P5444. Palo Alto, Calif.: Electric Power Research Institute.

Gellings, C., and J. Chamberlin 1986. *Demand-Side Management: Concepts and Methods.* Fairmont Press.

Heslin, J., and B. Hobbs 1989. "A Multiobjective Production Costing Model for Analyzing Emissions Dispatching and Fuel Switching." *IEEE Transactions on Power Systems* 4(3): 838–43.

——— 1991. "A Probabilistic Production Costing Analysis of SO_2 Reduction Strategies for Ohio: Effectiveness, Costs, and Regional Economic Impacts." *Journal of the Air and Waste Management Association* 41, Issue #7.

Heslin, J., B. Hobbs, and K. Sullenberger 1989. *Alternative SO_2 Emission Reduction Strategies for Ohio: Effectiveness, Costs, and Regional Economic Impacts.* Cleveland: Center for Regional Economic Issues, Case Western Reserve University.

Hirst, E., and M. Schweitzer 1989. "Uncertainty: A Critical Element of Integrated Resource Planning." *Electricity Journal* 2(6):16–27.

Hobbs, B. 1990. "Testimony on Acid Rain Special Topic Information." *In the Matter of the 1990 Long-Term Forecast Report of Ohio Power Company* and *In the Matter of the 1990 Long-Term Forecast Report of Columbus Southern Power Company.* Before the Public Utilities Commission of Ohio, Columbus, Ohio. Cases No. 90-659-EL-FOR and 90-660-EL-FOR, September 28.

——— 1991. "The 'Most Value' Criterion: Economic Evaluation of Utility Demand-Side Management Programs Considering Customer Value." *The Energy Journal* (in press).

Hobbs, B., and J. Heslin 1990. "Evaluation of Conservation for SO_2 Emissions Reduction Using a Multiobjective Electric Power Production Costing Model." In *Proceedings of the ACEEE 1990 Summer Study on Energy Efficiency in Buildings.* Vol. 4. Washington, D.C.: American Council for an Energy-Efficient Economy.

Hobbs, B., J. Heslin, and W. Huang 1991. *Comprehensive Emissions Reduction Model (CERM): User's Guide and Documentation.*

302 — Chapter Fifteen

Cleveland: Department of Systems Engineering, Case Western Reserve University.

Nixon, L., and C. Neme 1989. *An Efficient Approach to Reducing Acid Rain: The Environmental Benefits of Energy Conservation.* Washington, D.C.: Center for Clean Air Policy.

Ohio Environmental Protection Agency 1986. *Analysis of the Cost of H.R. 4567 to Electric Utilities in Ohio.* Columbus Division of Air Pollution Control, Ohio Environmental Protection Agency, Columbus, Ohio.

PEI Associates, Inc. 1989. *Evaluation of Acid Rain Controls for Ohio Electric Utility Coal-Fired Power Plants.* Cincinnati: PEI Associates, Inc.

South, D., *et al.* 1990. *Technologies and Other Measures for Controlling Emissions: Performance, Costs, and Applicability.* National Acid Precipitation Assessment Program Report 25 (Draft). Argonne, Ill.: Argonne National Laboratory.

Stone and Webster, Inc. 1986. *Feasibility Investigation of the Benefits and Constraints of a Power Pooling System Among the Ohio Regulated Electric Companies, Segment II Report.* Denver: Stone and Webster.

Stremel, J., R. Jenkins, R. Babb, *et al.* 1980. "Production Costing Using the Cumulant Method of Representing the Equivalent Load Curve." *IEEE Transactions on Power Apparatus and Systems* 99:1947–54.

Talukdar, S., and F. Wu 1981. "Computer-Aided Dispatch for Electric Power Systems." *Proceedings of the IEEE* 69: 1212.

U.S. Department of Energy 1986. *Cost and Quality of Fuels for Electric Utility Plants, 1986.* DOE/EIA-0191(86). Washington, D.C.: U.S. Department of Energy.

White, D. 1981. *Strategic Planning for Electric Energy in the 1980's for New York City and Westchester County.* Report MIT-EL-81-008. Cambridge: Energy Laboratory, Massachusetts Institute of Technology.

Yamayee, Z. 1985. "Production Simulation for Power System Studies." *IEEE Transactions on Power Apparatus and Systems* 104: 3376–81.

Yokoyama, R., *et al.* 1988. "Multiobjective Optimal Generation Dispatch Based on Probability Security Criteria." *IEEE Transactions on Power Apparatus and Systems* 3(1):317–22.

Benjamin F. Hobbs is an associate professor of Systems Engineering and Civil Engineering at Case Western Reserve University. Previously, he held research positions at Brookhaven and Oak Ridge National Laboratories. He received an M.S. from the State University of New York College of Environmental Science and Forestry and a Ph.D. from the Cornell University's School of Civil and Environmental Engineering.

James S. Heslin is senior consultant with Energy Management Associates, Inc. His responsibilities include developing new methodologies for full integrated resource planning as well as better screening of DSM programs. He received B.S. and M.S. degrees in Systems Engineering from Case Western Reserve University.

Chapter **16**

Building Energy Consumption and the Environment: What Past, Present, and Future Commercial Buildings Energy Consumption Surveys Can Tell Us About Chlorofluorocarbons

Julia D. Oliver and Eugene M. Burns, *U.S. Department of Energy, Energy Information Administration*[1]

Introduction

Chlorofluorocarbons (CFCs) and halons (hydrocarbons containing fluorine and bromine and used as fire extinguishers) are used extensively throughout the commercial, manufacturing, residential, and transportation sectors of the economy for both energy and nonenergy uses. These families of organic compounds have been ideal for many different uses because they are nontoxic, safe, nonflammable, and long-lived. Unfortunately, once released into the atmosphere, CFCs rise to the stratosphere and destroy the ozone molecules that block most of the

---

[1] The opinions and conclusions expressed here are solely those of the authors and should not be construed as representing the opinions or policy of any agency of the United States Government.

305

306 — Chapter Sixteen

sun's harmful ultraviolet radiation. This degradation of the ozone layer may eventually let in radiation severe enough to cause increased levels of skin cancer and cataracts and to cause a serious reduction in agricultural productivity. By the mid-1980s, enough scientific evidence had linked ozone depletion to CFC usage to create a worldwide consensus on the need to reduce the use of these chemical compounds.

As a result of this consensus, delegates from more than 90 governments met in Montreal, Canada, in September 1987 to sign an international treaty: the Montreal Protocol on Substances That Deplete the Ozone Layer, now known as the Montreal Protocol. This treaty identifies those compounds with the highest ozone-depleting potential (ODP)—CFC-11, CFC-12, CFC-113, CFC-114, CFC-115, and three halons—and freezes production of these substances at their 1986 production levels. The protocol also requires that by 1998, production of these compounds is to be reduced to 50% of their 1986 production levels.

Related substances known as hydrochlorofluorocarbons (HCFCs), seen by many as an interim solution to the phase-out of CFC refrigerants, have also come under scrutiny. The attachment of a hydrogen atom to the CFC molecule weakens the molecular bonds and allows more rapid dissipation in the stratosphere, thus lowering the ODP. For example, HCFC-22, the most widely used HCFC, has an ODP of 0.05 compared to an ODP of 0.9 for CFC-12 and 1.0 for CFC-11. Even though HCFC-22 has a very small ODP compared to the banned substances, some atmospheric scientists and environmental groups want to extend the list of regulated/banned substances to include the HCFCs.

In addition to depleting the atmospheric ozone, CFCs and halons are greenhouse gases that contribute to the global warming trend. Thus, control of CFCs and halons to prevent stratospheric ozone depletion will also help mitigate global climate change.

As shown in Figure 16-1, the main energy-related uses in the United States of the CFCs covered by the Montreal Protocol are for foam products, mobile air-conditioning (transportation), and refrigeration (including building air-conditioning). Chlorofluorocarbons provide most of the best refrigerant fluids available, and they constitute the foaming agents in low-density insulating materials that have greatly improved the energy efficiency of both buildings and appliances. In addition, others are used as solvents and cleaners that have been described as almost indispensable in the production of energy-conserving electronics.

The commercial sector, the fastest-growing energy-consuming

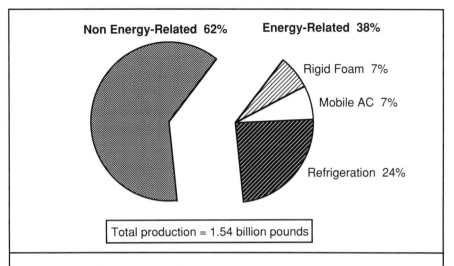

Figure 16-1. Sources of U.S. Production of Ozone-Depleting Substances in 1985 (Statt, T. 1988)

sector of our economy, is a major user of CFCs. During the 1990s, meeting the energy needs of commercial buildings while minimizing the environmental impacts will provide policymakers in the public and private sectors alike with a difficult challenge. To meet this challenge, analysts will need reliable data on commercial applications of CFCs and the potential effects of switching to other substances, and the potential for recycling and recapturing CFCs in commercial buildings. The Commercial Buildings Energy Consumption Survey (CBECS), which is conducted by the U.S. Department of Energy (DOE), will be an important source of information for DOE policymakers on the contribution to ozone depletion made by the commercial buildings sector. As the federal agency responsible for promoting energy supply and end-use options for the future, DOE will play a major role in developing and implementing this country's energy response to all climate change issues.

This chapter will use data available from the 1986 CBECS to quantify energy-related CFC usage in HVAC and non-HVAC equipment and in CFC-containing insulation in the commercial buildings population. We will also analyze the age of CFC-using equipment because turnover rates of existing equipment stock have implications for how to manage CFCs in existing and future equipment.

308 — Chapter Sixteen

The CBECS

The CBECS, formerly called the Nonresidential Buildings Energy Consumption Survey (NBECS), is conducted triennially by the Energy Information Administration (EIA), the independent statistical and analytical agency within DOE.

The purpose of the survey is to provide national and census-region estimates both of energy consumption and expenditures in commercial buildings and of building characteristics affecting energy consumption patterns. The definition of "building" is rather broad. A building is defined as a structure totally enclosed by walls extending from the foundation to the roof and intended for human occupancy. All commercial buildings over 1,000 square feet are included in the survey population. The definition of "commercial" is also broad and includes any building with more than 50% of its floor space used for commercial activities. Commercial buildings include, but are not limited to, stores, offices, schools, churches, gymnasiums, libraries, museums, hospitals, clinics, warehouses, and jails. Excluded are buildings in which more than 50% of the floor space is used for industrial, agricultural, or residential purposes.

The CBECS is the only publicly available survey of the current U.S. building stock that is statistically representative of the building population. Previous surveys have been conducted in 1979, 1983, and 1986. Data for the 1989 CBECS are currently being prepared for publication. At a minimum, each cycle of the CBECS has produced both national and census-region estimates of the number, square footage, energy consumption, and energy expenditures for the population of commercial buildings and for subpopulations defined by the following building-specific characteristics: use, location, energy sources, energy end uses, and conservation features such as roof and wall insulation and heating and cooling equipment and practices. The questions about the heating and cooling equipment specifically address chillers, individual room air conditioners, packaged units, and other types of heating and cooling equipment.[2]

---

[2] A chiller is a centrally located air-conditioning system that produces chilled water in order to cool air. The chilled water or cold air is then distributed throughout the building using pipes or air ducts, or both. A packaged unit is any unit built and assembled at a factory and installed as a self-contained unit to heat or cool all or portions of a building. Packaged units are in contrast to engineer-specified units built up from individual components for use in a given building.

What the 1986 CBECS Can Tell Us About CFC Usage in Buildings

CFC-Using Cooling Equipment

The 1986 CBECS provides a different perspective than generally referenced literature on CFC usage in commercial buildings cooling equipment. CBECS, because of its design, provides building-level information on the presence of applicable equipment rather than equipment-based data on production, sales, or installations.

Information of this type is important to trade associations, public policymakers, and public interest groups charged with implementing CFC-reduction strategies. These data—including the number of buildings that would be affected, the number of tons of air-conditioning involved, and the quantity of CFC by building type—provide policymakers not only with just the gross national production of CFCs but also with information necessary to design recycling, recapturing, incentive, and retrofit programs tailored to meet the differing needs of the building owners and utilities that may be required to reduce CFC usage. These data also provide rough estimates of the use of the equipment in buildings: buildings constructed in the 1960s and 1970s are candidates for equipment turnover, while those constructed in the 1980s should contain equipment with a useful life well into the next century.

An important example of the uncertainty and disparity caused by not having building-based measures occurs for the population of commercial chillers, which are widely used for space conditioning. Selected sources illustrate the range of estimates:

- EIA 1988: 347,000 buildings (10,000 square feet or larger) with central cooling (chillers) in 1986;

- Fischer and Creswick 1989: 50,000 "chillers in use";

- Hammitt *et al.* 1986: 30,000 "domestically installed central chillers between 1976 and 1984"; and

- EPA 1989: approximately 175,000 "domestically consumed chillers between 1955 and 1985."

The inconsistencies in these estimates are mostly the result of differing sources, definitions, assumptions, and methodologies used in collecting and analyzing the data. The Fischer and Creswick (1989), Hammitt *et al.* (1986) and EPA numbers are the result of various analyses of chiller production numbers, mostly taken from the U.S.

310 — Chapter Sixteen

Bureau of Census's industrial reports and from various industry sources.

In contrast, the 1986 CBECS gathered the chiller information at the building site. However, the 1986 CBECS asked if the building had "central cooling (for example, chillers)" as opposed to asking specifically about the presence of central chillers. Many of the 347,000 buildings with central cooling reported on the 1986 survey could actually have had packaged units or another type of system that delivers centrally conditioned air rather than custom-designed systems based on central chillers. The 1989 CBECS used a tighter definition and approach and should provide a more accurate and lower estimate of the number of buildings with chillers.

However, even the improved estimates from the 1989 survey will not provide a complete national estimate of chiller penetration. Neither the 1986 or 1989 survey asked for data on the number or types of chillers present, information that would provide a more complete description of the chiller population in commercial buildings. Furthermore, chillers installed in buildings that are predominantly industrial, agricultural, or residential are outside the scope of the CBECS and also would not be counted.

Using the CBECS data on cooling equipment, a rough estimate of the number of tons of air-conditioning and the number of pounds of CFC-11, CFC-12, and HCFC-22 in use in commercial buildings in 1986 can be calculated (Table 16-1). Three parameters were used to calculate these estimates: (1) tons of cooling capacity based on industry averages of square feet of floor are per ton of cooling capacity (TRW 1982; Hammitt et al. 1986); (2) working fluid characteristics as a function of equipment type (Fischer and Creswick 1989); and (3) refrigerant charge per ton of cooling capacity (Little 1989).

A four-step estimation process was followed:

1. The most frequently used working fluids were assigned to the CBECS equipment types (central systems, packaged cooling systems, individual room air conditioners, and heat pumps). CFC-11 and CFC-12 are usually found in chillers used in central air-conditioning systems in buildings over 10,000 square feet; very large central systems with 5,000–10,000-ton chillers use a mixture of CFC-500 and HCFC-22 (Fischer and Creswick 1989). HCFC-22 is used as the working fluid in packaged cooling systems, in room air conditioners, in very large central systems (as mentioned above), and in small central systems (capacities of 25 tons or less).

2. On the basis of industry averages for the square foot of floor area per ton of air-conditioning capacity in different building types,

Building Energy Consumption and the Environment — 311

Table 16-1. Distribution of Air-Conditioning Tonnage and Refrigerant Type by Principal Building Activity and Year Building Was Constructed

Table 16-1A. Cooling Equipment Assumed to Be Using CFC-11 or -12

Building Characteristic	Number of Buildings (000s)	Estimated Sq Ft/Ton	Cooled Area (MM sq ft)	Tons of Air-Conditioning (MM)[a]		Pounds of Refrigerant (MM lb)
				Centrifugal	Reciprocating	
All buildings	1,111	NA	9,333	22,445	6,555	125,147
Principal activity						
Assembly	196	330	1,574	2,765	1,870	20,488
Education	53	260	1,331	4,470	372	20,733
Food sales/ service	93	315	362	631	486	4,998
Health care	21	260	566	1,925	89	8,416
Lodging	29	300	410	1,136	158	5,584
Mercantile/ service	314	325	1,358	2,449	1,524	17,289
Office	282	320	2,864	6,958	1,388	35,494
Warehouse	58	350	348	964	129	4,681
Other	34	320	305	782	203	4,345
Vacant	32	350	214	364	336	3,120
Year constructed						
1960 or before	465	NA	3,464	5,378	5,537	48,051
1961 to 1979	472	NA	4,522	13,027	1,018	60,032
1980 to 1986	174	NA	1,347	4,040	1	17,064

Note:
[a] MM = Million.

Source: Unpublished data from 1986 CBECS; EIA 1988.

working fluids were assigned according to building size. For example, HCFC-22 was assumed to be the cooling medium for equipment in buildings of 10,000 or fewer square feet.

3. A typical refrigerant charge per ton of cooling capacity was assigned to the working fluid.

4. CBECS data on square footage of air-conditioned space were converted to tons of cooling capacity and, by extension, to estimates of total required refrigerant charge.

312 — Chapter Sixteen

Table 16-1B. Cooling Equipment Assumed to Be Using HCFC-22

Building Characteristic	Number of Buildings (000s)	Estimated Sq Ft/Ton	Cooled Area (MM sq ft)	Tons of Air-Conditioning (MM)[a]				Pounds of Refrigerant (MM lb)
				Packaged Units	Individual A/C Units	Reciprocating Units	Heat Pumps	
All buildings	2,122	NA	17,278	28,418	14,757	5,890	6,853	194,723
Principal activity								
Assembly	265	330	2,032	2,430	1,909	1,055	716	21,226
Education	123	260	1,726	3,582	2,292	303	443	23,111
Food sales/ service	212	315	929	1,687	551	448	238	10,148
Health care	37	260	1,063	2,559	1,253	52	221	14,289
Lodging	85	300	1,190	782	2,303	298	569	13,781
Mercantile/ service	661	325	4,707	8,662	2,977	1,196	1,584	50,253
Office	433	320	3,962	5,925	2,080	1,957	2,317	42,643
Warehouse	172	350	880	1,610	652	211	453	10,213
Other	69	320	411	594	364	223	174	4,704
Vacant	65	350	378	586	378	149	Q[b]	4,355
Year constructed								
1960 or before	826	NA	4,884	6,863	7,740	—[c]	1,278	55,586
1961 to 1979	900	NA	7,978	13,256	5,643	3,717	3,139	89,297
1980 to 1986	396	NA	4,417	8,299	1,374	2,174	2,436	49,840

Notes:
a MM = Million.
b Q = Withheld because relative standard error is ≥ 50%.
c — = Assumed to be zero.

Source: Unpublished data from 1986 CBECS; EIA 1988.

Using these factors and assumptions, we estimated that 56 million tons of cooling capacity in commercial buildings in 1986 were using HCFC-22. A typical refrigerant charge of 3.5 (3 to 4) lb/ton of cooling capacity (A. D. Little 1989) would mean that 195 million pounds of HCFC-22 were used in commercial buildings in 1986.

To calculate the amount of CFC-11 and CFC-12 in commercial buildings, the same basic four-step estimation procedure was followed except that a refrigerant charge of 4.5 (3 to 6) lb/ton of cooling capacity (A. D. Little 1989) was used. This procedure produces an estimate of 9 billion square feet of floor space in chiller-cooled buildings of more than 10,000 square feet. These commercial buildings contain approximately 29 million tons of cooling capacity using 125 million pounds of CFC-11 and CFC-12. In a separate, production-based study, EIA estimated that there were 97 million pounds of coolant in use for commercial air-conditioning (EIA 1989). Given the difficulties in assumptions and methods used to derive these estimates, and the confusion in the 1986 CBECS between central cooling with packaged units and central chillers, these two estimates are in reasonable accord.

Preliminary data from the 1989 CBECS on the relative prevalence of central chillers and packaged cooling units suggest that many 1986 respondents reporting "central cooling" were actually employing packaged cooling systems. Thus, the 1986 central cooling number overestimates the number of central chillers. This overestimation would be most serious for the smaller buildings, but it appears to have affected even moderately large buildings (50,000 to 200,000 sq. ft). The overestimation would modify the analysis results presented in Table 16-1 in three ways: (1) by reducing the number of chillers (to a figure closer to the EPA estimate); (2) by producing a large drop in the tonnage of reciprocating chillers, a smaller drop in the tonnage of centrifugal chillers, and a corresponding rise in the tonnage of packaged units; and (3) by shifting the estimated refrigerant amounts from CFC-11 and CFC-12 to HCFC-22. The amount of CFC-11 and CFC-12 in commercial buildings may be as low as one-half the amount shown in Table 16-1, while the amount of HCFC-22 may be one-quarter larger.

Commercial Refrigeration

Very little information about the stock of commercial cold-storage refrigeration equipment exists. Cold-storage refrigeration equipment used in the storage and distribution of meat, produce, dairy products, and other types of perishable goods is concentrated mainly in refrigerated warehouses, supermarkets, and restaurants. Schools, hospitals, laboratories, and retail activities also use CFCs but in less significant amounts. Thus, rough estimates of cold-storage refrigeration equipment containing CFCs can be obtained by analyzing the principal building activity, square footage, and measure of size information in the 1986 CBECS.

Of the CBECS building types, refrigerated warehouses consume most of the CFCs associated with refrigeration equipment (principally, CFC-12 and CFC-502). The 1986 CBECS estimated that there were 25,000 warehouses comprising 474 million square feet. By contrast, the U.S. Department of Agriculture (USDA)—the only other nation-

314 — Chapter Sixteen

al source of information on the capital stock of cold-storage ware-houses—has estimated that in 1985 there were 3,198 refrigerated warehouses containing roughly 74.8 million square feet (USDA 1986). However, the USDA survey does not include warehouses oper-ated by wholesale distributors, grocery chains, or other businesses that store food products less than 30 days, whereas the CBECS definition includes all buildings where 50% of the floor space is artificially cooled to 50°F or less. (The actual USDA estimate was 1.7 million cubic feet, which was converted to square feet by dividing by 22.7 feet, the average height of a refrigerated warehouse [EPA 1989].)

For refrigerated warehouses, the 1986 CBECS gathered informa-tion about heating and cooling equipment but not about refrigeration equipment. Without knowing the specific types of refrigeration equip-ment, we could not estimate CFC usage in refrigerated warehouses. In addition, there are widely varying estimates of the CFC charge required per ton of capacity (EPA 1989). If such estimates are deemed important, questions may be added to future rounds of the survey.

Insulation Blowing Agents

Every cycle of CBECS has gathered information on the presence of roof and wall insulation but not on their type and thickness; respon-dents were usually unable to provide the latter information. Table 16-2 shows previously unpublished CBECS data used to estimate square footage of insulated roof and ceiling in commercial buildings.

CFCs were not used in foam insulation until the early 1960s (Hammitt *et al.* 1986). By 1986, virtually all insulation being installed in commercial buildings contained CFCs (EIA 1989). Most of the roof and wall insulation in commercial buildings appears in the most recently constructed buildings and tends to be installed at the time of construction. Not surprisingly, only a very small fraction of the wall space had insulation added after construction. Most post-construction installation occurred from 1980 to 1986 and can be assumed to contain CFCs.

CBECS data on roof and wall insulation, floor space, and number of floors were used to derive estimates of insulated roof and wall area. Roof or ceiling areas were assumed to equal floor areas and thus could be calculated by dividing the total square footage by the number of floors. An insulated roof or ceiling was assumed to be completely rather than partially covered by insulation. The estimate for wall area was calculated assuming equal-area, square, 12-foot-high walls, which, if insulated, contained insulation throughout their nonglass area. Based on these assumptions, at least 14.7% (1,094 billion square

Building Energy Consumption and the Environment — 315

Table 16-2. Roof or Ceiling and Wall Insulation by Year Building Was Constructed

Year Constructed	Wall Insulation			Roof or Ceiling Insulation		
	Buildings (000s)	MM sq ft	Wall area[a] (MM sq ft)	Buildings (000s)	MM sq ft	Floor area[b] (MM sq ft)
Installed at Time of Construction						
All buildings	1,548	24,078	7,480	1,930	30,950	19,116
1900 or before	19	266	116	23	370	123
1901 to 1920	27	524	183	35	869	358
1921 to 1945	85	1,626	533	119	2,161	962
1946 to 1960	210	2,495	863	306	3,872	2,605
1961 to 1970	299	4,687	1,485	406	6,490	4,184
1971 to 1979	438	7,029	2,088	517	8,571	5,464
1980 to 1986	469	7,451	2,212	524	8,619	5,421
Added Since Construction—Before 1980						
All buildings	155	1,817	777	309	3,419	2,017
1900 or before	17	276	134	40	604	211
1901 to 1920	21	402	148	38	683	337
1921 to 1945	49	500	216	99	976	600
1946 to 1960	34	332	144	69	634	522
1961 to 1970	22	173	83	35	286	180
1971 to 1979	12	110	46	27	212	142
1980 to 1986	2	25	7	2	25	25
Added Since Construction—1980 to 1986						
All buildings	306	3,338	1,381	518	7,958	4,539
1900 or before	30	334	186	41	473	140
1901 to 1920	28	200	123	45	436	213
1921 to 1945	71	894	364	113	1,763	909
1946 to 1960	88	765	334	134	1,827	1,356
1961 to 1970	37	453	153	82	1,681	970
1971 to 1979	44	573	180	80	1,559	823
1980 to 1986	8	118	39	22	219	127

Notes:
[a] The wall area was calculated as

      Wall area = 12 ft (ceiling height) × 4 walls

           × square root (total sq ft/# of floors)

           × proportion nonglass exterior walls.

[b] The floor area was calculated (assuming equal size floors) as

      Floor area = total sq ft/# of floors.

Source: Unpublished data from 1986 CBECS.

316 — Chapter Sixteen

feet) of the wall area and an equal percentage of the roof or ceiling area in commercial buildings can be assumed to contain CFCs.

The presence of CFCs in insulation board is not really a vital energy issue. Researchers have determined that there is a minimal energy consumption penalty with the substitution of new types of non-CFC insulation board (Fischer and Creswick 1989). And, because CFCs, unlike asbestos, do not constitute a health hazard, it is not necessary to remove existing insulation from buildings. This information is of most use to climate modelers and others trying to project CFC emissions that will be released to the atmosphere from the existing building stock after the production of CFC-11 and CFC-12 have been banned.

What the 1989 CBECS Can Tell Us

In response to DOE's interest in the impacts of the Montreal Protocol, EIA made changes to the 1989 CBECS that would enhance the survey's ability to analyze and evaluate the impact of CFC curbs. The timing of the 1989 CBECS (at the start of the movement to curb CFCs) and its status as the only national survey of the existing U.S. building stock make it an excellent vehicle for the collection of baseline measurements on these issues.

As previously mentioned, on the 1989 CBECS, EIA clarified the categorization of space-conditioning equipment so that chillers in particular could be more reliably identified. EIA also added a question on the age of the chillers and of packaged air-conditioning equipment. A question was also added to determine the presence of other refrigeration units, including commercial refrigeration units, commercial freezers, residential-type refrigerators, residential-type freezers, ice-making machines, refrigerated vending machines, and water coolers.

Second, EIA expanded the scope of the survey to collect system-wide energy inputs and outputs from nonutility central plants supplying electricity, steam, and district heating and cooling to any commercial building falling within the sample. Additional information from the survey's facility form on the number and size of the chillers used on facilities, campuses, or complexes will complete the picture of the contribution of commercial buildings to CFC usage in cooling equipment.

In addition to the regular tables on heating and cooling equipment used in the building, the 1989 building characteristics report will include (subject to the statistical reliability of the data) at least the following new information useful in the analyses of CFC-related issues: (1) the ages of the chillers and packaged units cross-tabulated by the

Building Energy Consumption and the Environment — 317

year of construction and by the square footage of the building and (2) the types of commercial refrigeration equipment cross-tabulated by the building's principal activity, year of construction, and square footage.

Furthermore, the information on equipment age can be used to produce a special report on equipment life cycles. The 1989 data will only yield information on the age of existing equipment and, thus, would not be sufficient to describe equipment life cycles. However, the 1989 data would be a valuable complement to the new equipment production data already available from other sources. Beginning in 1992 the CBECS longitudinal panels (see below) will be another source of information on equipment turnover.

What Future CBECS May Tell Us

The timing of the CBECS on three-year cycles offers both a longitudinal and cross-sectional view of changes in the building population over time.

Beginning with the 1986 CBECS, the sample was designed with a longitudinal component—a set of buildings that will be included in the survey twice—to allow the tracking of energy-efficiency improvements in buildings. The buildings participating in the 1986 survey will be resurveyed in 1992 regarding replacements in equipment and other conservation steps taken since the 1986 survey. Demolition rates will be calculated and should prove useful in the development and testing of equipment life-cycle models. There is also a possibility for longitudinal analysis for the 1989–1995 and 1998–2003 surveys.

The 1992 sample will consist of the set of buildings that were surveyed in 1986 and a set of buildings constructed since the 1986 survey. The resurveyed buildings will be assessed for changes both in general building characteristics, such as size, occupancy, and operational patterns, and in energy-related characteristics, such as changes in fuel use, replacements of equipment, and conservation steps.

Although analysis of the 1989 data is just starting, the triennial survey schedule requires the planning of the content of the 1992 survey. Suggestions on the collection of additional data have already been received through extensive interaction with the various working groups and individuals using the CBECS data in support of DOE's work on the CFC and global-warming issues and on the National Energy Strategy (NES). There is a general consensus that EIA needs to collect more data on sector-specific consumption of energy; on the market-penetration of alternative fuels such as solar, wood, geothermal, and biomass; on the market penetration of energy-consuming equipment; and on consumer decision making with regard to equipment purchases.

318 — Chapter Sixteen

Summary and Conclusions

This chapter illustrates the usefulness of data from the CBECS for understanding the scope of the problem of CFCs in commercial buildings. Using data on building characteristics from the 1986 CBECS, we derived estimates of the total amount of CFC-11, CFC-12, CFC-113, CFC-114, and HCFC-22 used in commercial buildings. These estimates were found to be in reasonable accord with other production-based estimates. Less detail was available from the 1986 CBECS relative to CFCs for insulation and refrigerants, but it was possible to provide some estimate of the possible extent of CFC usage.

The 1989 CBECS, currently being published, will provide some data on these issues, and there is still an opportunity to make the 1992 CBECS more useful for analysis of CFCs and of building-related environmental problems in general.

Acknowledgments

The authors gratefully acknowledge the assistance of Dwight French and Arthur Rypinski of the EIA in the preparation of this paper.

References

A. D. Little 1989. "Impact of CFC Regulations on Electrically Powered Cooling Equipment." Prepared for Electric Power Research Institute, Palo Alto, Calif.: Unpublished draft.

EIA. See Energy Information Administration.

Energy Information Administration 1988. *Nonresidential Buildings Energy Consumption Survey: Characteristics of Commercial Buildings, 1986*. DOE-EIA-0246(86). Washington, D.C.: U.S. Department of Energy.

———— 1989. *Potential Costs of Restricting Chlorofluorocarbon Use*. SR/ESD/89-01. Washington, D.C.: Energy Information Administration.

Environmental Protection Agency 1989. "Documentation of Engineering and Cost Data Used in the Vintaging Analysis." Draft Report. Washington, D.C.: Environmental Protection Agency.

EPA. See Environmental Protection Agency.

Fischer, S., and F. Creswick 1989. *Energy-Use Impact of Chlorofluorocarbon Alternatives*. ORNL/CON-273. Oak Ridge, Tenn.: Oak Ridge National Laboratory.

Hammitt, J., K. Wolf, F. Camm, W. Mooz, T. Quinn, and A. Bamezai 1986. *Product Uses and Market Trends for Potential Ozone-*

Depleting Substances, 1985–2000. Rand/R-3386-EPA. Santa Monica, Calif.: Rand Corporation.

Statt, T. 1988. "Use of Chlorofluorocarbons in Refrigeration, Insulation and Mobile Air Conditioning in the USA." In *International Journal of Refrigeration* 11(7): 224–28.

TRW 1982. "RD and D Opportunities for Large Air Conditioning and Heat Pump Systems." ORNL 80-13817. Oak Ridge, Tenn.: Oak Ridge National Laboratory.

USDA. See U.S. Department of Agriculture.

U.S. Department of Agriculture 1986. *Biannual Survey of the Capacity of Warehouses in the U.S.* Washington, D.C.: U.S. Department of Agriculture, Agriculture Statistics Board.

Julia D. Oliver is the survey manager for the Commercial Buildings Energy Consumption Survey. She is also the co-chair of the Measured Energy Performance Subcommittee of TC 9.6, Systems Energy Utilization of the American Society of Heating, Refrigerating and Air-Conditioning Engineers.

Eugene M. Burns is a senior analyst for the Commercial Buildings Energy Consumption Survey. He received a Ph.D. from Cornell University.

Chapter 17

Measured Cooling Savings from Vegetative Landscaping

Alan K. Meier, *Lawrence Berkeley Laboratory*

Introduction

Planting trees, shrubs, and ivy around buildings is a familiar strategy to limit solar gain and to create a comfortable environment within (Olgyay 1963). If the building is air-conditioned, then vegetation will also provide economic benefits. Since many electric utilities in warmer regions experience their peak demand on summer afternoons as a result of widespread air conditioner operation, a reduction in air conditioner electricity demand can avoid construction of expensive power generation capacity.

The value of plantings as a technique to reduce air-conditioning loads has been neither well documented nor incorporated into analytical procedures. Building standards and design tools generally assume that buildings rest on treeless, exposed surfaces. Architects and engineers will thus favor technologically complex solutions over simpler, cheaper designs based on the use of trees and other plants. Partly as a result of this information gap, the energy and comfort benefits of vegetation have not been fully exploited.

This chapter reviews research measuring the cooling savings from the judicious use and siting of vegetation. The major pathways by which plants can reduce heat gains are described, as are attempts to simulate these processes. Several case studies are then summarized. Both a compilation of surface temperature reductions and energy savings and recommendations for further research are presented.

322 — Chapter Seventeen

Heat Gain Paths Influenced by Vegetative Landscaping

Trees, shrubs, and vines affect air-conditioning electricity use via five physically different paths. These paths are (1) direct gain through windows, (2) conduction gain through opaque surfaces, (3) latent heat from infiltrating air, (4) sensible heat from infiltrating air, and (5) air-conditioning system performance.

The relative importance of these paths depends on the vegetation being used, the climate, the building structure and orientation, and type of air-conditioning system. Shading of direct solar gain is typically considered the factor most influenced by trees and shrubs and has therefore received the most attention. In humid locations, however, infiltrating air can be responsible for as much as 50% of the peak cooling load (Roseme *et al.* 1979; Steen *et al.* 1976). Thus, the impact of vegetative landscaping on infiltration is also a major consideration. Proper shading of the air conditioner unit, especially the exterior condenser, can also lower energy use. This effect has been little researched; Parker (1983) claims that shading and evaporative cooling of the air surrounding the condenser lowers supply air by about 4°C. This decrease would improve the unit's coefficient of performance (COP) by as much as 10%.

Widespread plantings may ameliorate the urban heat island phenomenon and thereby reduce air-conditioning loads of nearby buildings, even though no specific landscape treatments have been applied (Akbari *et al.* 1990). The impact of vegetation on heat islands is discussed in chapter 20 by Huang, Davis and Akbari.

Simulations to Estimate Cooling Savings

The mechanism by which vegetation reduces air-conditioning energy use is extremely complex and therefore difficult to accurately predict. In general, current computer simulations appear to be most accurate in modeling the impact of shading by trees (assuming that accurate transmissivities are available) and less accurate in modeling wind shielding. The simulations are not yet capable of modeling evapotranspiration: one of the major drawbacks is the absence of field data from which to develop algorithms.

With regard to the optical properties, the simplest procedure is to treat the vegetation like a screen with limited transmittance. This method yields a kind of shading coefficient. Nayak *et al.* (1982) used this assumption to compare the relative performance of different shading strategies. They found that a vine-covered pergola (or trellis), combined with a rooftop water film, was the most effective means of reducing heat gain through the roof in a North Indian climate.

In a more sophisticated approach, McPherson *et al.* (1988) simulated the energy savings from tree shade using the MICROPAS hourly simulation model. Assumptions were made about the percent reduction in solar gain and infiltration that would be derived from various types and locations of plantings. The model was used to predict energy and dollars saved for four U.S. cities.

Holm (1989) adapted the DEROB hourly building simulation program to model the thermal effects of deciduous and evergreen vegetation on an external wall. The modifications included an outer surface with an absorptance spectrum similar to total leaf cover, multiple air spaces, ventilation, thermal mass, and thermal resistances. The physical values were calibrated through measurements of actual vegetation. The reduction in inside temperature due to a leaf cover on the external walls was then calculated and used to calculate air-conditioning savings due to reduced heat gain. This model did not attempt to include the change in energy flows caused by plant transpiration.

Huang *et al.* (1990) sought to simulate the impact of trees on heating and cooling energy. These simulations adjusted for the direct light transmissivities of the trees and for the diminished infiltration due to the plant windbreaks. The simulation model (DOE-2.1D) treated the shading effects of the trees as exterior building shades, whose transmissivities were determined by earlier work. The model included the impact of reduced diffuse light by adjusting the sky- and ground-form factors. DOE-2 does not have the means to calculate the savings from evapotranspiration, although the authors concluded from an earlier investigation (Huang *et al.* 1978) that its impact was greater than that from wind shielding or shading. The model could not include the tree-caused changes in long-wave radiation, which the authors believed to be small.

Detailed vegetation models exist (Terjung and O'Rourke 1980; Halvorson *et al.* 1980), but they are poorly linked to the building energy models. As a result, even the most sophisticated models cannot completely simulate some of the key processes associated with the effects of vegetation on heat gains (Huang *et al.* 1990; Holm 1989). Several phenomena resist easy simulation:

1. The heterogeneous optical characteristics of the plants. These short-wave characteristics—including the fractions of light that penetrate, reflect, or are absorbed by the plant—affect direct gain and sol-air calculations.

324 — Chapter Seventeen

2. Long-wave radiation energy exchanges between the building's surfaces and the surrounding surfaces (including nearby buildings).

3. The microclimate established in the area between the building surfaces and adjoining plants. The cooling caused by the vegetation shading, by moisture released through evapotranspiration, and by the cooler ground modifies the microclimate surrounding parts of the building by lowering temperatures and increasing humidity. The microclimate may lower building surface temperatures, induce convection currents, and protect building thermal mass.

4. The infiltration reductions caused by wind shielding. Plants can be used to create a barrier to wind and, therefore, to reduce air pressure differences on the building surfaces. Specific effects are determined by site configuration and by the wind speed and direction.

The dynamic and heterogeneous phenomena of vegetative landscaping are thus challenging, but not impossible, to simulate. Simplifying models could no doubt address specific situations; however, the chief obstacle remains the lack of measured data to validate algorithms and assumptions.

Case Studies

Likewise, air-conditioning energy savings caused by vegetative landscaping have rarely been directly measured. Such experiments are difficult because plants take many years to grow, and few researchers can wait so long for the desired environmental conditions to develop. To circumvent this problem, researchers have created temporary environments, which closely resemble a building surrounded by planted landscape, by moving fully grown plants to the site; have focused on plants that grow quickly, such as vines; and are attempting to locate serendipitous comparison groups (for example, the energy use of buildings with extensive plantings might be compared to a similar group without plants). Six measurement studies are summarized below.

Parker (1981, 1983; Meier 1987) investigated the cooling savings due to vegetative landscaping around a double-wide mobile home (used as a nursery school) in Miami, Florida. The hot/humid climate and high internal loads required air-conditioning for over half of the year. The building was originally situated on a clear site. Parker planted a multilayer canopy of fully grown shrubs and trees (2–8 m high) around the building. These plants were intended to shade windows and walls, create a cooler microclimate adjoining the walls, and shield the building from warm afternoon breezes. Parker com-

Measured Cooling Savings from Vegetative Landscaping — 325

pared air conditioner energy use before and after the trees and shrubs were planted for two days with similar conditions. Measured air-conditioning savings exceeded 50% for comparable hot days, but long-term savings were about 25%. The savings would have been higher, but the occupants selected a lower inside temperature after the trees were planted. In addition, the air conditioner was undersized; the occupants would have used more air-conditioning energy if the unit had more capacity. As a result, some of the potential energy savings were converted into lower indoor temperatures—that is, greater comfort—during the warmest periods.

Hoyano (1988) conducted a series of experiments to measure the cooling effects of vines and trees. These experiments mostly took place near Tokyo, whose summer climate is hot and humid. In one experiment, Hoyano placed a trellis planted with vines over a veranda. A neighboring, identical veranda without a trellis served as the control. In a second experiment, vines were grown over a west-facing wall of a residence. During the summer, the heat flows were compared on units with vines and without. Hoyano also tested the impact of trees placed in front of a west-facing wall. The trees were kept in containers so that the distances between trees and between the trees and the wall could be varied. Even for widely spaced trees, the heat gain through the wall was cut by over 50%.

McPherson *et al.* (1989) constructed 1/4-scale model homes and measured their cooling energy consumption with different landscaping. This experiment took place in Arizona, a desert climate where water is expensive. Therefore, a major goal was to find a landscaping strategy that reduced cooling energy but did not require a lot of water. One building served as a control; its "yard" consisted of a layer of decomposed granite about 5 cm deep. Two landscape treatments were compared: a traditional turf lawn and a selection of low-water-use shrubs. The turf landscape cut air-conditioning energy use by about 25% without increasing shade on the structure. Slightly greater savings were achieved by the shrubs, judiciously placed around the building (the remaining ground was left covered with decomposed granite).

The physical processes by which the two landscapes saved energy appeared to differ. The shrubs cut cooling energy by direct shading and modifying the microclimate around the building (less than a few meters wide). The turf cut cooling loads through extensive evaporative cooling, which decreased air temperatures over the whole yard and around the house. (The turf also used about ten times more water than the shrubs.) The turf probably also reduced long-wave radiation gains and increased reflection. The use of scale models introduces some uncertainty because physical processes scale differently (Meier 1989). The results from

326 — Chapter Seventeen

models may overstate the importance of effects that depend on the building's surface area.

Parker (1990) investigated the impact of vegetation around 25 air-conditioned houses in Florida and rated these impacts on a scale from 0 (no shading) to 3 (heavily shaded in all directions). He then inferred the impact of plants by regressing total energy use against level (0 to 3) of landscape and several other variables. The variables explained about 75% of the actual variation in total energy use. The shading class was a statistically significant determinant of air-conditioning use. Houses with moderate or heavy shading (shading class 2 or 3) used about 34% less air-conditioning energy than houses with no shading (class 0). This percentage corresponded to roughly 15 kWh/m²/d or 3,300 kWh/yr for the average house in the study.

The greatest electricity savings—80%—were reported by DeWalle et al. (1983). This project consisted of moving small, air-conditioned trailers between a forested and open site. The experiment took place in central Pennsylvania, whose summer is not particularly harsh: the initial average power use of the air conditioner was only about 300 watts. As a result, the air-conditioning loads could be nearly eliminated once solar gains were blocked. Nevertheless, the experiment demonstrated that forest vegetation can greatly reduce air-conditioning energy use.

The largest measured reduction in heat gain was reported by Harazono et al. (1989). Half of a roof was covered with vegetation (in trays grown with hydroponics). Essentially all of the heat gain into the top floor of the building was eliminated (and, at times, the direction of heat gain was reversed).

Measured Temperature Reductions in Exterior Surfaces

Strategically placed vegetation will lower temperatures on the building surfaces, but reduced surface temperature is only one effect of vegetative landscaping on building air-conditioning use; other effects include reduced infiltration and altered microclimate. Nevertheless, exterior surface temperature data are useful for calibrating simulation models, predicting energy savings in other conditions or locations, and calculating crude thermal resistances of the vegetation barrier.

Several researchers have measured exterior surface temperature reductions due to vegetative landscaping. These temperature reductions are presented in Figure 17-1, and some experimental details are listed in Table 17-1. All of the experiments reported considerably more information than is presented in the table, so only the results for similar conditions—a west wall at about 3 PM—are generally listed. Three PM is usually the time at which maximum temperature reductions are

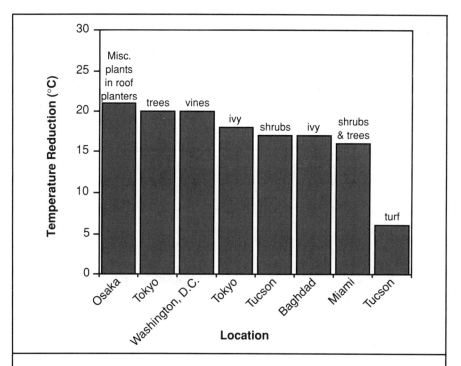

Figure 17-1. Surface Temperature Reductions Due to Vegetation.
Notes: Most temperatures were measured on west-facing walls, with and without vegetation. The principal form of planting is listed above the result. The reductions are peak values and generally occurred around 3 PM

found, except in the case of rooftop vegetation, whose maximum solar load occurs around midday.

The results must be interpreted with caution because each researcher used somewhat different experimental setups and procedures. For example, each measured west-wall temperature differently, so it is probably coincidental that the vegetation consistently lowered the wall surface temperature by about 17°C. However, the reduction is significant in all cases. The temperature reduction occurred with several types of vegetation, from thick ivy to strategically placed shrubs and trees. Moreover, the reduction occurred in both hot/dry climates such as Iraq's and in hot/humid climates such as Florida's. Surrounding the house with turf (McPherson 1989) caused only a 6°C wall temperature drop. This difference is not surprising because the turf added no shad-

328 — Chapter Seventeen

Table 17-1. Surface Temperature Reductions from Vegetative Landscaping

Author & Year	Location & Climate	Type of Planting	Wall-Veg. Distance	Difference Measured	ΔT	Notes
J. Parker 1981	Miami, Fla. hot/humid	shrubs and trees	shrubs < 1 m trees < 10 m	wall with & without plants	16°C	west wall, 5 PM, maximum value; about 1 month apart
Hoyano 1988	Tokyo, Japan hot/humid	ivy covering	touching	wall with & without ivy	18°C	west wall, 3 PM, maximum value; 1 year apart
Hoyano 1988	Tokyo, Japan hot/humid	dense canopy evergreens (Kaizuka hort)	0.2–0.6 m	wall & inside plant surface	5–20°C	west wall, 3 PM, parallel measurements
McPherson 1989	Tucson, Ariz. hot/dry desert	18 shrubs and 5-cm decomposed granite	0.5 m	wall with shrubs & no shrubs	17°C	west wall, 3 PM, different buildings
McPherson 1989	Tucson, Ariz. hot/dry desert	turf, extending about 5 m from structure	surrounding building	wall with turf vs. decomposed granite	6°C	west wall, 3 PM, different buildings
Makzoumi & Jaff 1987	Baghdad, Iraq hot/dry desert	vine (luffa cylindrica) on trellis	0.1–0.4 m	wall with & without vines	17°C	southwest wall, 3 PM, max. value, different buildings
Harazono 1989	Osaka, Japan hot/humid	rooftop hydroponic using lightweight planting substrates and mixed plants	0.1 m	half of roof with, half without	21°C	average for 10AM–6PM on clear August day
Halvorson 1984	Pullman, Wash. temperate	vertical vine canopy	n.a.	wall with & without vine	20°C	

ing, but the result suggests that the creation of a more humid microclimate around a building can alone reduce wall temperatures; this effect has not yet been incorporated into building simulations.

Measurements of Energy Savings

Measured energy savings from vegetative landscaping are difficult to standardize because researchers used diverse methods and measured

Measured Cooling Savings from Vegetative Landscaping — 329

different kinds of savings. Results are given in Table 2 and summarized in Figures 2 and 3. Some researchers measured the reduction in heat gain through a wall (Figure 2), while others measured the air-conditioning energy required to maintain a pre-set indoor temperature (Figure 3). The period of monitoring likewise varied. Heat gain was typically measured for a few hours, while air-conditioning energy use might have been monitored for several weeks. Measurements of heat gain were generally made on a wall section or test cell, while air-conditioning electricity use was measured in whole structures. For these reasons, both the absolute amounts and the fractions of energy saved (expressed in terms of average power, watts) are listed.

Measured air-conditioning savings from vegetative landscaping ranged from 25% to 80%. The variation in absolute savings was much greater but simply reflected the range in experimental conditions— from a small trailer in a temperate climate to a double-wide mobile home in a hot, humid climate. The measurements by McPherson *et al.* (1989) demonstrate the complexity of the relationship between landscaping and air-conditioning loads. Roughly 25% savings were achieved by two different physical processes. In one case, the savings were obtained by careful selection and siting of shrubs around the structure. The principal causes were probably direct shading, the creation of a narrow cooler microclimate surrounding the structure, and addition of a small amount of evaporative cooling. When the yard was covered with turf, the savings appeared to be due to the turf re-radiating less long-wave energy and creating a larger evaporatively cooled microclimate around the house.

Parker's (1981) and McPherson's (personal communication, June 18, 1989) results also indicate that vegetative landscaping affects more than just solar gains. This conclusion is supported by the extent to which energy was saved before the sun was even shining on the building. Moreover, this phenomenon was observed in two dramatically different climates.

Conclusions

Only a few attempts have been made to measure the air-conditioning savings due to plantings. However, consistently large savings were achieved: careful application of shrubs, trees, and vines could probably reduce cooling electricity use 25–50%.

These savings were achieved in a variety of climates and using greatly different landscape treatments. Reductions in air-conditioning energy were obtained even in humid climates. Large savings were also

330 — Chapter Seventeen

Table 17-2. Energy Savings from Vegetative Landscaping

Author & Year	Location & Climate	Type of Planting	Energy Measurment	ΔE (watts)	ΔE (%)	Notes
Air-Conditioning Savings						
DeWalle *et al.* 1983	Central Penn. temperate	forest site vs. clear site	AC electricity for identical mobile homes	230	80%	37-day test period
J. Parker 1983	Miami, Fla. hot/humid	shrubs & trees	AC electricity with & without landscaping	5,000	58% 24%	6-hr (afternoon) test period 10-day periods
McPherson *et al.* 1989	Tucson, Ariz. hot/dry desert	shrubs surrounding model house	AC electricity with & without shrubs	104	27%	2-week period
McPherson *et al.* 1989	Tucson, Ariz. hot/dry desert	turf surrounding model house	AC electricity with & without turf	100	25%	2-week period
D. Parker 1990	Palm Beach, Fla. hot/humid	misc. trees and shrubs	annual electricity for whole house	1.8 w/m²	34%	inferred from regression of 25 houses, from land-scape class 0→2, 3
Heat Gain						
Hoyano 1988	Tokyo, Japan hot/humid	vine-covered wall	heat gain with & without vines	175/m²	75%	peak value at 4 PM on west wall
Hoyano 1988	Tokyo, Japan hot/humid	row of evergreens next to wall	heat gain through wall	>60/m²	>50%	peak value at 4 PM on west wall for widely spaced trees
Harazono 1989	Osaka, Japan hot/humid	rooftop vegetation	half of roof with, other half without	130/m²	90%	average from 10 AM to 4 PM

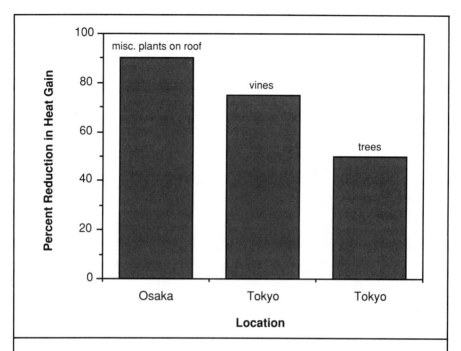

Figure 17-2. Heat-gain Reductions Due to Vegetation (Two Studies, Heat Gain Measured Through Wall)

obtained in a dry climate simply by planting grass around the building. These results suggest that vegetation interacts with heat gain through many physically different processes and that several combinations of vegetation and siting may yield similar savings.

In spite of the impressive savings achieved, the studies reported here are generally poorly documented and use widely differing measurement techniques. It is difficult to confidently apply these results to simulation models or other situations. Further research is needed to create a broader base of measurements. This research should include analysis of more buildings, different combinations of plantings, and careful monitoring of the temperature and energy use. The most valuable type of experiment appears to be one in which the same building can be compared with and without vegetative landscaping. In addition, special attention should be directed towards any maintenance problems resulting from the use of intensive planting around the structure.

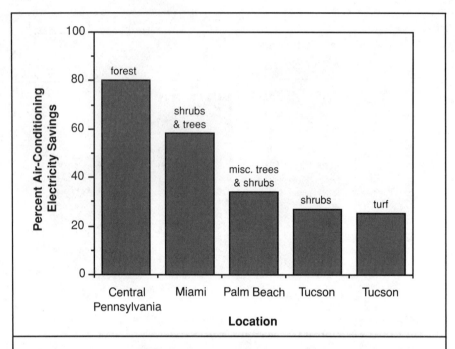

Figure 17-3. Air-conditioning Electricity Savings Due to Vegetation (Four Studies)

Acknowledgments

I am particularly grateful for the assistance by the members of the Heat Island Project at the Lawrence Berkeley Laboratory (Hashem Akbari, Phil Martien, Ron Ritschard, and Haider Taha). Several researchers have provided me details about their research and made important comments on earlier drafts; these researchers include Gordon Heisler, D. Holm, Akira Hoyano, Ken-ichi Kimura, Greg McPherson, Danny Parker, and John Parker. Peggy Sand provided many valuable comments.

References

Akbari, H., A. Rosenfeld, and H. Taha 1990. "Summer Heat Islands, Urban Trees, and White Surfaces." *ASHRAE Transactions* 96.

DeWalle, D., G. Heisler, and R. Jacobs 1983. "Forest Home Sites Influence Heating and Cooling Energy." *Journal of Forestry*, February, 84–88.

Halverson, H., J. Mawson, and B. Payne 1980. *A Computer Program to Map Tree Crown Shadows in the Urban Forest*. General Technical Report NE-59. Broomall, Penn.: U.S. Forest Service.

Halvorson, J. 1984. "Vine Canopy Effects on Wall Surface Temperature and Energy Fluxes." Unpublished master's thesis. Washington State University, Program in Environmental Science, Pullman, Wash.

Harazono, Y., S. Teraoka, I. Nakase, and H. Ikeda 1990/91. "Effects of Rooftop Vegetation Using Artificial Substrates on the Urban Climate and the Thermal Load of Buildings." *Energy and Buildings* 15–16: 435–42.

Holm, D. 1989. "Thermal Improvement by Means of Leaf-Cover on External Walls—A Computerized Simulation Model." *Energy and Buildings* 14: 19–30.

Hoyano, A. 1988. "Climatological Uses of Plants for Solar Control and the Effects on the Thermal Environment of a Building." *Energy and Buildings* 11: 181–99.

Huang, Y., H. Akbari, and A. Rosenfeld 1987. "The Potential of Vegetation in Reducing Summer Cooling Loads in Residential Buildings." LBL-21291. Berkeley: Lawrence Berkeley Laboratory.

Huang, Y., H. Akbari, and H. Taha 1990. "The Wind-Shielding and Shading Effects of Trees on Residential Heating and Cooling Requirements." *ASHRAE Transactions* 96: 1403–11.

Makhzoumi, J., and A. Jaff 1987. "Application of Trellises in Retrofitting Buildings in Hot Dry Climates." In Proceedings of the Third International Conference on Building Energy Management. A. Faist, E. Fernandes, and B. Sageldorf, eds. Lausanne, Switzerland: Ecole Polytechnique Féderale de Lausanne.

McPherson, E., L. Herrington, and G. Heisler 1988. "Impacts of Vegetation on Residential Heating and Cooling." *Energy and Buildings* 12: 41–51.

McPherson, E., J. Simpson, and M. Livingston 1989. "Effects of Three Landscapes on Residential Energy and Water Use in Tucson, Arizona." *Energy and Buildings* 13: 127–38.

Meier, A. 1989. "Using Water to Save Energy." *Home Energy Magazine*, July/August, 23–27.

Meier, A., and J. Friesen 1987. "Strategic Planting," *Energy Auditor and Retrofitter Magazine*, July/August, 7–12.

Nayak, J., A. Srivastava, U. Singh, and M. Sodha 1982. "The Rela-

334 — Chapter Seventeen

tive Performance of Different Approaches to the Passive Cooling of Roofs." *Building and Environment* 17(2): 145–61.

Olgyay, V. 1963. *Design with Climate: Bioclimatic Approach to Architectural Regionalism*. Princeton, N.J.: Princeton University Press.

Parker, D. 1990. "Monitored Residential Space Cooling Electricity Consumption in a Hot-Humid Climate." In *Proceedings of the 1990 ACEEE Summer Study on Energy Efficiency in Buildings*. Vol. 9. Washington, D.C.: American Council for an Energy-Efficient Economy.

Parker, J. 1981. *Uses of Landscaping for Energy Conservation, Department of Physical Sciences*. Miami: Florida International University.

————— 1983. "Landscaping to Reduce the Energy Used in Cooling Buildings." *Journal of Forestry* 81 (February): 82–84.

Roseme, G., C. Hollowell, A. Meier, A. Rosenfeld, and I. Turiel 1979. *Air-to-Air Heat Exchangers: Saving Energy and Improving Indoor Air Quality*. LBL 9381. Berkeley: Lawrence Berkeley Laboratory.

Steen, J., W. Shrode, and E. Stuart 1976. *Basis for Development of a Viable Energy Policy for Florida Residents*. Tallahassee: Florida State Energy Office.

Terjung, W., and P. O'Rourke 1980. "An Economical Canopy Model for Use in Urban Climatology." *International Journal of Biometerology* 24: 281–91.

Alan K. Meier is a staff scientist in the Center for Building Science at the Lawrence Berkeley Laboratory. He leads the Buildings Energy Data Group, which compiles measured data on the performance of energy efficient buildings and equipment in addition to measured savings from retrofit measures. He has a Ph.D. in Energy and Resources from the University of California, Berkeley.

Chapter **18**

Simulating Effects of Turf Landscaping on Building Energy Use

James R. Simpson, *Department of Soil and Water Science, University of Arizona*

Introduction

Vegetation can have an important role in the amelioration of urban climates. This fact has far-reaching implications for important issues such as municipal water use, electrical energy consumption for space cooling, and generation of greenhouse gases. It is important that both the benefits and costs of vegetation be considered when evaluating its place in the urban environment.

To date, the use of vegetation, or landscaping in general, to reduce air-conditioning loads has not been well documented. While the effects of shading by trees or shrubs has received some attention, treatment of nonshading vegetation, such as turfgrass, which cools the air by evaporation, thus reducing heat gain, has received little attention (see chapter 17). Partially due to the lack of information regarding benefits of vegetation and of the tendency, in arid regions, to focus on dwindling water reserves rather than escalating energy demands, conservation efforts in these regions have centered on reducing water use. As a result, vegetation removal tends to be encouraged without regard to the benefits that may simultaneously be eliminated.

This situation prompted a research program to better account for vegetation effects on cooling load and water use, especially in arid climates. Initial experimental work, described by McPherson, *et al.* (1989), used scale-model houses (one-quarter scale) situated in three representative landscape treatments: (1) Bermuda grass TURF with no shade, (2) rock mulch with SHADE from shrubs (no turf), and

336 — Chapter Eighteen

(3) ROCK mulch with neither turf nor shade. These treatments are referred to subsequently as simply TURF, SHADE, and ROCK. It was concluded that the house in the ROCK landscape used from 20% to 30% more energy for cooling than either TURF or SHADE houses. These energy savings were large enough to pay irrigation costs for plants using low and moderate amounts of water, but not for turf. However, simulation modeling of turf effects on full-sized buildings has shown that in arid climates, air-conditioning savings of 25% can exceed water costs, depending on the relative costs of electricity and power (McPherson 1990).

The fact that the TURF treatment reduced energy use almost as much as SHADE indicates that reduction in the temperatures of exterior air, building exterior, and surrounding ground surface by turf is nearly as effective in reducing heat gain as heavily shading the building's walls from the direct sun. To better understand the mechanisms by which turf landscaping influences cooling load, this chapter, combining the results of the aforementioned scale-model study with computer simulation analysis, predicts how contrasting turf and rock mulch ground covers, through their influence on microclimate, will affect building cooling load. (See appendix for explanation of methods used.)

Results and Discussion

Building Microclimates

Large temperature differences were observed between TURF and ROCK treatments due to the evaporative cooling effect of turf. The ROCK treatment was always warmer than TURF, with maximum ground temperature differences of about 20°C, and maximum air temperature differences, averaged over model height, of about 6°C (Figure 18-1). Maximum exterior wall temperature differences ranged from about 5°C for the west-facing walls to as high as 13°C for the south-facing walls. In contrast, total incident solar radiation between treatments agreed within 1% at times of peak irradiance, differing by less than 10 W/m², since measured ground surface reflectances were similar (23% for turf and 21% for rock ground covers). Winds were light (1–2 m/s) and easterly until 1 PM, when a shift to the west was accompanied by a wind speed increase to 3–4 m/s. Very light east winds returned after 7 PM. Since treatment exposures were nearly identical, the same wind speed was used for each treatment.

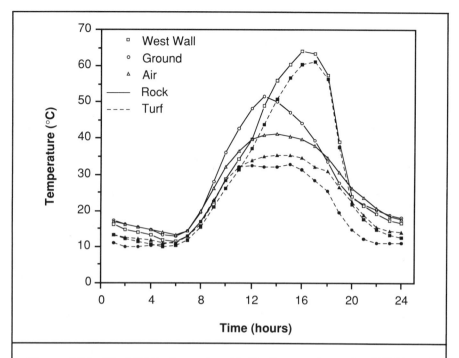

Figure 18-1. West Wall, Ground, and Air Temperatures for TURF and ROCK Treatments
Note:
TURF = dashed lines
ROCK = solid lines

Magnitude of Heat Gain Components

Glazed solar loads accounted for at most 2% of the hourly (Figure 18-2) or 3% of the total (Table 18-1) difference in energy use, due to the similarity in total solar radiation between treatments. However, glazed solar represents 58% of the TURF cooling load but only 43% of the ROCK. Consequently, on a percentage basis, reducing the temperature of the air and the surroundings has the effect of making cooling load more sensitive to solar radiation.

Infiltration gain accounted for 37% of the total difference in energy use (Table 18-1). The primary determinant of infiltration differences between treatments was air temperature, since the same wind speed was used for each treatment (equation [5]). Calculated infiltra-

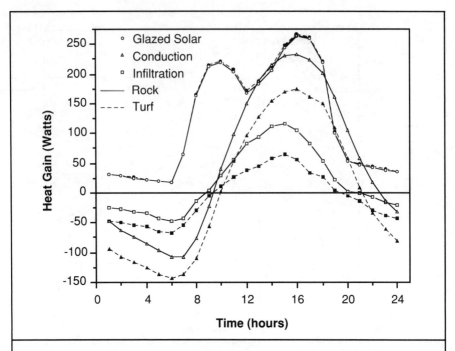

Figure 18-2. Comparisons of Selected Components of Space Cooling Electrical Use for TURF and ROCK Treatments

Note:
TURF = dashed lines
ROCK = solid lines

tion rates ranged from 0.5 to 3 air changes per hour, which, given the large surface-to-area ratio of the models, compares favorably with the value of 2 air changes per hour given for a full-sized building of similar configuration under average conditions (ASHRAE 1985). Heat gain from infiltration is generally estimated to be 40% or more of the total conduction gain for well-insulated buildings (Goldschmidt 1986). For this experiment, the magnitude of infiltration gain from 8 AM to 8 PM was 29% and 41% of total conduction gain for TURF and ROCK, respectively. Measured relative humidities were similar for both treatments during daylight hours, ranging from 8% to 30% for TURF and from 5% to 22% for ROCK. Consequently, the same latent heat load fraction (20%) was used for both treatments in calculating infiltration gain.

Simulating Effects of Turf Landscaping on Building Energy Use — 339

Table 18-1. Air-Conditioning Electrical Energy Use by Heat Gain Pathway

Electrical Use			% of ROCK	% of total ROCK-TURF	Cooling Load Component
TURF (kWh)	ROCK (kWh)	Difference (kWh)	(Total)	(Difference)	
40	278	238	6	24	Wall conduction
186	261	145	2	8	Roof conduction
− 194	145	339	9	34	Glazed conduction
1,658	1,630	− 29	− 1	− 3	Glazed solar
− 107	257	364	10	37	Infiltration
1,256	1,256	0	0	0	Internal
2,838	3,826	988	26	100	Simulation totals
2,923	4,253	1,330	31		Measured totals

Opaque conduction accounted for 32% and glazed conduction for 34% of the cooling load difference (Table 18-1). For glazed conduction, as in the case of infiltration, this difference was entirely due to air temperature, since air temperature and not sol-air temperature is used in the calculations. Opaque conduction, however, is a function of not only air temperature, but also solar radiation, and is the only term that depends on longwave radiation, through its effect on sol-air temperature (equation [4]).

To estimate the proportion of between-treatment *differences* in cooling load due to longwave radiation, ROCK treatment opaque conduction for the 24-hour period was calculated substituting h_o and ΔR values for TURF into ROCK sol-air temperatures. The resulting values depended only upon air temperature and solar radiation. The difference in opaque conduction as a percentage of total opaque conduction using these and the unmodified sol-air temperatures showed that about 11% of wall/door conduction and about 6% of roof conduction were due to longwave radiation. Based on this analysis, longwave radiation accounted for about 3% of the total cooling load differences.

This procedure was repeated using first T_o and then I_T determined for TURF to calculate ROCK sol-air temperatures, with all other terms remaining the same. As a result, it was found that air temperature was responsible for 68% of wall/door conduction and 94% of roof conduction. There was virtually no difference in roof conduction due to solar

340 — Chapter Eighteen

radiation, since the roofs received almost no reflected radiation due to their small pitch (12°). Total incident solar radiation was responsible for the remaining 21% of wall/door conduction, which was about 5% of total cooling load.

Comparison of Measurements and Simulations

Overall, the ROCK treatment used about 26% more energy for cooling than did TURF as a percentage of total simulated ROCK energy use based on simulations, and 31% more based on measurements (Table 18-1), which is in substantial agreement with earlier results (McPherson et al. 1989). Modeled electrical usage is in substantial agreement with measured values (Figure 18-3). For the 24-hour period, total load is underestimated by 3% for TURF and 10% for ROCK. On an hourly basis, maximum disagreement between modeled and measured load is better for the TURF treatment than is ROCK, and generally less than 30 W for both treatments. Parker (1983) indicates that an about 4°C increase in air conditioner ambient operating temperature can decrease operating efficiency by as much as 10%. Hence, higher ambient temperature of the ROCK air conditioning unit (Figure 18-1) may partially explain the larger difference between measured and simulated results for ROCK compared to TURF.

Conclusions

Microclimatic differences in this study were found to be predominantly due to air temperature, and, to a lesser degree, to longwave and shortwave radiation. Solar radiation and wind speed were similar for each treatment, so that their effects on microclimate differences were small. Air temperature and longwave radiation differences resulted from contrasts in building and ground surface temperature, which were in turn due primarily to the evaporative cooling effect of TURF. Air temperature was responsible for about 95% of treatment differences through its effect on conduction and infiltration gain, with longwave and solar radiation responsible for the remainder.

The actual air temperature, and hence cooling load, reduction obtainable from turf landscaping in an arid climate is expected to be somewhat less than found here. This difference is due to the diminishing effects of immediate ground surface conditions and to the growing influence of upwind surfaces on air temperature as height increases. A large part of the change in air temperature over a cool surface occurs close to that surface. In addition, dry, hot surfaces such as paved areas,

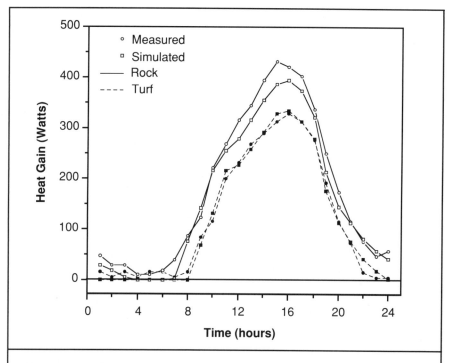

Figure 18-3. Comparisons of Measured and Simulated Total Electrical Use for Space Cooling for ROCK and TURF
Note:
TURF = dashed lines
ROCK = solid lines

bare soil, and other buildings upwind tend to dilute the localized evaporative cooling effect of a vegetated area. Consequently, the cooling potential of turf on full-sized buildings will depend on building height, horizontal extent of the vegetated area, and surface composition upwind of the vegetated area.

Under different circumstances, other microclimatic factors may assume more importance. For example, a 2% difference in ground albedo and the resulting effect on reflected solar radiation accounted for approximately 8% of the cooling load difference in this study through its effects on glazed and opaque conduction gains. Since ground albedos of 10% to 35% are not unrealistic for urban areas (Taha et al. 1988), and a potentially wider range is possible for building sur-

342 — Chapter Eighteen

faces, albedo differences may substantially impact cooling load. However, while increased ground reflectance will increase radiation reflected to building surfaces, it will at the same time tend to reduce ground and air temperatures, since less radiation is absorbed. More research is needed to evaluate these complex and sometimes competing effects on microclimate and building cooling loads.

Longwave radiation is responsible for a relatively small portion of the treatment difference in this analysis, despite the striking surface temperature contrasts, partially because longwave radiation affects only opaque conduction, which in the present case limits its effects on treatment differences to less than 32%. However, if it is assumed that air temperature differences between treatments would be reduced for full-sized buildings due to the scaling considerations just discussed, while surface temperature contrasts remain the same, it follows that longwave radiative effects will become a larger proportion of the total difference. Hence, although these results indicate that differences in longwave radiation between different landscapes is likely to be of little consequence, more study using full-sized buildings is indicated.

The smaller surface-area-to-volume ratio of full-sized buildings will tend to reduce the magnitude of their conduction and infiltration gain with respect to scale models, since the surface area for conduction and leaks for full-sized buildings decreases (by a factor of four in this study) relative to the volume of air to be cooled. However, conduction and infiltration, the predominant terms in this study, scale in a similar fashion. Hence, it is expected that model results regarding the relative impact of microclimate on cooling load in terms of air temperature effects are reasonably representative of full-sized buildings.

Some uncertainty results from the empiricism and assumptions involved in simulating building cooling load—for example, in the descriptions of infiltration and convective heat transfer. However, while this uncertainty may affect the absolute value of the results, it is expected that differences between treatments were more closely represented, since any constant errors tend to cancel out. In this study, the simulation model predicted measured electrical use to within 10%. Improved data sets and model formulations hold the potential for much better model performance. Consequently, this methodology seems promising for further use in predicting the effects of microclimate on building energy use. Additional research is necessary to improve our understanding of the complex microclimatic factors that influence building cooling load, with the goal of developing simulation modeling procedures applicable to a wide range of environments.

Appendix: Methods

Measurements

Data were gathered on September 18, 1988, a clear, warm day in Tucson, Arizona. Building characteristics, building operating conditions, and landscape treatments described in McPherson et al. (1989) apply to the current study, so only salient points or differences are described here. The experimental site was located at the University of Arizona Campus Agricultural Center (30.3°N, 111.0°W). The model houses, constructed on concrete slabs, were centered in 15.3 m × 15.3 m plots arranged in a row with azimuth 74°. Buildings were oriented 16° clockwise from the cardinal direction, so that, for example, when referring to the south wall, it is understood that its azimuth is actually 164°. With the exception of a greenhouse 10 m north (downwind) of the ROCK model, plots were surrounded by about 12 m of bare soil. Beyond this, upwind surroundings consisted of a mixture of short crops and bare soil extending 100, 30, and 45 m to the west, south, and east, respectively. Wood frame walls were 3.7 m wide and 0.7 m tall on north and south sides, with 19-cm overhangs, and 3 m wide and 0.7 to 1 m tall on east and west sides, with 2-cm overhangs. Plywood siding was painted light gray.

Measurements for each treatment consisted of hourly averaged interior and exterior air temperature and building electrical use. Net all-wave (Micromet Systems model Q3) and reflected (Weathertronics model 3020) solar radiation were measured over turf and rock ground covers. Global and diffuse solar radiation were measured between ROCK and SHADE models with pyranometers (LICOR model LI200S), the latter being mounted in a shadow band. Global solar radiation, air temperature, relative humidity, and wind speed and direction were measured at the nearby Arizona Meteorological Network (AZMET) weather station (360 m to the west). Measurements next to the model buildings prior to the experiment confirmed that AZMET wind and radiation data were suitable for characterizing conditions at the study site. Data were sampled at 15- to 60-second intervals, using a combination of Campbell Scientific CR21 and 21X data loggers, and stored as either 15-minute or hourly averages.

Wall albedos (reflectances) were determined from a vertically oriented pyranometer (Weathertronics model 3020), which was alternately turned to face either the wall or the surroundings for a range of solar azimuths and altitudes. Hourly surface temperature measurements of building walls, roofs and surroundings were recorded manually from 6 AM to 11 PM using a hand-held infrared thermome-

344 — Chapter Eighteen

ter (IRT) (Everest Interscience Model 112, 15° field of view) oriented within 20° of the surface normal. Surface temperature results reported prior to 6 AM are estimates based on measurements from similar days when the IRT was mounted to view a single surface for a 24-hr period. Reported ground temperatures are averages of two IRT measurements of sunlit areas, one on the north and one on the south side of each model.

Building Heat Gain Components

Building microclimate is characterized by solar radiation, net-long wave (terrestrial) radiation between a structure and its surroundings, air temperature, relative humidity, and wind speed. Vegetation modifies the microclimate through processes such as shading, evaporation with its resulting cooling effect, and wind speed modification. These processes directly influence the primary paths for heat transfer between buildings and the environment: conduction through walls, roof, doors (opaque conduction) and windows (glazed conduction); solar radiation through windows (glazed solar); sensible and latent heat gain from air leakage (infiltration); and internal gain from sources such as lights and electrical appliances. In the analysis that follows, the emphasis will be on evaluating how differences between TURF and ROCK microclimates influence heat transfer to produce the observed differences in cooling load.

Conduction. The total energy incident on an opaque exterior building surface available for conduction into the interior (q_{to}) is the sum of shortwave radiation, convection, and longwave radiation, respectively, written as

$$q_{to} = \alpha I_T + h_o(T_o - T_s) - \varepsilon_s \Delta R, \tag{1}$$

where

α = absorptance for solar radiation,
I_T = total incident solar (shortwave) radiation,
h_o = the convective heat transfer coefficient,
T_o and T_s = ambient air and building exterior surface temperature,
and ε_s = surface emissivity.

Net longwave radiation for the building surface radiating as a black body at air temperature (ΔR), is given by

$$\Delta R = L_{in} - \sigma T_o^4, \tag{2}$$

where

L_{in} = longwave irradiance from the surroundings,
and σ = the Stefan-Boltzmann constant.

Due to the temperature contrasts expected between treatments, h_o was modified to account for the resulting differences in longwave radiation regimes. This step was necessary because a portion of h_o is actually the result of heat transfer by longwave radiation due to the way in which heat transfer coefficients were empirically determined (Walton 1983).

Sol-air temperature (T_e) is used to represent outdoor conditions for calculation of opaque conduction. T_e incorporates the effects of both solar and longwave radiation into an effective air temperature that allows the radiative components of heat transfer to opaque building surfaces to be treated as convective transfer, so that equation (1) can be written as

$$q_{to} = h_o(T_e - T_s). \tag{3}$$

Combining (1) and (3), T_e can be expressed in terms of the surface energy balance as

$$T_e = T_o + \frac{\alpha I_T}{h_o} - \frac{\varepsilon_s \Delta R}{h_o}. \tag{4}$$

This formulation allows a simplified analysis of the transient heat flow equation. Separate values for T_e must be computed for each building surface for each hour, since I_T is a function of the position of the sun in relation to the orientation of each surface. Air temperature is used for determination of glazed conduction, since transmission and absorptance of solar radiation are treated separately for glazing.

Infiltration. Infiltration rate (I, air changes per hour) was calculated using the empirical expression (ASHRAE 1985)

$$I = k_1 + k_2(T_o - T_i) + k_3 U, \tag{5}$$

where

T_i = indoor air temperature,
U = windspeed,
and k_i's = constants chosen here for loose construction.

Infiltration rate from equation (5) is for full-sized buildings, and since surface-to-volume ratio for the scale models was four times that of full-scale buildings, infiltration rate was scaled up by a factor of four. Latent heat load was estimated to be a constant fraction of sensible heat load, a value of 20% being appropriate for an arid climate (ASHRAE 1985).

346 — Chapter Eighteen

Glazed Solar and Internal Loads. Solar radiation transmitted and absorbed by glazing were calculated according to ASHRAE (1985). A standard absorptance of 0.06 for single-strength glass was used, while transmittance was estimated from vertically oriented pyranometer measurements immediately inside and outside the windows. Internal loads were the same for each treatment, consisting of single 100-W light bulbs.

Cooling Load Simulation

Cooling load was determined by combining heat gain components using the transfer function method (ASHRAE 1985; McQuiston and Parker 1988). This two-step procedure uses transfer functions to represent the effect of thermal storage on heat gain and cooling load. In the first step, heat gain at interior surfaces is computed, while in the second the transfer of this heat to the room air as cooling load is determined. This method is strictly limited to cases in which T_e is calculated with a constant h_o of 17 $Wm^{-2}\,°C^{-1}$. However, the bulk of the longwave effect on T_e is contained in the ΔR terms, and calculated h_o's were close to 17, so that final results were minimally affected by the choice of h_o.

The various energy budget calculations and the transfer function models for heat gain and cooling load were implemented using Lotus 1-2-3 spreadsheets, one of which computes the sol-air temperatures for each surface of each building for input into the second spreadsheet, where heat gain and cooling load are calculated using the transfer function method. Energy use comparisons were done in terms of electrical use for air-conditioning, which was determined from the measured kWhs by subtracting known power used by other loads, in this case lights and air conditioner fans. No other electrical loads were present. Cooling load computed from the simulations was converted to electrical load using the manufacturer's supplied energy efficiency ratio (EER) of 7.2, EER for the TURF air conditioner was reduced by approximately 8% based on comparisons of energy use when landscape treatments were identical (rock ground cover, no shade).

Simulated energy use in early morning hours was slightly negative for both treatments, especially for TURF, from heat loss due to the inside temperature falling slightly below the thermostat setpoint. Since the model assumes constant inside temperature, cooling represented by these low temperatures was treated as storage of cooling energy and subtracted from the modeled cooling load once inside temperatures rose back to the setpoint later in the morning.

References

ASHRAE 1985. *ASHRAE Handbook,* Vol. on *Fundamentals.* Atlanta: American Society of Heating, Refrigerating and Air-Conditioning Engineers.

Goldschmidt, V. 1986. "Average Infiltration Rates in Residences: Comparison of Electric and Combustion Heating Systems." In *Measured Air Leakage of Buildings,* edited by H. Trechsel and P. Lagus. Philadelphia: American Society for Testing and Materials, 70–98.

McPherson, E. 1990. "Modeling Residential Landscape Water and Energy Use to Evaluate Water Conservation Policies." *Landscape Journal* 9(2): 122–34.

McPherson, E., J. Simpson, and M. Livingston 1989. "Effects of Three Landscape Treatments on Residential Energy and Water Use in Tucson, Arizona." *Energy and Buildings* 13: 127–38.

McQuiston, F., and J. Parker 1988. *Heating, Ventilating, and Air Conditioning Analysis and Design.* 3rd edition. New York: Wiley.

Parker, J. 1983. "Landscaping to Reduce the Energy Used in Cooling Buildings." *Journal of Forestry* 81(2): 82–84.

Taha, H., H. Akbari, A. Rosenfeld, and J. Huang 1988. "Residential Cooling Loads and the Urban Heat Island—The Effects of Albedo." *Building and Environment* 23: 271–83.

Walton, G. 1983. *Thermal Analysis Research Program Reference Manual.* National Bureau of Standards report NBSIR 83-2655. Springfield, Va.: National Technical Information Service.

James R. Simpson is an assistant professor in the Department of Soil and Water Science at the University of Arizona. His research interests include effects of microclimate on urban water and energy conservation, and determination of plant water use. He received a B.S. in Physics at Pacific Lutheran University and a Ph.D. in Forest Meteorology from the University of Washington.

Chapter 19

Economic Modeling for Large-Scale Urban Tree Plantings

E. Gregory McPherson, *Northeastern Forest Experiment Station, U.S. Department of Agriculture Forest Service*

Introduction

Citizens in communities throughout the United States are beginning to plan and implement large-scale reforestation projects. The goal of the American Forestry Association's Global ReLeaf program is to plant 100 million trees in U.S. cities. The momentum generated by this campaign is expected to be accelerated by President Bush's America the Beautiful Community Tree Program. Tree plantings are advocated as a means to conserve energy and improve environmental quality. However, relatively little data exist to evaluate the economic and ecologic implications of different planting strategies. This chapter describes an economic-ecologic modeling approach applied to the Trees for Tucson/ Global ReLeaf (TFT/GR) reforestation program, which proposes planting 500,000 desert-adapted trees before 1996. This model estimates selected urban forest benefits and costs; it can be used by any community to evaluate a proposed investment in tree planting.

Urban forest valuation techniques emphasize the current capital asset value of the stock of standing biomass without including management costs (Franks and Reeves 1988; Neely 1988). However, to evaluate the economic and ecologic impacts of different planting scenarios over time, one must account for changes in tree management costs and for benefits accrued over the time period covered by the scenario. Miller (1988) modeled the costs and benefits of street trees using a computer program wherein specified management actions directly impact the condition and value of trees.

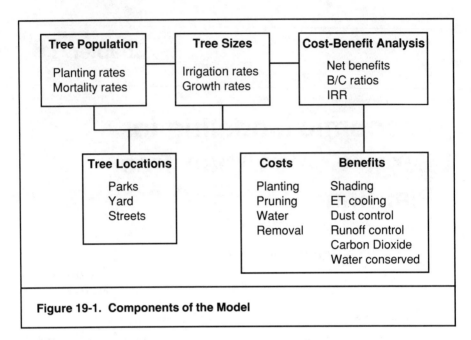

Figure 19-1. Components of the Model

The present study presents a model that incorporates specific urban forest benefits associated with off-street as well as on-street trees. This model assumes that urban trees can substitute for technology in cities by providing air cooling, carbon dioxide reduction, and rainfall and dust interception (Merriam and Gilliland 1981; Coughlin and Strong 1983; Rowntree 1986).

A microcomputer spreadsheet program is used to project average annual benefits and costs for five-year time periods that span a 40-year planning period (1990–2030). Benefits and costs are calculated for the midpoint of each five-year period and assumed to be uniform throughout each period. All trees are assumed to be planted during the first five years and not to be replaced if lost.

Components of and Inputs to the Model

As shown in Figure 19-1, the model has five components: tree population, tree size, tree location, costs/benefits, and cost-benefit (c-b) analysis. Inputs to the first four components are used to produce results that then serve as inputs to the fifth component, the c-b analysis.

Tree Location and Population

Tucson, Arizona, is a rapidly urbanizing city of 156 square miles and 404,000 residents. It is located within the hot, arid Sonoran Desert and receives an average of 11 inches of rain a year. Vegetation cover has diminished in recent years because of citywide efforts to conserve water (McPherson and Haip 1989). Although desert landscapes reduce water use, the overall reduction in vegetation reduces community attractiveness and may accentuate urban warming and other environmental problems. To improve environmental quality and conserve natural resources, the community-based Trees for Tucson/Global ReLeaf (TFT/GR) program helps citizens reforest neighborhoods with desert-adapted trees. It is estimated that planting 500,000 trees by 1996 will increase tree numbers from 1.25 million to 1.75 million and tree canopy cover from 20% to 30% when the trees reach maturity.

Planting and management costs, growth rates of the trees, and benefits from this planting will depend on where trees are planted within Tucson. Benefits and costs are simulated for three tree locations. The model also assumed that 25% of the trees would be planted in areas that receive professional care, such as parks, schools, and commercial landscapes. The highest survival and growth rates are expected for these trees, hereafter referred to as park trees.

The second location is in residential yards. Because TFT/GR encourages homeowners to plant their own shade trees, 60% of all trees are assumed to be planted in yards. Yard trees are expected to receive less intensive maintenance than park trees.

TFT/GR is also working with neighborhoods to plant trees along residential streets. Hence, 15% of all trees are assumed to be for roadsides. Slowest growth rates and highest mortality are anticipated for these trees because the city currently prohibits irrigation systems along roadsides. Maintenance of the trees is left to residents adjacent to them, and residents seeking planting permits are required to sign a maintenance agreement.

Tree population is calculated from the number of trees cited above at each location during each five-year period based on planting rates and expected tree mortality.

Mortality. Vandalism, damage from vehicles, improper planting and maintenance, and storm damage are examples of factors likely to influence life span and loss rates for trees in Tucson. Therefore, the assumed life span of the mesquite has been reduced from over 100 years in the desert to 60 years in the city.

Three types of mortality are projected for trees at each location:

352 — Chapter Nineteen

type A—transplanting-related losses of young trees; type B—age-independent losses, constant over time, due to weather, site modifications, and other factors; type C—losses caused by aging (Richards 1979, Table 1).

Tree Size

The tree size component calculates total leaf area for each location and time period using data on tree population and projected growth rates.

Tree Type and Leaf Area. All planted trees are assumed to be similar to the "typical tree," a native velvet mesquite (*Prosopis velutina*), popular because of its rapid growth, drought tolerance, and moderately dense shade. Mature crown size is assumed to be 25 feet tall and wide. A leaf area index (LAI) of 3 is assumed, based on preliminary research data from an open-grown mesquite tree in a Tucson park. Leaf area (LA) is calculated using a ground projection (GP) term, where GP is the area under the tree crown dripline:

$$LA = LAI \times GP$$

A unique aspect of this study is linking benefits and costs to leaf area because many benefits and costs increase as leaf surface area increases. The dollar value of each benefit and cost for a mature mesquite tree is divided by the total leaf area (1,473 sq ft) to derive values per square foot. Benefits and costs are assumed to be linearly related to leaf area, which may not always be true (for example, removal costs may increase nonlinearly when more expensive equipment is required to remove larger trees).

Growth Rates and Irrigation Water Costs. Calculated growth rates for trees in each location depend upon estimated irrigation rates. Potential evapotranspiration rates (PET) for low-water-use plants are modified to account for the anticipated effects of deficit irrigation (Sacamano undated). Although desert trees such as mesquite require ample irrigation during their first years, they can perform reasonably well with little supplemental irrigation after establishment. However, growth rates will slow as drought stress increases.

Reductions in water use are assumed to be least for park trees, which are irrigated regularly. Reductions are assumed to be greatest for street trees because most will be infrequently watered with hoses by neighborhood residents. The situation for yard trees is intermediate. Trees receiving ample irrigation (100% of PET) are assumed to grow 3 ft/yr horizontally and vertically, while trees receiving 50% and 15% of PET are projected to have annual growth rates of 2 ft/yr and 1 ft/yr, respectively. Assumed irrigation rates and tree sizes are shown in

Economic Modeling for Large-Scale Urban Tree Plantings — 353

Table 19-1 for each location and time period. Annual water use and cost are projected using crown diameter, irrigation rate, and local water price ($0.002/gal) with a model previously developed at the University of Arizona (1976).

Costs

Planting, pruning, and removal costs are estimated from information obtained from local landscape professionals. Costs for disease and pest control are not included because most desert-adapted trees are resistant to these problems. Similarly, other tree care costs are omitted because of infrequent use and limited cost. Although some liability and public-health costs—such as property damage and allergies from pollen—accrue as a result of tree planting, these types of expenditures are difficult to quantify and are therefore not considered.

Unit Planting Costs. This model assumes that 500,000 trees in five-gallon containers are planted between 1990 and 1996. Average planting costs per tree range from $12 to $20 depending on location (Table 19-2). It is assumed that all yard and street trees are planted by residents.

Unit Pruning Costs. Pruning costs are based on the anticipated pruning frequency and costs for a mature mesquite. Pruning frequency refers to the percentage of trees expected to be pruned once by a paid professional during the 40-year planning period. Half of all park trees are assumed to be pruned by a professional arborist at an average cost of $250 for a mature mesquite (Table 19-2). Pruning frequencies are assumed to be less for yard and street trees. Pruning costs are assumed to be greater for yard and street trees due to a higher probability of contact with vehicles, power lines, and buildings.

Unit Removal Costs. Removal costs are estimated based on mortality rates and the assumed costs of removal of mature mesquites in different locations. Mature-tree removal costs reflect anticipated location-related conflicts (Table 19-2).

Quantifiable Local Benefits

Numerous benefits are claimed for urban trees. Some of these can be quantified, others cannot. Some benefits accrue on-site to the land owner; others accrue to the local community; others are global (McPherson and Woodard 1989). If a local policymaker is deciding whether to endorse urban reforestation, then arguably only local, quantifiable benefits should be compared with local, quantifiable costs. Thus, the monetary benefits such as reduced global atmo-

354 — Chapter Nineteen

Table 19-1. Projected Average Annual Irrigation Rates, Tree Crown Sizes, and Mortality Rates for Each Five-Year Time Period

Years	1990–94	95–99	00–04	05–09	10–14	15–19	20–24	25–29
Park								
% PET[a]	100	40	20	20	20	20	20	20
Size (ft)	12	24	25	25	25	25	25	25
% Mortality[b]								
Type A	1.0							
Type B	0.5	0.5	0.5	0.5	0.5	0.5	0.5	0.5
Type C							0.5	1.0
% Loss Rate	1.5	0.5	0.5	0.5	0.5	0.5	1.0	1.5
Yard								
% PET	75	26	15	15	15	15	15	15
Size (ft)	11	20	25	25	25	25	25	25
% Mortality								
Type A	1.0							
Type B	0.6	0.6	0.6	0.6	0.6	0.6	0.6	0.6
Type C						0.5	1.0	2.0
% Loss Rate	1.6	0.6	0.6	0.6	0.6	1.1	1.6	2.6
Street								
% PET	50	15	15	15	15	15	15	15
Size (ft)	10	16	21	25	25	25	25	25
% Mortality								
Type A	4.0							
Type B	1.0	1.0	1.0	1.0	1.0	1.0	1.0	1.0
Type C					0.5	1.0	2.0	3.0
% Loss Rate	5.0	1.0	1.0	1.0	1.5	2.0	3.0	4.0

Notes:
[a] PET = percent of potential evapotranspiration rate
[b] Type A = Transplanting-related losses
Type B = Age-independent losses
Type C = Aging-related losses

Economic Modeling for Large-Scale Urban Tree Plantings — 355

Table 19-2. Location-Specific Assumptions for Modeling Costs and Benefits

Planting Location	Planting Cost ($/tree)	Pruning Cost ($/tree)	Pruning Frequency %	Removal Cost ($/tree)	% Bldgs w/AC	% Shade Effic.
Park	20	250	50	450	100	25
Yard	12	350	25	550	50	66
Street	15	300	15	350	50	25

spheric carbon dioxide and reduced water consumed at off-site power plants are not considered because these benefits are remote. Improved aesthetics, increased urban wildlife habitat, reduced human stress, and increased leasibility of commercial property are not considered because of valuation problems. Effects of trees on property values and the value of sales of trees by local nurseries also are not considered because of problems of double-counting. Excluding these costs and benefits, monetary benefits are then estimated for cooling energy savings due to shade on buildings and reduced air temperatures; implied values for dust control and storm water runoff control are also calculated.

Air-conditioning Energy Savings from Direct Shading and Evapotranspirational Cooling. Estimates of air-conditioning (AC) energy savings incorporate the direct effects of tree shade on buildings and the indirect effects of evapotranspirational (ET) cooling from trees on air temperatures (Huang *et al.* 1987).

Direct cooling savings are projected by estimating the potential AC energy savings from a mature mesquite tree, then applying reduction factors to account for less than maximum shading and for shading buildings with and without AC. The potential annual AC savings from a mature mesquite shading the west wall of a well-insulated Tucson home is calculated as 250 kWh (about 5% of a total AC costs) based on previous computer simulation results (McPherson 1990). Assumed reduction factors for the percentage of homes with air-conditioning (% AC) and percentage shading efficiency (% SE) are shown in Table 19-2. Direct energy savings are calculated using the 1988 electricity sales price to residential customers of $0.072/kWh.

Indirect effects of ET cooling are calculated by first multiplying Tucson Electric Power's (TEP) 1989 electricity sales for each end-use sector by the percentage of sales used for AC:

356 — Chapter Nineteen

Residential = 2,000 GWh × 17% AC = 340 GWh
Small commercial = 1,193 GWh × 35% AC = 418 GWh
Large users = 1,678 GWh × 17% AC = 285 GWh
(Jon Guenther, TEP, personal communication, Oct. 9, 1989)

We then estimated the impact of trees on drybulb temperature depression. Tucson's afternoon summertime temperatures appear to be increasing at a rate typical for many U.S. cities, about 1°F per decade (Balling and Brazel 1987). Studies indicate that increasing tree canopy cover by 10% reduces drybulb temperatures by as much as 4–6°F (Myrup 1969; McGinn 1982). Planting 500,000 trees will increase the Tucson canopy cover by approximately 10%, and it was conservatively assumed that this increase would reduce urban heat island warming by 3°F. Computer simulations for typical residential buildings in Tucson indicate that this 3°F temperature reduction may lower annual air-conditioning energy consumption by 21–25% (846–1,263 kWh) compared to a no-planting scenario. Thus, a conservative potential AC energy savings of 20% is assumed for the residential sector. Values of 12% and 5% are applied for small commercial and large users based on values used for a similar analysis (Akbari et al. 1988). Using these figures, the maximum potential indirect AC energy savings from 500,000 trees is calculated by sector to be 68 GWh, 50 GWh, and 14 GWh for residential, small commercial, and large users, respectively.

Implied Values for Reducing Airborne Particulates. Programs aimed at reducing airborne particulates in the Tucson area include paving dirt roads and switching from diesel to compressed natural gas in buses. Paving one million square yards of unpaved roads within the city limits will cost about $0.78/sq yd/yr, when paving and maintenance costs are averaged over the 40-year planning period (Mary Lou Arbaugh, City of Tucson Transportation Dept., personal communication, Dec. 11, 1989). However, paving roads generates benefits other than dust control; assigning 80% of the paving costs to dust suppression gives an average annual paving cost of $0.63/sq yd. Annual dust control costs through paving are $0.12/lb because each square yard of unpaved road produces about 5.2 lb of particulates (PAG 1988).

The annual mass of particulates that trees remove from the air is estimated to be between 42 lb and 400 lb per tree (Johnson and Baker 1990). Data on particulate removal by mature desert trees have yet to be developed, so a conservative annual removal rate of 40 lb per mature mesquite tree is adopted.

Implied Values for Reducing Storm Water Runoff. Urbanization increases the land area that is covered with impermeable surfaces, which increases the incidence and severity of flooding. One means of

Economic Modeling for Large-Scale Urban Tree Plantings — 357

controlling storm water runoff is to construct basins that detain runoff and thus reduce stream flows and flooding potential. The county in which Tucson is located requires construction of on-site detention basins for new development to ensure that off-site flow does not exceed predevelopment rates. It costs about $67,000 to purchase land and to construct and landscape a basin to store one acre-foot of runoff (Tom Nunn, Pima County Dept. of Transportation and Flood Control, personal communication, Sept. 28, 1989). The annualized cost of detention basins is $0.0025/gal when construction and maintenance costs are averaged over 40 years. The canopy of a mature mesquite tree can store about 3 gal of rainwater, which ultimately evaporates (Aston 1979). More significantly, trees planted in accordance with principles of rainwater harvesting provide miniature catchment basins. Trees planted in 8-ft-wide basins 4 inches deep with runoff directed into them provide about 125 gal of storage. Basins of 8-ft, 6-ft, and 4-ft widths are assumed for park, yard, and street trees, respectively.

Non-Community Benefits

Carbon Dioxide Reduction. Urban trees can reduce atmospheric carbon dioxide directly by assimilating it during photosynthesis and indirectly by reducing carbon dioxide emissions from power plants (Akbari *et al.* 1988). It is estimated that mature mesquite trees sequester 13 lb of carbon per year. TEP power plants produce about 0.9 lb of carbon in the form of carbon dioxide per kWh of power produced. Hence, total conserved carbon is calculated as the sum of avoided power plant emissions and carbon dioxide sequestered in tree biomass.

Water Conserved. Approximately 0.6 gal of water is used at TEP power plants to produce 1 kWh of electricity (pers. comm., Jon Guenther, TEP, Oct. 9, 1989). Trees that reduce power production thus reduce water consumption. However, TEP's marginal power supply is generated in plants in northern Arizona and New Mexico, making the water savings insignificant to the local community. Nonetheless, the extent to which this conserved water offsets the water consumed by evapotranspiration is of interest.

Cost-Benefit Analysis

Several types of cost-benefit analysis are used to evaluate the proposed planting project. Net benefits are calculated for the entire 40-year planning period. Benefit-cost ratios are calculated for each location and time period to compare the temporal and spatial aspects of the proposed investment. For instance, many of the tree planting and management costs are incurred early on, while the benefits grow with the

358 — Chapter Nineteen

trees. Because benefit-cost ratios do not incorporate the time value of money, an internal rate of return (IRR) is calculated for the community investment in trees planted at each location. The IRR is the interest rate that equates the present value of the cash flow series to the initial investment. Net present values are not calculated because selecting a discount rate is problematic when public, private, and corporate entities are involved.

Results

Projected Tree Numbers and Leaf Area

Tree numbers are projected to increase rapidly but never reach 500,000 due to transplanting-related losses of about 10% per year (Figure 19-2). Tree numbers decline at a slow and steady rate from 1955 to 2020 because no replacement planting is assumed. Loss rates increase during the fourth decade due to increased age-related mortality. Forty-three percent (roughly 215,000 trees) of the 500,000 trees planted are projected to die by the year 2030. Total leaf area is projected to increase more slowly than tree numbers because 10–15 years are required for trees to reach full size. Once all trees reach full size, leaf area gradually decreases due to mortality.

Projected Management Costs

Planting costs averaged over the 40-year period range from $0.30 to $0.50/tree/yr (Figure 19-3). Pruning and water costs range from about $1 to $3 per tree per year. The projected average annual water use is estimated to be 1,071 gal/tree, or $2.14, about the same amount of water used inside the home by a single person for ten days. Projected water and pruning expenses for the more intensively managed park trees are nearly twice as great as projected expenses for the street trees maintained by adjacent homeowners. Tree removal is the most significant expense, with average annual costs ranging from $3.11 to $6.63 per tree. Per tree removal costs are greater for yard and street trees than park trees due to higher mortality rates for full-sized trees.

Average annual costs for each five-year time period depict how the demand for management resources are expected to vary with time (Figure 19-4). Average annual expenses during the first five years are projected to be about $3.5 million, primarily due to large one-time planting costs. Total average annual expenses drop to about $3 million for the next 20 years, but are projected to increase rapidly to over $5 million annually by the year 2028 due to increased removal costs.

Because pruning costs are assumed to be directly linked to leaf

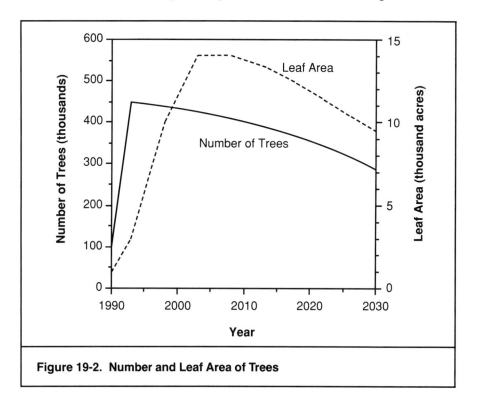

Figure 19-2. Number and Leaf Area of Trees

area, projected expenses (Figure 19-4) mirror the leaf area curve shown in Figure 19-2. Removal costs gradually increase with time as trees grow larger. Increased mortality of mature trees accounts for higher projected expenditures for removal during the last 15 years. Water costs follow a pattern similar to that projected for pruning, except initial costs are higher. High irrigation rates are projected for the establishment period, and offset the effect of small tree size on total water demand. Although irrigation rates will diminish from 1995 to 1999, rapid increases in tree size are projected to increase total water costs compared to the previous five-year period. Water costs will gradually decrease during the remaining 30 years as leaf area diminishes and irrigation rates remain constant.

The projected costs per year of managing trees in the three types of locations are shown in Figure 19-5. Annual management costs for park trees range from $8 to $10 per tree during the first 20 years and are greater than costs for yard or street trees due to substantial expenses for pruning and water. Relatively greater mortality rates for

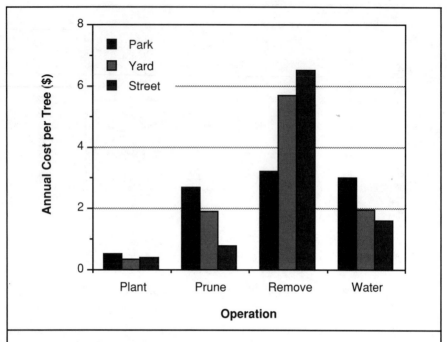

Figure 19-3. Average Annual Cost per Tree over 40-Year Period (by cost component and location)

mature yard and street trees during the last 20 years result in annual management costs as high as $20 per tree.

These data reflect the modeling assumption that funds spent initially to promote tree establishment, rapid growth, and strong crown structure can prolong the serviceable life of a tree. Annual management costs averaged for the entire 40-year period are smallest for park trees ($9.28) and greatest for yard trees ($9.87), with an overall average of $9.62 per tree (Table 19-3).

Projected Energy Savings and Environmental Benefits

Average annual cooling energy savings from direct shade for all trees is 61 kWh per tree, and average annual benefits are projected to range from $1.74 to $5.07 per tree depending on location (Figure 19-6). Yard trees provide the most shade to buildings and hence the greatest savings. The greatest cooling energy benefits result from evapotranspirational cooling effects, and average 227 kWh per year per tree. ET cooling benefits ranged from $15 to $17 and are about three times

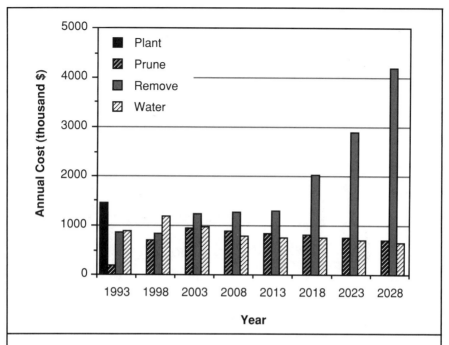

Figure 19-4. Average Annual Costs for All Trees at Five-Year Intervals (by cost component)

greater than direct energy savings from shade. This finding agrees with results from another computer simulation study for a single-family residence in Phoenix (Huang et al. 1987). In that study, 80% of the total cooling energy savings are attributed to ET cooling. Park trees provide the greatest ET cooling benefits due to high irrigation rates and rapid growth. Total average annual cooling energy savings for all trees is projected to be 288 kWh ($20.74 per tree, with yard and park trees providing average annual benefits exceeding $21 per tree).

Projected average annual particulate control benefits range from $3.81 to $4.35 per tree, and storm water control benefits range from $0.06 to $0.29 per tree (Figure 19-6). For all trees, the average annual implied value for dust and storm water control is $4.16 (34.7 lb) and $0.18 (73 gal) per tree. Implied values for dust and storm water runoff control vary little across locations.

Carbon savings averaged 408 lb annually per tree. Mature trees are estimated to each conserve 477 lb of source carbon per year. Hence, 97% (464 lb) of the total carbon conserved by a mature mesquite tree

362 — Chapter Nineteen

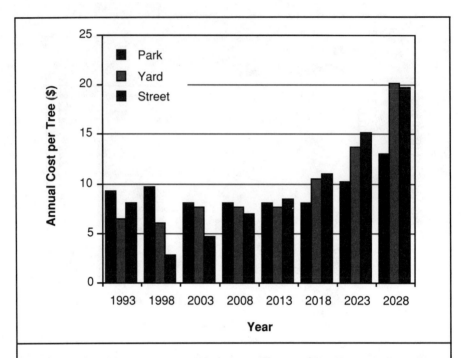

Figure 19-5. Average Annual Costs per Tree at Five-Year Intervals (by location)

Table 19-3. Projected Benefits and Costs

Location	Costs (MM$) Annual/Tree	40-Yr Total	Benefits (MM$) Annual/Tree	40-Yr Total	Net (MM$) Benefits	B/C	IRR %
Park	9.28	38.9	25.68	106.3	67.4	2.74	5.47
Yard	9.87	91.4	25.69	241.9	150.6	2.65	14.43
Street	9.54	15.7	20.59	34.3	18.6	2.18	2.00
All Trees	9.61	145.9	25.09	382.5	236.6	2.62	7.11

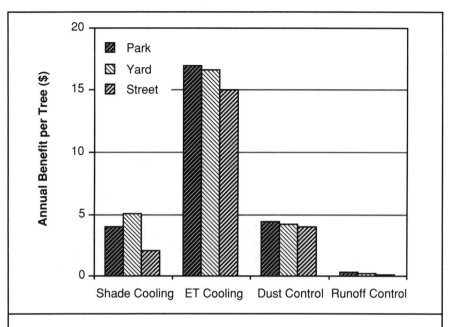

Figure 19-6. Average Annual Benefits per Tree over 40-Year Period (by component and location)

is attributed to reduced power plant emissions resulting from tree shade and ET cooling. Water conserved at the power plant due to reduced electricity demand is calculated to average 171 gal annually per tree, or 16% of each tree's average annual water consumption.

The stream of savings associated with each functional benefit (Figure 19-7) follows the trend of rapidly increasing and then gradually decreasing leaf area shown in Figure 19-2. Total annual benefits are projected to exceed $10 million from the years 2000 to 2025.

Projected annual benefits per tree range from $4.22 for young street trees to $30 for mature yard trees (Figure 19-8). Rapid growing park trees provide the greatest benefits during the first decade. Once trees mature, greater savings are projected from yard trees because of increased building shade and air-conditioning energy savings. Street trees provide less benefit than yard or park trees initially, because deficit irrigation is assumed to result in slower growth rates and less leaf area. Annual monetary benefits from mature street trees are projected to average $26.25 per tree. This figure is less than for mature yard and

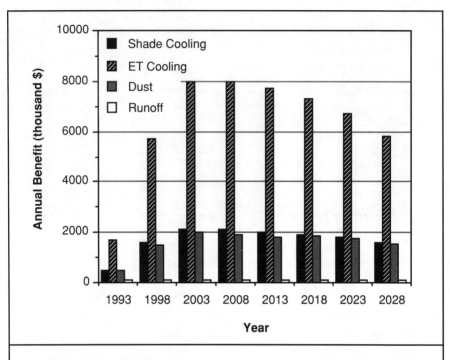

Figure 19-7. Average Annual Benefits for All Trees at Five-Year Intervals (by component)

park trees because of assumed locational differences in direct shade on buildings. When annual benefits per tree are averaged over the entire 40-year period, yard and park trees provide the largest benefit ($25.69, $25.68), street trees provide the least benefit ($20.59), with an overall average benefit of $25.09 per tree (Table 19-3).

Projected Benefit-Cost Ratios and Internal Rate of Return

Projected total benefits exceed total costs by $236.5 million for the 40-year period (Table 19-3). Sixty-four percent ($150.6 million) of net benefits are projected for trees in yards, where 60% of all trees are assumed to be planted. The ratio of benefits to costs for all trees is 2.62, indicating that benefits are over 2.5 times greater than costs. The benefit-cost ratio is largest for park trees (2.75) and smallest for street trees (2.18).

An internal rate of return (IRR) of 7.1% is calculated for all trees (Table 19-3). The IRR for yard trees is 14.4%, largely because

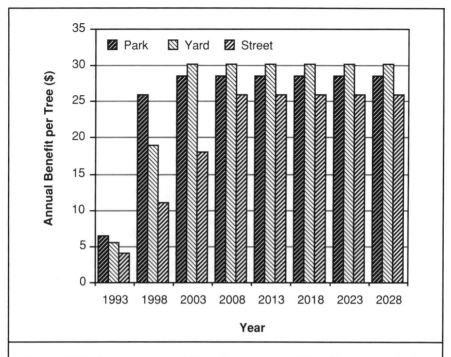

Figure 19-8. Average Annual Benefits per Tree at Five-Year Intervals (by location)

of the relatively small initial investment in planting costs on a per tree basis ($12). The IRR for street trees is only 2% because high establishment-related mortality and slow growth rates yield small functional benefits, despite relatively large initial expenditures for planting and management.

The projected annual stream of benefits and costs shows that costs exceed benefits during the first five years, largely due to one-time planting expenses (Figure 19-9). However, for the next 25 years, projected benefits are three or more times greater than costs. During the last decade, costs begin to catch up with benefits as the end of the serviceable life of the trees grows near. If one extrapolates this trend for the next decade, costs will begin to exceed benefits.

Conclusions

Modeling of selected benefits and costs associated with Trees for Tucson/Global ReLeaf program suggests that energy savings, dust con-

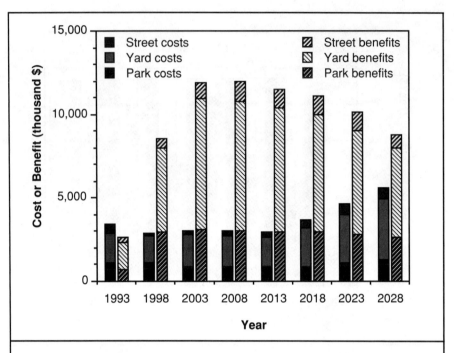

Figure 19-9. Average Annual Benefits and Costs at Five-Year Intervals (by location)

trol, and storm runoff detention benefits may outweigh tree planting and maintenance costs. Although the homeowner can obtain substantial cooling energy savings from direct building shade, greater benefits accrue to the community as a whole due to cooling caused by evapotranspiration. Public investment in tree planting may be warranted because economic, health, and aesthetic benefits extend beyond the site where individual trees are planted. Greatest net benefits can be expected from trees planted in parks and yards. Finally, substantial global benefits accrue as the trees reduce atmospheric carbon dioxide directly through the tree biomass and indirectly by reducing fossil fuel consumption. The estimated indirect effects are over 30 times the direct effects. Thus, planting trees in urban areas, though expensive, may be highly effective at reducing atmospheric carbon dioxide.

Modeling results are only as reliable as the data used to generate the findings. The research basis for this type of economic-ecologic modeling is paltry. Modeling limitations can be reduced through research in the following areas:

1. Developing rapid and accurate means to estimate leaf area of trees and explore relations between leaf area and functional benefits.

2. Obtaining more data on tree mortality rates and how these rates vary with location.

3. Developing better tree growth models that account for site conditions, water use, irrigation practices, and other factors.

4. Developing more accurate estimates of relations between local climate, vegetation structure, and building cooling energy, and landscape water use.

5. Measuring dust, solar radiation, and rainfall interception rates for commonly used landscape plants.

6. Considering incorporating other important costs (liability and increased winter shade) and benefits (effects on property values, peak energy use, and water demand effects) into the model.

7. Incorporating benefits and costs for existing trees and replacement plantings into the model.

Despite the lack of a well-researched data base for modeling, the approach described here offers decision-makers a timely and relatively sophisticated tool for evaluating the economic and environmental implications of proposed tree plantings. To improve this tool, studies should begin to monitor and document the effects of new tree plantings.

References

Akbari, H., J. Huang, P. Martien, L. Rainer, A. Rosenfeld, and H. Taha 1988. "The Impact of Summer Heat Islands on Cooling Energy Consumption and CO_2 Emissions." In *Proceedings from the ACEEE 1988 Summer Study on Energy Efficiency in Buildings*. Washington, D.C.: American Council for an Energy-Efficient Economy.

Aston, A. 1979. "Rainfall Interception by Eight Small Trees." *Journal of Hydrology* 42: 383–96.

Balling, R., and S. Brazel 1987. "Temporal Variations in Tucson, Arizona, Summertime Atmospheric Moisture, Temperature, and Weather Stress Levels." *Journal of Climate and Applied Meteorology* 26: 995–99.

Coughlin, R., and A. Strong 1983. *Forests, Fields, and Urban Development: Planning as Though Vegetation Really Mattered.*

Research Report Series No. 2. Philadelphia: University of Pennsylvania, Department of City and Regional Planning.

Franks, E., and J. Reeves 1988. "A Formula for Assessing the Ecological Value of Trees." *Journal of Arboriculture* 14(10): 255–59.

Huang, Y., H. Akbari, H. Taha, and A. Rosenfeld 1987. "The Potential of Vegetation in Reducing Summer Cooling Loads in Residential Buildings." *Journal of Climate and Applied Meteorology* 26(a): 1103–16.

Johnson, C., and F. Baker 1990. *Urban and Community Forestry*. Ogden, Utah: USDA Forest Service.

McGinn, C. 1982. "Microclimate and Energy Use in Suburban Tree Canopies." Unpublished Ph.D. thesis. Davis: University of California.

McPherson, E. 1990. "Modeling Residential Landscape Water and Energy Use to Evaluate Water Conservation Policies." *Landscape Journal* 9(2): 122–34.

———, and R. Haip 1989. "Emerging Desert Landscape in Tucson." *Geological Review* 9(4): 435–49.

———, and G. Woodard 1989. "The Case for Urban ReLeaf: Tree Planting Pays." *Arizona's Economy*, December, 1–4.

Merriam, A., and M. Gilliland 1981. "How Do We Define 'Appropriate' in Appropriate Technology Applications?" *A Basic Science of the System of Humanity and Nature*. Gainesville: University of Florida, Center for Wetlands.

Miller, R. 1988. *Urban Forestry, Planning and Managing Urban Greenspaces*. Englewood Cliffs, N.J.: Prentice Hall.

Myrup, L. 1969. "A Numerical Model of the Urban Heat Island." *Journal of Applied Meteorology* 8: 896–907.

Neely, D. (ed.) 1988. *Valuation of Landscape Trees, Shrubs, and Other Plants*. Urbana, Ill.: International Society of Arboriculture.

PAG. See Pima Association of Governments.

Pima Association of Governments 1988. *The Group II Committal State Implementation Plan for Particulate Matter—10 (PM-10) for the Tucson Planning Area*. Tucson: Pima Association of Governments.

Richards, N. 1979. "Modeling Survival and Consequent Replacement Needs in a Street Tree Population." *Journal of Arboriculture* 5(11): 251–55.

Rowntree, R. 1986. "Ecology of the Urban Forest—Introduction to Part II." *Urban Ecology* 9(1/2): 229–43.

Sacamano, C. undated. "Estimating Consumptive Water Use by Arid Landscape Plants." Unpublished technical report. University of Arizona, Tucson.

TEP. See Tucson Electric Power Company.

Tucson Electric Power Company 1989. *Tucson Electric Power Company 1988 Annual Report*. Tucson: Tucson Electric Power Company.

University of Arizona 1976. *Water Conservation for Domestic Users*. Tucson: Tucson Water.

E. Gregory McPherson is an urban forest researcher with the USDA Forest Service, Northeastern Forest Experiment Station. He is lead scientist on the Chicago Urban Forest Climate Project. He received a B.G.S. from the University of Michigan, an M.L.A. from Utah State University and a Ph.D. from the State University of New York, College of Environmental Science and Forestry in Syracuse.

Chapter 20

A Synopsis of *Cooling Our Communities: The Guidebook on Tree Planting and Light-Colored Surfaces*

Joe Huang, Susan Davis, and Hashem Akbari, *Lawrence Berkeley Laboratory*

Introduction

In all but a few exceptional cases, cities throughout the world have created heat islands with temperatures noticeably higher than those in the surrounding countryside. This phenomenon is a moderate asset for cities in cold climates, where the heat island raises wintertime temperatures and lowers heating bills. But for cities in temperate or cooling-dominant climates, the heat island increases air-conditioning usage, adds to human discomfort, and exacerbates smog and urban pollution.

Although the characteristics of heat islands are still being investigated by scientists, the main causes are well established. Human-made heat from cars, heating equipment, and machinery plays a significant role in making the winter heat island. But summer heat islands are caused primarily by alterations that change the urban landscape's response to solar radiation.

These alterations fall into four broad categories:

1. Hard surfaces have replaced vegetation and soil, thereby diminishing the city's capacity to cool via evaporation or plant transpiration.

2. The replacement of soil and vegetation with dark buildings has reduced the ability of city surfaces to reflect incoming radiation.

372 — Chapter Twenty

3. The high thermal mass of buildings and pavement have resulted in increased heat retention throughout modern cities.

4. The increase in the number of buildings has lowered wind speeds and thus the rate at which surface heat gain is carried away.

The combination of these effects has resulted in urban temperatures that are, on average, 2–9°F hotter than those in surrounding countrysides. Until recent years, urban developers have not considered the negative effects of summer heat islands on energy use and environmental quality. Recent estimations of energy costs for this human-caused climate change, however, have highlighted the necessity—and possibility—of mitigating it. In the U.S., researchers estimate, each °F of warming may increase the amount of cooling electricity consumed by as much as 1–2%, and the production of low-level ozone by as much as 5%. Together, these factors cost individual cities hundreds of thousands of dollars a year. In Los Angeles, for example, the additional cost of cooling energy caused by heat islands is $150,000 an hour on hot summer afternoons. In Washington, D.C., the added cooling costs reach $40,000 an hour, or $40 million a year (Akbari *et al.* 1989).

Some of the causes of summer heat islands, such as increased thermal mass or surface roughness, are inevitable. Their elimination would require drastic changes in the way cities are built. Other factors, such as the loss of vegetative cover and reduced albedo, are correctable if urban planners and city residents can be convinced of the importance and the benefits of proposed mitigation strategies.

Cooling Our Communities: The Guidebook on Tree Planting and Light-Colored Surfaces is being produced by the Heat Island Project at Lawrence Berkeley Laboratory (LBL/HIP), with contributing authors from around the nation. The project is supported by the Environmental Protection Agency (EPA), Electric Power Research Institute (EPRI), the Department of Energy (DOE), California Institute of Energy Efficiency (CIEE), Los Angeles Department of Water and Power (LADWP), and the University-wide Energy Research Program at the University of California (UERG/UC). The guidebook will bring together the findings to date on this citywide conservation issue and will answer some of the more frequently raised questions.

Background of Project

Cooling Our Communities is a direct outcome of a heat island workshop held in Berkeley in February 1989. This workshop, the first ever devoted solely to the topic of urban heat islands and their energy impacts, brought together more than 80 attendees, representing federal

Cooling Our Communities: The Heat Island Reduction — 373

Table 20-1. Guidebook Contents	
Chapter 1	Causes and Effects of Urban Heat Islands
Chapter 2	Using Trees to Cool Our Communities
Chapter 3	Using Light-Colored Surfaces to Cool Our Communities
Chapter 4	Potential Problems
Chapter 5	Developing Tree Programs
Chapter 6	Planting and Light-Colored Surfaces for Energy Conservation
Chapter 7	Developing Ordinances for Cooling Communities
Chapter 8	The Future Face of Our Cities
	Technical Appendices Resource Guide

and state energy offices, utility planners, meteorologists, foresters, landscape architects, and community activists (Garbesi *et al.* 1989). Such broad-based and enthusiastic participation demonstrated the increased recognition that the summer urban heat island is an energy issue and that public support is growing for citywide campaigns like tree planting.

Papers at the workshop fell into three topic areas: meteorological studies of urban heat island characteristics; field study and computer-simulated analysis of mitigation strategies; and descriptions of actual citywide implementation efforts, such as the proposed planting of a million trees in Los Angeles.

At the close of the workshop, attendees generally agreed that while considerable research still needs to be done, enough exists to support a guidebook for public or private interest groups wishing to embark on programs in this area. The contents of this guidebook are listed in Table 20-1.

Heat Island Mitigation Strategies

The guidebook focuses on two potential strategies for mitigating summer heat islands: (1) increasing the number of urban trees and (2) increasing the urban albedo[1] through the use of light colors on surfaces

---

[1] Albedo is a measure of the ability of a surface to reflect incoming radiation. Albedo differs from "reflectivity" in that it is measured across all wavelengths, rather than just the visible spectrum.

374 — Chapter Twenty

such as roofs, parking lots, and streets. The first measure increases the amount of shading and cooling through plant evapotranspiration. The second measure helps keep urban surfaces cool by reflecting most of the heat from the sun.

Each of these strategies has a direct and an indirect effect. "Direct" refers to those effects that modify the heat exchange of buildings and their surroundings. "Indirect" refers to those that change the surrounding urban conditions themselves. Direct effects accrue only to the immediate area where the strategy is implemented—a single building, for example, or a shaded plot. These benefits are independent of the city conditions as a whole and can be viewed as ways to counteract the heat island by keeping oneself cool. Indirect effects accrue to larger areas—for example, a neighborhood or even a whole city—through the widespread implementation of the same heat island mitigation strategies. These benefits require concerted efforts by a community and can be viewed as ways to reduce or eliminate the heat island effect.

The LBL/HIP has used computer simulations to estimate the potential cooling energy savings from heat island mitigation strategies (Akbari *et al.* 1988). If cooling measures are implemented on a scale large enough to affect urban climate, the potential energy savings from the indirect effects are estimated to be roughly equal to those from the direct effects. Researchers also estimate nationwide savings of 30% in residential, 16% in small commercial, and 5% in large commercial building energy use if these strategies are used in cities throughout the nation. The yearly savings in electricity are estimated at half a quad (quad $= 10^{15}$ Btu).

Increased Number of Urban Trees

The guidebook discusses the location of trees around a house to maximize the shading benefits during the summer, while minimizing unwanted shading during the winter. In most hot climates, it is most effective to place shade trees on the south and west sides of a building and to shade the air conditioner. As a general rule, tall trees should be placed farther away, and short trees close to a building to create the most optimal shade patterns. Trees can also be planted in such a way as to funnel wind and create cooling breezes.

Buildings are the primary targets for tree shading, but parking lots and streets also benefit from this strategy. These trees do not provide direct energy savings, but they contribute indirectly to cooling the entire city and to producing more comfortable outdoor environments. The best planting strategy in these public areas is to create islands of trees rather than to scatter individual plants. Trees clumped together share soil, help keep each other cool, and create a broader shade shape than do individual trees. To reduce maintenance costs, trees that drip or drop sticky berries and leaves should be avoided. To achieve maximum indirect effects, trees should also be planted in parks, along streets and sidewalks, or wherever they can thrive.

Albedo

Although there is no record of any concerted efforts to modify urban albedos, the direct cooling energy savings for houses with light-colored roofs and walls are well documented through field measurements and computer simulations (Griggs *et al.* 1989; Martien *et al.* 1989). Common sense tells us that the primary targets for urban albedo efforts are building roofs, walls, parking lots, and possibly streets. Indeed, the Greeks have been cooling their cities for centuries by whitewashing these surfaces.

Building surfaces are the simplest and cheapest surfaces for albedo modifications. Since many buildings are routinely painted every ten years or so, changing the albedo of walls is simply a matter of substituting light paint for dark. Changing the albedo of roofs and paved areas can be more expensive, because it may require using different materials—a concrete tile is more expensive than a shingle, for instance, and asphalt with light-colored aggregate can be more expensive than the original. Similarly, repainting or resurfacing is quite expensive when done outside the regular maintenance schedule for the sake of energy conservation.

Implementation Issues

Although computer analyses suggest that heat island mitigation strategies have very large energy-saving potentials, many practical questions remain about their implementation and possible pitfalls. The reasons cited for the large potential benefits for the indirect effects also confound their implementation. That is, they are effective only if carried out on a scale large enough to affect the urban climate itself.

Several chapters in the guidebook address the practical aspects of instituting a heat island mitigation program. Almost all these chapters concern tree planting, with very little on albedo modification. Although tree planting programs have the added complexity of working with living plants, the lack of discussion on implementing albedo programs mainly reflects the current state of knowledge and experience. Increasing the urban albedo has been hampered both by the lack of public awareness and by the absence of pilot programs that can be used for developing guidelines. In comparison, the chapters on developing tree planting programs benefit from the experiences of community

376 — Chapter Twenty

groups, the American Forestry Association, and review of over 20 existing programs in different cities.

Approach

A few general rules can be developed, however. Both trees and albedo modification will depend on public education programs and legislation. Public education campaigns can range from Tree Days, Energy Days, or perhaps Light Surfaces Days, to corporate sponsorship, including mailings with utility bills, to broad-based media campaigns. Legislation includes municipal ordinances that would both mandate or encourage tree planting and albedo changes and provide legal authority and support for public education campaigns. An ideal ordinance is flexible in giving property owners leeway in choosing tree species and light colors, for example, and is positive instead of punitive. That is, citizens respond more willingly when the ordinance mandates tree planting rather than levies fines for no trees. State and national legislation is a possibility as well, although political and climatic differences throughout these larger regions may make passage and enforcement of an ordinance difficult.

Public Information

Both public and private agencies have written extensively about developing and sustaining tree planting programs. The American Forestry Association, the U.S. Forest Service, TreePeople, and dozens of local groups are examples of the number and diversity of organizations involved in tree planting. The number of organizations involved, however, does not mean that tree planting programs are easy to implement or sustain. Andy Lipkis, president of TreePeople, says that the most important element in the success of a tree planting program is public recognition that a program is a long-term commitment:

> Because the solution appears somewhat simple, people are rushing to embrace massive planting proposals. But there is a catch to this "solution." The "technology" is a living one requiring extensive ongoing care if it is to work. That is, tree planting is not a technical "fix" that will handle a problem regardless of human action. It mandates an ongoing partnership between people and their environment.

To some degree, this level of public awareness is also needed for other conservation strategies. For example, researchers have found that people who purchase energy-efficient appliances or buildings without adopting an energy-saving lifestyle will continue to be heavy energy users. Tree plantings need even more public involvement because trees are living things that will die from neglect or abuse. Therefore, tree planting programs need to strongly emphasize maintenance after the thrill of planting is gone.

In general, a number of program leaders have found that community participation is key to assuring longevity of urban trees, simply because participation inspires a caretaking attitude. Other factors in successful programs include developing paid professional staff rather than relying solely on volunteers; using computerized tree inventories, which help both with maintenance schedules and budget requests; and using corporate sponsorship to market and publicize the program.

> The American Forestry Association's Global ReLeaf Utility Program is successfully convincing utility companies to plant trees for energy conservation. The Utility Program invites companies to sponsor Global ReLeaf as part of a customer education and community and public relations program. Individual customers learn how to plant trees to save energy and money. Communities are encouraged to develop tree-planting projects—and to support them. The program also gives utility employees background on the savings potentials of trees and strategic landscaping methods, and coaches utility companies on corporate outreach and the development of citizen-based environmental activities.

Although there are no recorded efforts to implement urban albedo measures, similar general rules should also apply. The question of volunteer versus staff may be unimportant, as most surface-changing programs will be implemented by either property owners or local government. But the idea of using press coverage and corporate sponsorship for public education still applies.

Potential Conflicts

For tree planting programs, potential problems areas are water usage and yard waste. The guidebook Appendix C, "Water Constraints in Arid Landscapes," addresses the water usage issue and shows a method for estimating water use in different landscape scenarios. The use of non-native vegetation, especially lawns, in hot, arid locations will greatly boost water consumption, but the use of native trees, shrubs, and ground covers can keep water requirements low while still providing cooling benefits from shading. The indirect effects from plant evapotranspiration in such landscaping scenarios will of course be reduced. A shorter section in Chapter 4, "Trees and Landfills," looks at the problem of increased yard debris from trees and other veg-

378 — Chapter Twenty

etation. The section concludes that composting will prevent the amount of solid waste in landfills from increasing, although the cost of waste management will probably increase.

Albedo modifications present issues not only of cost-benefit but also of aesthetics and safety. Some people simply won't want white houses or white streets. And glare on streets or buildings can impede vision. Here the issue of energy conservation collides with issues of public property, aesthetics, and free choice. Public education on the potential for energy saving may help in this case.

Conclusion

The guidebook represents a compendium of current knowledge about summer urban heat islands and possible mitigation strategies. The primary objective for the book is to heighten public awareness of the heat island issue and to help foster positive citywide programs to combat it. Since the intended audiences are city planners, officials, and community groups, most of the book addresses practical implementation issues.

In the area of urban tree planting, there are numerous active programs and established organizations, both professional and community-based, that can provide guidance and experience. In the area of increasing urban albedo, such information sources do not exist, so that any attempted program must be based on research results and common sense.

Although the two strategies encompass very different "technologies," their implementations both require the involvement of large numbers of people in order to have an impact on the heat island problem. Experiences with tree planting programs suggest that such large-scale efforts must have volunteer community support, professional staffing, and a stable institutional structure.

Numerous technical questions need further research. From a conservation point of view, the energy savings from the indirect effects of heat island mitigation strategies are based on theoretical studies using relatively crude computer models. These effects need to be better assessed and, ideally, corroborated with actual data. The durability of and effects of urban wear-and-tear on high-albedo products and surfaces also need to be monitored and evaluated. The numerous issues surrounding tree survival, water use, and conflicts with other urban requirements constitute an entire research area that falls in the domain of urban foresters and landscape architects.

References

Akbari, H., J. Huang, P. Martien, L. Rainer, A. Rosenfeld, and H. Taha 1988. "The Impact of Summer Heat Islands on Cooling Energy Consumption and CO_2 Emissions." In *Proceedings of the ACEEE 1988 Summer Study on Energy Efficiency in Buildings.* Washington, D.C.: American Council for an Energy-Efficient Economy.

Akbari, H., A. Rosenfeld, and H. Taha 1989. "Recent Developments in Heat Island Studies: Technical and Policy." In *Controlling Summer Heat Islands.* Washington, D.C.: Dept. of Energy CONF-8902142.

Garbesi, K., H. Akbari, and P. Martien 1989. *Controlling Summer Heat Islands—Proceedings of the Workshop on Saving Energy and Reducing Atmospheric Pollution by Controlling Summer Heat Islands.* LBL 27872. Berkeley: Lawrence Berkeley Laboratory.

Griggs, E., T. Sharp, and J. MacDonald 1989. *Guide for Estimating Differences in Building Heating and Cooling Energy Due to Changes in Solar Reflectance of a Low-Sloped Roof.* ORNL-6527. Oak Ridge, Tenn.: Oak Ridge National Laboratory.

Martien, P., H. Akbari, and A. Rosenfeld 1989. *Light-Colored Surfaces to Reduce Summertime Urban Temperatures: Benefits, Costs, and Implementation Issues.* Berkeley: Lawrence Berkeley Laboratory.

Joe Huang is a staff scientist with the Energy Analysis Program at Lawrence Berkeley Laboratory. He was one of the founding members of the LBL Heat Island Project, which has been investigating the potentials of conserving building energy use through ameliorating urban climates since 1985. He has a B.S. from Stanford University and an M.A. from the University of California, Berkeley.

Susan Davis is an independent writer and editor. She edited Cooling Our Communities: The Guidebook on Tree Planting and Light-Colored Surfaces. *She has authored articles on energy, the environment and geography that have appeared in nationwide publications.*

Hashem Akbari is a staff scientist and principal investigator in the Applied Sciences Division at Lawrence Berkeley Laboratory. He is the leader of the LBL Heat Island Project. He received a Ph.D. in Engineering from the University of California, Berkeley.

Index

Abatement costs. *See* Cost-of-control method; Environmental abatement costs
ACEEE. *See* American Council for an Energy-Efficient Economy (ACEEE)
ACEEE 1990 Summer Study on Energy Efficiency in Buildings, 1, 6–7
Acid rain
 and the Clean Air Act, 3, 185
 energy conservation and reducing SO_2 emissions, 289–303
 and energy efficiency, 174
 evaluating control options, 5
 and fish loss in Lake Michigan, 241
 gases contributing to, 252
 impacts of demand-side management programs on, 145
 implementing policies to address, 105
 NEES matrix for weighting and rating contributing factors, 234
 and nitrogen oxides, 225, 236–37
 reducing precursors of, 2
 See also Sulfur dioxide (SO_2)
Aerosols, CFCs in, 101
Aesthetics, NEES matrix for weighting and rating contributing factors, 235
Afforestation, and environmental abatement costs, 261–62. *See also* Reforestation
Africa, projected CO_2 emissions (chart), 14
Agriculture
 energy use by fuel type in the U.K. (chart), 91
 and the greenhouse effect, 72

greenhouse-gas emissions from, 130–31
 and ozone depletion, 306
Ahuja, D., 218–19
Air-conditioning
 CFC use in, 309–13
 chillers, 308n, 309–13, 316–17
 distribution of air-conditioning tonnage and refrigerant type (table), 311, 312
 effect of turf landscaping on, 338, 339, 341
 electricity savings due to vegetation (chart), 332
 energy demands of, 96
 energy-efficient technologies for, 80–81
 energy savings from direct shading and evapotranspirational cooling, 355–56
 halocarbon emissions from, 126
 impact of global climate change on, 172, 175
 impact of vegetative landscaping on, 321–34
 packaged units, 308n
 residential, efficiency in the United States, 81
 shading of air conditioners, 322
 in the U.K., 91, 96
 and urban heat islands, 371
 usage of HCFCs in, 310–13
Air pollution
 emission projection algorithms, 278–80
 emission projection model design considerations, 277–78
 estimating social costs related to, 221
 NEES matrix for weighting and rating contributing factors, 235

381

pollutants produced by fossil
fuels, 193
projection methodologies,
271–87
reducing through energy efficiency, 1–2
See also Pollution; Tropospheric
air pollution
Air quality scenarios project (of the
SCAQMD), 280–87
Air Resources Board (ARB), air pollution emissions projections,
272–74
Air shed models, for projecting air
pollution emissions, 272
Air transportation
estimating greenhouse gases produced by, 128
fuel efficiency improvements in,
135, 138
See also Transportation
Akbari, Hashem, ix, 322, 334, 371,
379
Alabama
electric capacity and energy savings from DSM programs,
149
incorporating environmental
externality costs (table),
202–3
Alaska
electric capacity and energy savings from DSM programs,
149
incorporating environmental
externality costs (table),
202–3
Albedo
defined, 373n
increasing to reduce urban heating, 373–78
Allowance and offset policies for
reducing CO_2 emissions,
164, 175
Allowance pooling, and compliance
with the Clean Air Act
Amendments, 185

Allowance trading. *See* Emissions
allowance trading
Alternative energy
collecting data on, 317
as a means to energy efficiency,
58
Alternative fuels. *See* Alternative
energy
American Council for an Energy-
Efficient Economy
(ACEEE), vii
American Electric Power System
analysis of, 294
benefits of conservation to, 299
American Forestry Association
Global ReLeaf program,
349, 376, 377
American Society of Heating,
Refrigerating and Air-
Conditioning Engineers
(ASHRAE) Standard 90.2,
recommended insulation levels, 107
America the Beautiful Community
Tree Program, 349
Ameritech Corporation, 300
Annual fuel utilization index
(AFUE), 108
Appliance rebates, and reducing CO_2
emissions, 3
Appliances. *See* Electric appliances
ARB/SCAQMD method for air pollution emissions projections,
272–74
Arizona
case study of vegetative landscaping and cooling savings,
325–26, 328, 330
electric capacity and energy savings from DSM programs,
149
incorporating environmental
externality costs (table),
202–3
Trees for Tucson/Global ReLeaf
(TFT/GR) reforestation program, 349–69

Arkansas
  electric capacity and energy savings from DSM programs, 149
  incorporating environmental externality costs (table), 202–3
Arsenic, and the Clean Air Act Amendments of 1990, 224n
ASHRAE Standard 90.2. *See* American Society of Heating, Refrigerating and Air-Conditioning Engineers (ASHRAE) Standard 90.2
Asia
  case studies of energy use and carbon emissions, 24
  projected CO_2 emissions (chart), 14
Atomic bomb testing, and weather change, 40
Australia
  case study of energy use and carbon emissions, 11, 19
  policy analysis of energy use and carbon emissions, 16
Automobile fuel efficiency, as a means to energy efficiency, 58, 61–62, 133, 135–36, 137, 138
Automobiles, tuning up, 133, 142
Avoided-cost consideration
  and demand-side management programs, 265–66
  and environmental externality costs, 206, 242
  and the environmental standards approach, 261–63

Backward linkages, in energy conservation, 298
Bernow, Stephen, ix, 5, 249, 269
Best available control technology (BACT), 177
Beyea, Jan, ix, 121, 124, 144
Bicycling, as a means to energy efficiency, 137

Bid evaluation consideration, and environmental externality costs, 207
Big Green Initiative (Proposition 128), 169
Biofuels, research and development programs on, 167
Blumstein, Carl, 6
Boilers
  efficiency of, 95–96
  tuning up to reduce greenhouse-gas emissions, 132, 141
Bolze, Dorene, ix, 121, 144
Bonneville Power Administration (BPA), 4, 6
  economic forecasts, 115
  incorporating environmental externality costs, 201
Brazil
  assumptions for unit energy consumption, 25
  case study of energy use and carbon emissions, 11, 23–24
  policy analysis of energy use and carbon emissions, 16
Breglio, Vince, 30n
BREHOMES model of energy use in the U.K. residential sector, 98
Brower, David, vii
Brundtland Report, 71
Brusger, Eileen, 245
Btus (British thermal units), measuring energy, 153
Buchanan, Shepard C., 211
Building heat gain components
  calculation of, 344–46
  magnitude of, 337–40
Building microclimates, effects of turf landscaping on, 336, 340–43
Building Research Establishment, 102
Buildings
  carbon dioxide emissions from energy use in the U.K., 93–94

CFC usage in, 309–18
commercial
  defined, 308
  energy consumption in, 305–19
cooling load simulation for, 346
distribution of air-conditioning
    tonnage and refrigerant type
    (table), 311, 312
effect of turf landscaping on
    energy use in, 335–47
energy efficiency in, 1, 71–103
  improvements in, 94–96, 317
energy requirements of, 89
energy use in the U.K., 90–91
  by fuel type (chart), 91
heat gain components of, 337–40,
    344–46
insulation requirements in Wash-
    ington State, 106–7
residential, space heating require-
    ments, 109
techniques for reducing heat loss,
    80, 95–96
and urban heat islands, 371–72
usage of insulation in, 314–16
See also Buildings sector
Buildings sector
  carbon emissions from, 77–79
  increasing energy efficiency in, 87
  technology options for, 79–83
  See also Buildings
Building standards
  for enforcing energy efficiency,
    85, 164
  and reducing CO_2 emissions, 3,
    164
  in the U.K., 94–95
  Washington State insulation stan-
    dards, 105–20
Burns, Eugene M., ix, 5, 305, 319
Bush, George, vii, 349
Byers, Richard, ix, 3, 105, 120

Cadmium, relative potency of, 219
California
  air pollution projection methodol-
    ogies, 271–87

collaborative consideration, 207
electric capacity and energy sav-
    ings from DSM programs,
    149
incorporating environmental
    externality costs, 202–3, 205
monetary values assigned to envi-
    ronmental externalities
    (table), 216
urban tree-planting program, 227
California Energy Commission
    (CEC)
  air quality scenarios project, 280–
    87
  emission projection algorithms,
    278–80
  emission projection history, 276
  emission projection model
    design considerations, 277–
    78
  energy demand forecasts, 274–76
    integrating with emission pro-
    jections, 5, 272, 274–80
  1990 Electricity Report, 280–81
  valuation of environmental exter-
    nalities, 227
Cambridge Reports, poll indicating
    public support for environ-
    mental protection, 45
Canada
  carbon emissions from the build-
    ings sector, 77
  case study of energy use and car-
    bon emissions, 11, 15, 19–
    20
  policy analysis of energy use and
    carbon emissions, 16
Canadian Department of Energy,
    Mines, and Resources, 15
Carbon dioxide (CO_2)
  atmospheric content of, 37, 72,
    122–23, 161–62
  contribution to global warming,
    124, 161–62
  environmental abatement costs
    (table), 262, 262n
  as a fuel-dependent emission, 73

global warming potential of,
262n
and nuclear power, 193
properties of, 9
quantifying emissions of, 3, 146–
58
quantity produced by gasoline,
128
reducing emissions of, 2–3
costs, 24–26, 113–14, 163,
169–72
through DSM programs, 148–
58
through electrification, 151–58
through energy efficiency, 71–
88, 96–99, 110–12
through individual actions,
131–41
legislative proposals for, 168–
69
through market "pull" mechanisms, 165–68
policy options for, 162–68
through recycling, 128
through regulatory "push"
mechanisms, 163–65
research, development, and
demonstration programs for,
167–68
SCAQMD region emission projections summary (table),
284
technology choices for, 153–58
through tree planting, 133, 142,
357, 361, 363
in the United Kingdom, 3, 96–
99
regulatory emissions standards,
164
relative potency of, 218–19
See also Carbon emissions
Carbon dioxide emissions. *See* Carbon emissions
Carbon emission factors of various
fuels (table), 76
Carbon emissions
annual per capita by region, 14

from building energy use in the
U.K., 93–94
from coal-fired electric plants, 111
country-specific policy analysis
results, 16
current trends in the U.K., 99–
101
emission factors for various fuels
(table), 76
by end-use categories, 77–79
from energy end-use sectors, 75–
77, 78
from energy use in U.K. buildings, 89–103
estimation of reductions achieved
by the Washington State
energy codes, 110–12
from fuel consumption in the
U.K., 91–93
global reference scenario, 11–26
of IEA countries in 1988 (table),
76
from the industrial sector, 77
international case studies of, 11
from the James Bay project, 254n
measured in calories, 129–30
projections by region (chart), 14
quantifying production by individuals, 121
reducing
cost-effective methods of, 96–
99
costs of, 24–26, 113–14, 163,
169–72
through electrification, 151–58
through energy efficiency, 96–
99
implementing policies for, 105
through individual actions,
121–44
legislative proposals for, 168–
69
through market "pull" mechanisms, 165–68
policy options for, 162–68
technically feasible methods
for, 96–99

through regulatory "push"
mechanisms, 163–65
technologies for, 153–58
reference scenario results (table),
12
regulatory emissions standards,
164
See also Carbon dioxide (CO_2)
Carbon monoxide (CO)
environmental abatement costs
(table), 262
global warming potential of, 262n
relative potency of, 218–19
Carbon taxes, 177, 177n
as a means to encourage energy
efficiency, 86
See also Energy taxes
Carpooling, as a means to reducing
CO_2 emissions, 133
Case Western Reserve University
Center for Regional Eco-
nomic Issues, 300
Catalytic converters, 225
Cataracts, and ozone depletion, 306
Cavanagh, Ralph, vii
Caverhill, Emily J., ix, 5, 211, 215,
228
CBECS. *See* Commercial Buildings
Energy Consumption Survey
(CBECS)
CEC. *See* California Energy Com-
mission (CEC)
Center for Global Change, ranking
systems for environmental
externality costs, 208
Centolella, Paul, 7, 294–95, 296,
297, 300
CERM (Comprehensive Emissions
Reduction Model), 299. *See
also* MODES (Multiple
Objective Dispatch Evalua-
tion System)
CFCs. *See* Chlorofluorocarbons
(CFCs)
CH_4. *See* Methane (CH_4)
Chandler, Bill, 17
Chang, Elaine, 286

Chernick, Paul L., ix, 5, 211, 215,
228
Chernobyl, 193
Children, as a justification for environmental protection, 50–56,
50n
Chillers, 308n, 309–13, 316–17.
See also Air-conditioning
China
assumptions for unit energy consumption, 25
case study of energy use and carbon emissions, 11, 23–24
policy analysis of energy use and
carbon emissions, 16
programs to reduce energy intensity, 13
Chlorofluorocarbons (CFCs)
and the Clean Air Act of 1990, 5
and commercial building energy
consumption, 305–19
contribution to global warming,
162
as greenhouse gases, 101, 152,
161–62, 306
and hydrochlorofluorocarbons
(HCFCs), 306
and the 1987 Montreal Protocol,
5, 306
and ozone depletion, 305–6
phasing out of, 5, 10, 306
properties of, 9, 305–6
quantifying energy-related usage
of, 307, 309–17
substitutes for, 101, 306
usage in air-conditioning, 309–13
usage in buildings, 309–16
usage in commercial refrigeration,
313–14
usage in insulation, 306, 314–16
See also Halocarbons
Chromium, relative potency of, 219
Clean Air Act of 1990
acid rain control provisions, 3
CFC control provisions, 5
and integrated resource planning,
177–88

and pollution control, 190
   See also Clean Air Act Amendments of 1990
Clean Air Act Amendments of 1990, 177–88
   "bonus" allowances, 299
   compliance options, 180–83
   credit for emissions reductions, 290
   and DSM programs, 182–83
   emissions caps, 177–78
   estimating costs of controlling greenhouse gases, 222
   exemptions given to electric utilities, 224n
   marketable permits, 166–67
   promoting energy efficiency, 4
   and SO_2 emissions, 177, 178–82, 290
   tradable emissions allowances, 163n, 179–82, 290
   See also Clean Air Act of 1990
Climate, distinguished from weather, 39, 44
Clothes dryers, energy consumption in the U.K. from, 100
Cloud seeding, lay opinions about, 41
CO_2. *See* Carbon dioxide (CO_2)
CO_2 Diet for a Greenhouse Planet, 121–44
   assessing emissions of greenhouse gases by individuals, 125–31
   CO_2 Diet example, 134–37
   importance of energy conservation, 140
   initial goals of, 123–24
   means for reducing greenhouse-gas emissions, 131–41
   policy complements to, 137–40
   reducing plan for a sample household, 136, 139
   ten-year goals, 137
   worksheet for, 127, 135
CO_2 emissions. *See* Carbon emissions
Coal

carbon emission factor, 76
CO_2 emissions from generating electricity with, 111, 153–54
construction costs of new coal-fired generating units, 296
cost-effectiveness of, 242–43
cost of low-sulfur coal, 184
residential/commercial energy use (chart), 75
See also Fuel switching; Solid fuels
Coal bed methane recovery, 170
Coal fluidized bed, environmental loadings coefficients (table), 255
Coal gasification, research and development programs on, 168
Cogeneration
   and damage risk valuation, 198
   environmental loadings coefficients, 254–56
   environmental performance of, 250
   QF-IPP scenarios, 254–56
   as a resource option, 251
Collaborative consideration, and environmental externality costs, 207
Colorado
   bid evaluation consideration, 207
   electric capacity and energy savings from DSM programs, 149
   incorporating environmental externality costs (table), 202–3
"Command and control" regulations, 163
   and emissions caps, 178
Commercial Buildings Energy Consumption Survey (CBECS), 308, 316–18
   and CFC use in buildings, 309–19
   information on CFC use, 5, 307, 310–19

1989 building characteristics
report, 316–17
Commercial refrigeration. *See*
Refrigeration
Commercial sector, technologies for
reducing CO_2 emissions,
154, 156–57
Compact fluorescent light bulbs,
133, 136–37
Component Testing Project (of the
University of Washington),
107
Conduction, in buildings, calculation of, 344–45
Connecticut
electric capacity and energy savings from DSM programs,
149
incorporating environmental
externality costs (table),
202–3
Conservation and Renewable Energy
Reserve, 183
Conservation programs
evaluating, 299
for reducing SO_2 emissions in
Ohio, 294–95
and tradable emissions allowances, 183
See also Energy conservation
Consumer-sector energy models,
projecting energy demand,
276
Contingent valuation, and damage
risk valuation, 198
Control costs. *See* Cost-of-control
method
Cook, James H., ix, 121, 124,
144
Cookers, reduction of CO_2 emissions possible through
improvements to, 98
Cooking
carbon emissions from, 79
fuels used in U.K. buildings
(chart), 92
Cooling, measured savings from

vegetative landscaping, 321–
34. *See also* Air-conditioning
*Cooling Our Communities: The
Guidebook on Tree Planting
and Light-Colored Surfaces*,
synopsis of, 371–79. *See
also* Heat Island Project;
Urban heat islands
Cooper-Synar Bill (H.R. 2663),
168
Cornell Carnegie-Mellon model, and
environmental dispatch, 208
Cost-of-control method
environmental abatement costs
(table), 262
of estimating values for environmental externalities, 222–26,
231, 245, 258–59
imperfections in the regulatory
system, 225–26
and marginal cost, 223–25
See also Shadow-pricing approach
Costs
benefits realized from energy conservation, 105–6, 115–17
cost-effectiveness of electric utility resource options, 242–43
including environmental externalities in, 229–30
of reducing carbon emissions,
24–26, 113–14, 163, 169–
72
of tree planting programs, 6, 227,
350, 353, 357–60
See also Environmental externality costs
Country-specific policy analysis of
energy use and carbon emissions, 16
Crawley, Drury, 7
Creswick, F., estimates of commercial chiller usage, 309–10

Daimler-Stiftung Foundation, 211
Damage costs. *See* Environmental
externality costs
Damage risk valuation, 196–200.

See also Environmental externality costs

Davis, Susan, ix, 322, 371, 379

DeCicco, John M., ix, 3, 121, 144

Deforestation
and the costs of reducing carbon emissions (chart), 170
and the greenhouse effect, 72
research programs, 169

Delaware
electric capacity and energy savings from DSM programs, 149
incorporating environmental externality costs (table), 202–3

Demand-side management (DSM) programs, 86–87, 145–60, 251, 251n
and avoided costs, 265–66
benefits of, 145, 146, 175, 215, 223, 226, 257, 265–66
capacity savings from, 148–51
combined with environmental dispatch, 182
and compliance with the Clean Air Act Amendments, 182–83, 187
and conservation programs, 183
cost-benefit ratios for demand-side options (table), 234–35
cost-effectiveness of demand-side resource options, 242–43
demand-side planning versus supply-side planning, 229, 240, 249
emission projections for a high-DSM scenario, 282–85
emissions impacts of, 145–60
estimating, 146–48
energy savings from, 148–51
environmental loadings coefficients, 255, 257
environmental performance of, 250, 266
and fuel switching, 257, 265–66

and global climate change, 161–76
impacts of global climate change on, 174–75
impetus for, 174
integrating emission projections with energy demand forecasts, 271–87
rate-of-return consideration, 205–6
results of the Vermont IRP study, 265–66
and SO_2 emissions, 182–83

Denmark
carbon emissions from the buildings sector, 77
enforcing energy efficiency in existing housing, 85

DEROB hourly building simulation program, 323

Descendants, preserving the environment for, 50–56, 51n

Developing countries
assumptions for unit energy consumption (table), 25
end-use model for, 13
growth in energy use and carbon emissions, 13–14

DeWalle, D., 326, 330

Diaz-Flores, Hebert, 286

Digest of U.K. Energy Statistics, 90, 93

Direct estimation of costs method of estimating values for environmental externalities, 220–22, 226–27

Disaggregation
in consumer-sector energy models (table), 276
and emission projection model design considerations, 277–78

Discount rates
and damage risk valuation, 198–200
in the Vermont IRP study, 263

390 — Index

Dishwashers, energy consumption in the U.K. from, 100
Distillate oil combustion turbines, 251
  environmental loadings coefficients (table), 255
District of Columbia, electric capacity and energy savings from DSM programs, 149
DOE (U.S. Department of Energy)
  and climate change issues, 307
  energy-efficiency initiatives, 10
  environmental externality valuation, 210–11
  environmental loadings coefficients, 253
  incorporating environmental considerations into energy efficiency programs, 5
Domestic Policy Council, working groups on climate issues, 4
DSM. *See* Demand-side management (DSM) programs
Dudek, Daniel, 190–91
Dunlap, R., and the environmental ethic, 47
Dutt, G., 134

Eastern Europe
  case studies of energy use and carbon emissions, 20–23
  projected CO_2 emissions (chart), 14
Eco-labels, 84
Economic growth
  in developing countries, 14
  and pollution control, 190
  predictions for, 11, 17
Economic instruments for encouraging energy efficiency, 85–87
Economic planning, integrated with energy and environmental planning, 4–5
ED. *See* Emissions dispatching (ED)
Edison Electric Institute survey, 148
EIA. *See* Energy Information Administration (EIA)

EIS. *See* Energy and Industry Subgroup (EIS)
Electric appliances
  carbon emissions from the use of, 81, 99–100
  in U.K. buildings, 93–94
  efficiency of, 133–34
  reduction of CO_2 emissions possible through improvements to, 98
  technology options for increasing energy efficiency, 81–83
Electric chillers, as a technology for reducing CO_2 emissions, 156. *See also* Chillers
Electric commercial cooking technologies, for reducing CO_2 emissions, 156–57
Electric heat pumps, as a technology for reducing CO_2 emissions, 154, 155
Electric heat pump water heaters, as a technology for reducing CO_2 emissions, 154, 155, 156–57
Electricity
  CEC demand forecasts, 274, 276
  efficiency of generation in the U.K., 93
  end uses in U.K. dwellings (chart), 92
  energy demand trends, 74–75
  energy use by sector in the U.K. (chart), 91
  greenhouse gases generated by the use of, 110–11, 113
  heating efficiency compared with natural gas heat, 109
  means of generating, 111
  residential/commercial energy use (chart), 75
  summary of emission projections for the SCAQMD region (table), 284
  See also Electric utilities
Electricity demand, impact of global climate change on, 172

Electricity generation
  carbon emission factor, 76–77
  effect of the load profile on fuel
    use, 77
Electric kettles, energy consumption
  in the U.K. from, 100
Electric Power Research Institute,
  energy demand forecasts,
  274
Electric resource evaluation, including environmental goals,
  249–69. *See also* Integrated
  resource planning (IRP);
  Resource selection (by electric utilities)
Electric utilities
  air pollutants produced by, 193
  allowance pooling, 185
  CO_2 emissions generated by, 152–
    58
    technology choices for reduc-
      ing, 153–54
  cost-effectiveness of resource
    options, 242–43
  damage risk valuation, 196–
    200
  and demand-side management
    (DSM) programs, 145–60,
    182–83
  emission projections, 282–85
  and emissions allowance trading,
    86, 179–82
  energy demand forecasts, 274
  energy-efficiency programs, 146–
    47
  energy-efficiency standards for,
    164–65
  environmental externalities
    costs of, 189–213, 215–28,
      230–47
    and energy conservation plan-
      ning, 215–28
    incorporating in integrated
      resource planning, 229–
      47
  environmental reserve margins,
    184–85

  impact of global climate change
    on, 172–74
  incorporating environmental goals
    in resource evaluation, 249–
    69
  information on reducing consumer
    energy use, 134
  least-cost utility planning, 165,
    229
  methodology for estimating power
    plant output and emissions,
    290–93
  proportion of emissions generated
    by, 283
  quantifying emissions reductions,
    4, 146–51, 282–85
  regulation of, 189–213
  resource selection, 189–213,
    250–52
  retirement of old generating units,
    295–96, 298
  risk management and compliance
    with the Clean Air Act
    Amendments, 183–85
  summary of emission projections
    for the SCAQMD region
    (table), 284
  See also Electricity; Integrated
    resource planning (IRP);
    Resource selection (by elec-
    tric utilities)
Electric vehicles (EV)
  and emissions reductions, 283–85
  impact on emission projections,
    282–85
  projected penetration of, 282
Electrification, reducing CO_2 emissions through, 151–58
Electromagnetic fields, damage
  from, 193
Electrostatic precipitators, 229
Electrotechnologies, in the industrial
  sector, 156, 158
Emission avoidance costs. *See*
  Avoided-cost consideration
Emission projections
  algorithms for, 278–80

and energy demand forecasting
models, 271–72
high-DSM scenario, 282–85
high-EV scenario, 282–85
history of, 276
industrial electrification scenario,
282–85
integrated energy/emission model,
279–80
integrating with energy demand
forecasts, 271–87
and level of disaggregation, 277–
78
model design considerations,
277–78
stationary source emissions, 281–
82
Emissions allowance system
of the Clean Air Act, 177–78
and emissions caps, 179, 290
trading allowances, 179–82,
290
Emissions allowance trading, 86,
179–82
and conservation programs,
183
and environmental organizations,
190–91
market price of allowances, 181–
82, 186
regulating, 186
Emissions caps
of the Clean Air Act Amendments
of 1990, 177–78
compliance options, 180
versus emissions rates or technology requirements,
179–80
formula for determining, 179
for SO_2, 224
Emissions coefficients. *See* Environmental loadings coefficients
Emissions control retrofit technologies, 296
Emissions dispatching (ED)
combined with fuel switching,
293, 297

and compliance with acid rain legislation, 289–90
cost-effectiveness of, 290, 297
and emissions control retrofit technologies, 296
and emissions reductions, 289
and energy conservation, 289–303
environmental benefits of, 300
estimating emissions and costs
with MODES, 291–93
Emissions standards, for reducing
CO_2 emissions, 164. *See also*
Emissions caps
Emissions taxes, for reducing CO_2
emissions, 166. *See also* Carbon taxes
Endangered Species Act, 52
Energy and Environment Policy
Overview (of the IEA), 71
Energy and Industry Subgroup (EIS)
(of the RSWD), international
case studies of energy use
and carbon emissions, 11–26
Energy conservation
benefits of, 215, 299–300
and compliance with acid rain legislation, 289–90
cost-effectiveness of, 115–17,
289–90, 294–99
and the costs of reducing carbon
emissions (chart), 170
economic justifications for, 105–6
and electric rate changes, 300
and emissions control retrofit technologies, 296
and emissions dispatching, 289–
303
and environmental externalities,
215–28
evaluating conservation programs,
299
incentives for, 167
and job losses for, 298–99
low-cost measures of, 170
price-driven, 174–75
reducing SO_2 emissions through,
289–303

reliable achievement of, 131
role in reducing greenhouse-gas
  emissions, 140, 175
strategic conservation programs,
  294–95
three categories of, 140
and tradable emissions allowances, 183, 290
See also Energy efficiency
Energy consumption
  CEC forecast of, 276
  in commercial buildings, 305–19
  in developing countries (table), 25
  and greenhouse gases, 2, 10, 122
  trends in, 73–75
  See also Energy production
Energy demand
  forecasting models, 271–72
  forecasts, 274–76
  integrating forecasts with emis-
    sion projections, 271–87
  and pollutant emissions, 72–79
  temperature effects on (chart), 173
  trends by economic sector (table),
    74
Energy efficiency
  and acid rain, 174
  barriers to, 83–84
  in buildings, 1, 71–103
  and CO_2 reduction, 71–88, 96–
    99, 110–12, 161–62
  costs of, 24–26, 169–72
  regulatory standards for, 164–
    65
  as a criteria for purchase deci-
    sions, 83–84
  economic instruments for encour-
    aging, 85–87
  economic justifications for, 105–
    6, 115–17
  efforts to improve in China, 13
  electric utility programs, 146–47
  and emission projections, 282–85
  and energy security, 71–72
  environmental benefits of, 105–20
  and environmental protection, vii
  feedback loop required for, 84

and global climate change, 3
and greenhouse-gas emission
  reductions, 2, 9–27, 71–88
improving in existing buildings,
  94–96
incentives for, 167
information programs for encouraging, 84–85
lay reactions to policy options,
  56–59
legislative proposals addressing,
  168–69
and the 1990 Clean Air Act
  Amendments, 4
and pollution prevention, 1–2
regulatory instruments for enforcing, 85
and retrofitting buildings, 79
technology options, 85
  in the buildings sector, 79–83
  in U.K. buildings, 89–103
*Energy Efficiency and the Environ-
  ment: Forging the Link*
  acknowledgments, 6–7
  issues addressed, 1–2
  as a resource for professionals, 6
Energy-efficient technologies
  availability of, 72
  importance of, 140
  See also Energy efficiency
Energy/environment target IRP
  method, 250
Energy Information Administration
  (EIA), 308
  estimates of commercial chiller
    usage, 309
Energy intensity
  defined, 73n
  evolution of, 73
  programs to reduce, 13, 17
Energy planning
  incorporating environmental
    goals, 249–69
  integrated with environmental and
    economic planning, 4–5
  See also Integrated resource plan-
    ning (IRP)

Energy production, and greenhouse gases, 2, 10. *See also* Energy consumption

Energy taxes
as a means to encourage energy efficiency, 61–62, 85–87, 166
as a means to finance research, 61–62
for reducing CO_2 emissions, 166
See also Carbon taxes

Energy use
in buildings, 90–91
effect of turf landscaping on, 335–47
and CO_2 emissions, 121–22
country-specific policy analysis results, 16
environmental impacts of, 71
ethical questions regarding, 3
global reference scenario, 11–26
reference scenario results (table), 12
savings from implementing the Washington State Residential Building Energy Code, 107–10
sources of information on reducing, 133–34
See also Energy efficiency

Environmental abatement costs (table), 262, 262n

Environmental benefits
of energy efficiency, 105–20, 226
of tree planting, 321–34, 350, 353–57, 360–64, 365–67

Environmental damage costs. *See* Environmental externality costs

Environmental Defense Fund, and emissions allowance trading, 190–91

Environmental dispatch
and DSM programs, 182, 187
and incorporation of environmental externality costs, 207–8, 210

Environmental ethic, 47. *See also* Environmental morals; Environmental values; Species preservation ethic

Environmental externalities
applying NEES methodology to integrated resource planning, 239–43
defined, 216–17
and energy conservation planning, 215–28
estimating costs of controlling, 222–26, 231, 245, 258–59
geographic scope of, 217
incorporating in integrated resource planning, 229–47
multiple effects of, 225
NEES cost penalties for environmental damage, 241–42
NEES matrix for weighting and rating, 233–39
valuation methods, 218–26, 231–39, 244–45, 258–61
control cost approach, 258–59
damage cost approach, 258
environmental goals approach, 260–61
hybrid approaches, 232, 244
qualitative approaches, 231
quantitative approaches, 231–32, 244, 249–69
See also Environmental externality costs

Environmental externality costs
avoided-cost consideration, 206, 242, 261–63
bid evaluation consideration, 207
collaborative consideration, 207
control cost approach, 258–59
control costs versus damage costs, 194–96, 231
damage cost approach, 258
damage risk valuation, 196–200
direct estimation of, 220–22
and electric utilities, 189–213
and environmental dispatch, 207–8, 210

environmental goals approach,
260–61
and environmental LCUP, 209,
210
incorporation methods, 207–10
internalizing, 191–92, 210
methods of evaluating, 191, 192–
96, 215–26, 230–39, 241–
42, 244–45, 249, 258–63
monetary values assigned to
(table), 216
NEES penalties, 241–42
planning consideration, 207
pollution mitigation funds, 209–
10
and ranking systems, 208
rate-of-return consideration, 205–
6, 210
recommendations for incorporation, 210
state incorporation of, 201–7
uncertainty in, 200
valuation methods, 218–26, 231–
39, 244–45
See also Environmental
externalities
Environmental goals
approach to utility resource planning, 260–61
including in electric resource evaluation, 249–69
Environmental impacts. *See* Environmental externalities
Environmental LCUP (Least Cost
Utility Planning), and incorporation of environmental
externality costs, 209, 210
Environmental loadings coefficients
for cogeneration facilities, 254–
56
for the NEPOOL margin, 256–57
table of, 255
in the Vermont study, 253–57
Environmental morals, 48. *See also*
Environmental ethic; Environmental values
Environmental organizations, and

pricing environmental
impacts, 190–91
Environmental planning, integrated
with energy and economic
planning, 4–5
Environmental protection
economic incentives for, 86
and energy efficiency, vii
as a parental responsibility, 50
preserving the environment for
our descendants, 50–56,
51n
public support for, 45–47
Environmental protection agencies,
at the state level, 4
Environmental regulations
market-based, 3–4
in the United States, 3
Environmental reserve margins, and
compliance with the Clean
Air Act Amendments, 184–
85
Environmental screening matrix, 5
Environmental standards approach,
260–61
derivation of emission avoidance
costs, 261–63
Environmental values, 47–56
EPA (Environmental Protection
Agency), 17
compliance options allowed by,
178
environmental externality valuation, 210–11
environmental loadings coefficients, 253
estimates of commercial chiller
usage, 309–10
Global Climate Change Program,
169
"Unfinished Business" reports,
233
Ethnographic interviewing methods,
30
European Economic Community,
computing environmental
externality costs, 192

Fax machines, as a technology for reducing CO_2 emissions, 156–57
Federal Energy Regulatory Commission (FERC), avoided-cost consideration, 206
Federal Republic of Germany. *See* Germany
Felix, Curtis S., ix, 145, 160
Financial inducements, as a means for encouraging energy efficiency, 85–87
Finland, case study of energy use and carbon emissions, 11
Fischer, S., estimates of commercial chiller usage, 309–10
Florida
    case studies of vegetative landscaping and cooling savings, 324–25, 326, 328, 330
    electric capacity and energy savings from DSM programs, 149
    incorporating environmental externality costs (table), 202–3
Florida Solar Energy Center, information on reducing energy use, 134
Flue gas desulfurization (FGD), 180
Forests, role in the atmospheric content of CO_2 and oxygen, 37
Forward linkages, in energy conservation, 298
Fossil fuels
    air pollutants produced by, 193
    as causes of global warming, 56–57
    CO_2 generated by, 152, 162
    dominance of, 13
    limiting the use of, 122, 174
France
    case study of energy use and carbon emissions, 11
    policy analysis of energy use and carbon emissions, 16

pollution taxes, 190
    program to reduce energy intensity, 13, 17–18
    reduction of CO_2 emissions, 17–18
Frantz, Stephen, 6
Fraunhofer Institut, 211
Freezers
    cost impact of CFC reductions, 175
    energy consumption in the U.K. from, 100
French, Dwight, 318
FRG. *See* Germany
Fuel switching
    combined with emissions dispatching, 293, 297
    and the costs of reducing carbon emissions (chart), 170
    and demand-side management, 257, 265–66
    and emission projections, 282–85
    environmental benefits of, 250, 257
    environmental loadings coefficients, 255, 257
    estimating emissions and costs with MODES, 291–93
    heat rates, 257
    and reducing SO_2 emissions, 293, 296–97
Furnaces
    high-efficiency, 154
    tuning up to reduce greenhouse-gas emissions, 132, 137–38, 141

Gas
    carbon emission factor, 76
    end uses in U.K. dwellings (chart), 92
    energy demand trends, 74
    energy use by sector in the U.K. (chart), 91
    residential/commercial energy use (chart), 75

See also Gasoline; Natural gas

Gasoline
  amount of CO_2 produced by, 128
  summary of emission projections
    for the SCAQMD region
    (table), 284
Geba, Vera B., ix, 145, 160
Georgia
  electric capacity and energy sav-
    ings from DSM programs,
    149
  incorporating environmental
    externality costs (table),
    202–3
Geothermal power, research and
    development programs on,
    167
German Marshall Fund of the United
    States, 211
Germany
  case study of energy use and car-
    bon emissions, 11, 15, 17–
    18
  emissions trading schemes, 86
  policy analysis of energy use and
    carbon emissions, 16
  pollution taxes, 190
  program to improve energy effi-
    ciency, 18
  program to reduce energy inten-
    sity, 13, 17
Global Change Research Act, 168
Global climate change, 2–3, 122,
    161–62
  consequences of, 161
  and demand-side management
    programs, 161–76
  and energy efficiency, 3
  and greenhouse gases, 89, 218–
    19
  impact on demand-side manage-
    ment programs, 174–75
  impact on electricity demand, 172
  impact on electric utilities, 172–
    75
  and landscaping, 5–6

lay perspectives on, 29–69
lay reactions to policy options,
    56–62
policy options for mitigating,
    162–68
scientific versus lay concerns
    about, 63–64
See also Global warming; Green-
    house effect
Global Climate Change Program,
    169
Global primary energy demand by
    energy type (chart), 15
Global reference scenario. *See* Ref-
    erence scenario
Global warming
  adapting to, 59–60
  balanced response to, 10
  biotic effects of, 51–52
  causes of, 2, 161, 218–19
  confused with experienced tem-
    perature variations, 33, 37–
    39
  confused with plant photosyn-
    thesis, 33, 36–37
  confused with stratospheric ozone
    depletion, 33, 34–35
  confused with tropospheric air
    pollution, 33, 35–36
  confused with warmer weather,
    38–39
  costs of, 165–66
  and energy taxes, 61–62, 166
  ethical questions regarding, 3,
    47–56
  individual actions for slowing,
    121–44
  lay reactions to adaptation without
    prevention, 59–60
  lay reactions to policy options,
    56–62
  legislative proposals addressing,
    168–69
  NEES matrix for weighting and
    rating contributing factors,
    234

policy options for mitigating,
162–68
public concern about, 65
relationship to greenhouse-gas
concentrations, 9, 218–19
See also Global climate change;
Greenhouse effect
Global warming potentials (GWPs)
of greenhouse gases, 218–
19. *See also* Global warming;
Greenhouse gases
Goldsmith, Marc W., ix, 145, 160
Granite State Electric. *See* New
England Electric System
(NEES)
Green consumerism, 84
Greenhouse effect
lay reactions to adaptation without
prevention, 59–60
lay reactions to policy options,
56–62
public awareness of, 33, 62–63
related to atmospheric CO_2 concentrations, 72
scientific versus lay concerns
about, 63–64
See also Global climate change;
Global warming
Greenhouse-gas emissions. *See* Carbon emissions; Greenhouse
gases
Greenhouse gases, 89, 124, 252
assessing individual contributions
to, 125–31
CFCs as, 101, 152, 161–62,
306
contribution to global warming,
9, 72, 110, 161, 218–19
emission rates from natural gas
and electricity (table), 113
emissions subject to individual
control, 125
emissions worldwide and in the
United States (table), 125
and energy production, 2
estimating costs of controlling,
222–26

food-related emissions, 130–31
generated by the use of electricity,
110–11, 113, 152–58
global warming potentials
(GWPs) of, 218–19
halocarbons as, 121, 124
halons as, 306
impact of Washington state residential energy codes on
emissions of, 105–20
increased emissions of, 2, 89
man-made emissions of, 124–25
methane as, 101, 110, 124, 152,
161–62
nitrous oxide as, 101, 124, 152
and population growth, 121
produced by electric utilities,
193
reducing emissions of, 2, 65n
costs, 24–26, 113–14, 163,
169–72, 222–26
through electrification, 151–
58
through energy efficiency, 9–
27, 71–88, 110–13, 161
through individual actions,
131–41
legislative proposals for, 168–
69
through market "pull" mechanisms, 165–68
policy options for, 162–68
regulatory emissions standards,
164
through regulatory "push"
mechanisms, 163–65
relative potencies of, 218–19
sources of (chart), 10
See also Carbon dioxide (CO_2)
Grimsrud, Dave, 6
Gross Domestic Product (GDP), and
oil consumption, 72

Halocarbons
in consumer items (table), 130
contribution to the greenhouse
effect, 124–25

emissions in the United States
(table), 126
estimating individual production
of, 129
as greenhouse gases, 121, 124
replacements for, 138
See also Chlorofluorocarbons
(CFCs)
Halons, 124
as greenhouse gases, 306
and ozone depletion, 305–6
Halvorson, J., 328
Hamilton Township, interviews on
global warming, 31
Hammitt, J., estimates of commercial chiller usage, 309–10
Harazono, Y., 326, 328, 330
Hardcastle, R., 90
Hawaii
electric capacity and energy savings from DSM programs,
149
incorporating environmental
externality costs (table),
202–3
Hayes, Mary Sharpe, ix, 229, 247
HCFCs. *See* Hydrochlorofluorocarbons (HCFCs)
Heat gain paths, and vegetative landscaping, 322
Heat-gain reductions due to vegetation (chart), 331
Heating
benefits of high-efficiency systems, 132, 141
comparison of efficiency of natural gas heat and electric heat,
109
electric heat pumps, 154, 155
energy-efficient technologies for,
80–81
impact of global climate change
on, 172
residential, efficiency in the
United States, 81
See also Space heating
Heat Island Project, 334, 372–79.

*See also Cooling Our Com-
munities: The Guidebook on
Tree Planting and Light-
Colored Surfaces*; Urban heat
islands
Heat islands. *See* Urban heat islands
Heat pump water heaters (HPWHs).
See Electric heat pump water
heaters
Heavy metals, relative potencies of,
219
Heberlein, T., and the environmental
ethic, 47
Hedonic pricing, and damage risk
valuation, 198
Heisler, Gordon, 334
Henderson, George, ix, 3, 89, 102–
3
Hendrey, G., 219
Herring, H., 99
Heslin, James S., ix, 5, 289, 297n,
298, 299, 303
Hicks, Elizabeth, 245
Hobbs, Benjamin F., ix, 5, 289,
297n, 298, 299, 302–3
Hohmeyer, O., 219
Holm, D., 323, 334
Home weatherization programs,
167
Household waste, estimating greenhouse gases produced by,
128–29
Hoyano, Akira, 325, 328, 330,
334
Huang, Joe, ix, 6, 322, 323, 371,
379
Hungary
case study of energy use and carbon emissions, 11, 22–23,
25–26
policy analysis of energy use and
carbon emissions, 16
Hybrid approaches, to evaluating
environmental externalities,
232, 244
Hydrochlorofluorocarbons (HCFCs),
306

usage in air-conditioning, 310–13
See also Chlorofluorocarbons
(CFCs)
Hydroelectric power
and "low CO_2 per kilowatt-hour"
electricity generation, 153
projected global demand, 15
Hydrogen fuel cells, research and
development programs on,
167
Hydro-Quebec Contract, 250, 251,
251n
environmental loadings coefficients (table), 255
results of the Vermont IRP study,
263–65

Idaho
electric capacity and energy savings from DSM programs,
149
incorporating environmental
externality costs (table),
202–3
rate-of-return consideration, 206
sources of electric power, 111n,
112
IEA. *See* International Energy
Agency (IEA)
Illinois
electric capacity and energy savings from DSM programs,
149
incorporating environmental
externality costs (table),
202–3
"Implied valuation" approach. *See*
Cost-of-control method
India
assumptions for unit energy consumption, 25
case study of energy use and carbon emissions, 11, 23–24
policy analysis of energy use and
carbon emissions, 16
reducing heat gain in buildings,
323

Indiana
electric capacity and energy savings from DSM programs,
149
incorporating environmental
externality costs (table),
202–3
Individual action for reducing greenhouse gases, 121–44
Indonesia
assumptions for unit energy consumption, 25
case study of energy use and carbon emissions, 11, 23–24
Industrial Revolution, and increased
atmospheric concentration of
CO_2, 162
Industrial sector
and carbon emissions, 77, 78,
152
energy use by fuel type in the
U.K. (chart), 91
technologies for reducing CO_2
emissions, 156, 158
Infiltration, in buildings, calculation
of, 345
Information programs for encouraging energy efficiency, 84–
85
Insects, importance of, 53n
Institutional barriers to energy efficiency, 83–84
Institutional fragmentation, 4
analytical efforts to bridge, 5
Insulation
benefit-cost ratio of, 113–14
blowing agents, 314, 316
cost-effective means of reducing
carbon emissions through, 98
and energy efficiency in buildings,
79
in existing buildings (table), 315
impact on construction costs and
energy savings, 107–10
regulatory standards for, 164
as required by the Washington
State Energy Code, 106–7

in U.K. buildings, 94–95
usage of CFCs in, 306, 314–16
for water heaters, 132
Integrated energy/emission model
limitations of, 280
schematic for, 279
See also Emission projections
Integrated resource planning (IRP)
application of NEES methodology
to, 239–43
and the Clean Air Act Amendments, 177–88
compliance with, 183–85
cost-effectiveness determination,
242–43
energy/environment target IRP
method, 250
incorporating environmental
externalities in, 229–47
incorporating environmental
goals, 249–69
valuation of environmental
impacts, 260
See also Resource selection (by
electric utilities)
Intergovernmental Panel on Climate
Change (IPCC)
international discussions on climate change, 10, 71
reducing greenhouse gases, 65n
International Energy Agency (IEA),
11, 71n
Energy and Environment Policy
Overview, 71
International Energy Studies Group,
Energy Analysis Program,
17
Interviewing methods, survey and
ethnographic, 30n
Investment, as a means of energy
conservation, 140
Investor-owned utilities (IOUs). *See*
Electric utilities
Iowa
electric capacity and energy savings from DSM programs,
149

incorporating environmental
externality costs (table),
202–3
IPCC. *See* Intergovernmental Panel
on Climate Change (IPCC)
Iraq, case study of vegetative landscaping and cooling savings,
327, 328
Irons, energy consumption in the
U.K. from, 100
IRP. *See* Integrated resource planning (IRP)

Jacobson, Bonnie B., x, 161, 176
Jaff, A., 328
James Bay project, 250, 252
carbon emissions from, 254n
land-use impacts, 253
Japan
case study of energy use and carbon emissions, 11, 19, 20,
26
case study of vegetative landscaping and cooling savings, 325,
328, 330
policy analysis of energy use and
carbon emissions, 16
Jaske, Michael R., ix, 5, 271, 287
Job losses, and energy conservation,
298–99

Kansas
electric capacity and energy savings from DSM programs,
149
incorporating environmental
externality costs (table),
202–3
rate-of-return consideration,
206
Kathan, David W., x, 4, 161,
176
Kempton, Willett, x, 3, 29, 69
Kentucky
electric capacity and energy savings from DSM programs,
149

incorporating environmental externality costs (table), 202–3
Kimura, Ken-ichi, 334
Krause, Florence, 123–24, 209
Krupnick, Alan, 211

Land ethic, 47
Landfill gas recovery, 170
Landscaping, and global climate change, 5–6. *See also* Turf landscaping; Vegetative landscaping
Land use
  environmental loadings coefficients, 253
  monetary values of land-use impacts, 260
Lang, Carolyn M., x, 4, 177, 188
Lashof, D., 218–19
Latin America
  case studies of energy use and carbon emissions, 24
  projected CO_2 emissions (chart), 14
Lawrence Berkeley Laboratory
  end-use model for developing countries, 13, 17
  environmental LCUP, 209
  Heat Island Project, 334, 372–79
Lay perspectives on global climate change, 29–69
  demographic and social data on informants interviewed (table), 32
  interview method used, 30
  See also Global climate change
LBL/HIP. *See* Heat Island Project
Leach, G., 99
Least-cost utility planning, 165, 229. *See also* Integrated resource planning (IRP); Resource selection (by electric utilities)
Legislative proposals for reducing CO_2 emissions, 168–69
Leopold, A., 47

Light bulbs
  compact fluorescent, 133, 136–37
  regulatory standards for, 164
Lighting
  carbon dioxide emissions from use in U.K. buildings, 93–94
  carbon emissions from, 77, 79, 81, 98–100
  commercial sector demands for energy for, 83
  compact fluorescent light bulbs, 133, 136–37
  energy consumption in the U.K. from, 100
  fuels used in U.K. buildings (chart), 92
  opportunities for efficiency improvements, 82
  share of electricity consumption from, 82
Lipkis, Andy, 376
Load shifting, 175
  and DSM programs, 182
Longwave radiation, effect of turf landscaping on, 339, 342
Louisiana
  electric capacity and energy savings from DSM programs, 149
  incorporating environmental externality costs (table), 202–3
Lowell, Jon, 245
Lower atmosphere, warming of, 2, 9. *See also* Global warming
Low-flow showerheads, for saving on water-heating, 132, 134, 141

McAuliffe, Patrick K., 286
McInnes, Genevieve, x, 3, 71, 88
MacLean, D., 51
McPherson, E. Gregory, x, 6, 323–36, 343, 349, 369
Maine
  electric capacity and energy sav-

ings from DSM programs, 149

incorporating environmental externality costs (table), 202–3

Makzoumi, J., 328

Management, as a means of energy conservation, 140

Manne, A., 59

Marginal costs
and the cost-of-control method, 223–25
defining the margin, 223–25

Markal Model, 13

Marketable permits, for reducing CO_2 emissions, 166–67

Market barriers to energy efficiency, 83–84, 87

Market-based environmental regulation, 3–4

Market prices, and damage risk valuation, 197

Market "pull" mechanisms (for reducing CO_2 emissions), 165–68
emissions taxes, 166
energy taxes, 166
incentives for efficiency and conservation, 167
marketable permits, 166–67
tax credits, 167

Marron, Donald, x, 5, 249, 269

Martien, Phil, 334

Maryland, electric capacity and energy savings from DSM programs, 149

Massachusetts
electric capacity and energy savings from DSM programs, 149
incorporating environmental externality costs, 202–3, 205
monetary values assigned to environmental externalities (table), 216
tree-planting program, 227

valuation of environmental externalities, 217, 227, 230–39

Massachusetts Department of Public Utilities
requiring utilities to consider environmental externalities, 230, 244–45
valuation of environmental externalities, 227, 230–39, 244–45

Massachusetts Electric. *See* New England Electric System (NEES)

Mass transit, as a solution to air pollution, 35–36

Meier, Alan K., x, 5, 321, 334

Methane (CH_4)
contribution to global warming, 162
environmental abatement costs (table), 262
global warming potential of, 262n
as a greenhouse gas, 101, 110, 124, 152, 161–62
natural gas production and increased emissions of, 174n
properties of, 9
relative potency of, 218–19

Mexico
assumptions for unit energy consumption, 25
case study of energy use and carbon emissions, 11, 23–24
policy analysis of energy use and carbon emissions, 16

Michigan
case study of CO_2 reduction strategies, 171–72
electric capacity and energy savings from DSM programs, 149
incorporating environmental externality costs (table), 202–3

Michigan, Lake, fish loss due to acid rain, 241

Michigan Electricity Options Study,

environmental loadings coefficients, 253

MICROPAS hourly simulation model, 323

Microwave ovens, as a technology for reducing CO_2 emissions, 154

Miller, R., 349

Minnesota
electric capacity and energy savings from DSM programs, 149
incorporating environmental externality costs (table), 202–3

Mission to Planet Earth research program, 169

Mississippi
electric capacity and energy savings from DSM programs, 149
incorporating environmental externality costs (table), 202–3

Missouri
electric capacity and energy savings from DSM programs, 149
incorporating environmental externality costs (table), 202–3

MODES (Multiple Objective Dispatch Evaluation System), 291–93, 299–300
purpose of, 291

Modified total-resource cost test, 242. *See also* Resource selection (by electric utilities)

Montana
electric capacity and energy savings from DSM programs, 150
incorporating environmental externality costs (table), 202–3
rate-of-return consideration, 206

sources of electric power, 111n, 112

Montreal Protocol, 306
impacts of, 316
and marketable permits, 166–67
reduction of CFCs, 5

Moskovitz, David, 209

Muir, John, vii

Multiobjective electric power production costing model, reducing SO_2 emissions, 289–303

Murray, Glee, 6

Narragansett Electric. *See* New England Electric System (NEES)

NARUC. *See* National Association of Regulatory Utility Commissioners (NARUC)

National Acid Precipitation Assessment Program (NAPAP), 211, 272, 298

National Allowance Data Base, and emissions caps, 179

National Ambient Air Quality Standards (of the Clean Air Act), 195

National Appliance Efficiency Act, 108

National Association of Regulatory Utility Commissioners (NARUC), 186
incorporating environmental externality costs, 211

National Audubon Society report, 122

National Energy Strategy, 168

National Science Foundation, 300

Natural gas
CEC demand forecasts, 274, 276
CO_2 emissions from generating electricity with, 154
domestic use in the U.K., 90
as the dominant stationary fuel, 274

environmental loadings coefficients (table), 255

greenhouse-gas emission rates, 113

heating efficiency compared with electric heat, 109

and increased methane emissions, 174n

projected global demand, 15

research and development programs on, 167

SCAQMD region emission projections (table), 284

as a transition fuel, 172, 174

See also Gas

Natural gas combined-cycle generating facilities, 251

Nature, religious opinions about protection of, 47

Nayak, J., 322–23

NBECS. *See* Nonresidential Buildings Energy Consumption Survey (NBECS)

Nebraska

electric capacity and energy savings from DSM programs, 150

incorporating environmental externality costs (table), 202–3

NEES. *See* New England Electric System (NEES)

NEES matrix for weighting and rating externalities, 233–39

applying to integrated resource planning, 239–43

limitations and planned revisions to, 243–44

See also New England Electric System (NEES)

NEESPLAN 1990, 232, 239–45.

See also New England Electric System (NEES)

NEPOOL. *See* New England Power Pool (NEPOOL)

NEPOOL margin, environmental

loadings coefficients, 255, 256–57

Netherlands

case study of energy use and carbon emissions, 11, 13, 18–19

policy analysis of energy use and carbon emissions, 16

pollution taxes, 190

Nevada

electric capacity and energy savings from DSM programs, 150

incorporating environmental externality costs (table), 202–3

monetary values assigned to environmental externalities (table), 216

sources of electric power, 111n, 112

New England Electric System (NEES)

applying NEES methodology to integrated resource planning, 239–43

comparison of NEES resource options (table), 240

cost penalties for environmental damage, 241–42

incorporating environmental externality costs, 230–47

limitations and planned revisions to the NEES methodology, 243–44

multiple approaches to rating and weighting, 233, 238

NEES matrix for weighting and rating externalities, 233–39

polling of experts, 233

valuation of environmental externalities, 230–39

evaluation methodology used, 232–39

worksheet for including environmental externalities, 220

406 — Index

See also NEES matrix for weighting and rating externalities

New England Power Company. *See* New England Electric System (NEES)

New England Power Pool (NEPOOL), 251, 251n
environmental loadings coefficients, 253

New Hampshire
electric capacity and energy savings from DSM programs, 150
incorporating environmental externality costs (table), 202–3

New Jersey
avoided-cost consideration, 206
bid evaluation consideration, 207
electric capacity and energy savings from DSM programs, 150
incorporating environmental externality costs (table), 202–3

New Mexico
electric capacity and energy savings from DSM programs, 150
incorporating environmental externality costs (table), 202–3

New Source Performance Standards (NSPS), 238n, 245
for reducing sulfur dioxide, 163, 204

New York
avoided-cost consideration, 206
bid evaluation consideration, 207
electric capacity and energy savings from DSM programs, 150
and environmental dispatch, 208
environmental regulations, 192n
hydroelectric generation in, 172
incorporating environmental externality costs, 201–4, 209–10
monetary values assigned to environmental externalities (table), 216
rate-of-return consideration, 206

New York Public Service Commission, valuation of environmental externalities, 227

New York Times, poll indicating public support for environmental protection, 45

Night setback, as a means for reducing greenhouse-gas emissions, 132, 134

Nisson, J., 134

Nitrogen (N_2), global warming potential, 262n

Nitrogen oxides (NO_x)
and air shed modeling, 272
emission inventory by sector and source (table), 275
environmental abatement costs (table), 262
as precursors of acid rain, 225
as precursors to ozone, 273
properties of, 9, 225
reducing emissions of, 2, 158
sources of, 273–74
summary of emission projections for the SCAQMD region (table), 284
See also Nitrous oxide (N_2O)

Nitrous oxide (N_2O)
environmental abatement costs (table), 262
as a greenhouse gas, 101, 124, 152
reducing emissions through DSM programs, 148n, 158
relative potency of, 218–19
See also Nitrogen oxides (NO_x)

Nonresidential Buildings Energy Consumption Survey (NBECS), 308. *See also* Commercial Buildings

Energy Consumption Survey (CBECS)

Nonutility generators (NUGs), 180, 251
  and emissions trading allowances, 181

Nordhaus, W., 116

North America, annual per capita carbon emissions, 14

North Carolina
  electric capacity and energy savings from DSM programs, 150
  incorporating environmental externality costs (table), 202–3

North Dakota
  electric capacity and energy savings from DSM programs, 150
  incorporating environmental externality costs (table), 202–3

Northwest Power Planning Council (NWPPC), 4
  economic forecasts, 115
  incorporating environmental externality costs, 201–3

Norway
  carbon emissions from the buildings sector, 77
  case study of energy use and carbon emissions, 11
  policy analysis of energy use and carbon emissions, 16
  pollution taxes, 190

Nowak, Z., 99

Nuclear energy
  and carbon dioxide emissions, 193
  dangers of radiation from, 193
  and "low CO_2 per kilowatt-hour" electricity generation, 153
  projected global demand, 15
  research and development programs on, 168
  risks of, 193

O&R rating and weighting scheme, 232, 236, 239

Oceanic weather effects, research programs on, 169

Ocean thermal power, research and development programs on, 167

ODP. See Ozone-depleting potential (ODP)

OECD. See Organization for Economic Cooperation and Development (OECD)

Office automation, commercial sector demands for energy for, 83

Office of Technology Assessment report, on CO_2 reduction goals, 162–63

Ohio
  conserving energy to reduce SO_2 emissions, 289–303
  electric capacity and energy savings from DSM programs, 150
  and environmental dispatch, 208
  evaluating conservation programs, 299
  incorporating environmental externality costs (table), 202–3
  incremental costs of reducing SO_2 emissions (chart), 293, 297–98
  projections for load growth, 294
  SO_2 emission reduction compliance targets, 293–94

Ohio Air Quality Development Authority, 300

Oil
  carbon emission factor, 76
  CO_2 emissions from generating electricity with, 154
  energy demand trends, 74
  projected global demand, 15
  residential/commercial energy use (chart), 75

See also Petroleum

Oil prices, predictions for, 11

Oklahoma
  electric capacity and energy savings from DSM programs, 150
  incorporating environmental externality costs (table), 202–3
  rate-of-return consideration, 206

Oliver, Julia D., x, 5, 305, 319

Ontario Hydro system, environmental loadings coefficients, 254

Orange and Rockland (O&R), 232

Oregon
  electric capacity and energy savings from DSM programs, 150
  incorporating environmental externality costs, 202–3, 205
  sources of electric power, 111n, 112

Oregon Public Service Commission, quantifying environmental externality costs, 195n

Organization for Economic Cooperation and Development (OECD), 3, 17, 71n
  projected CO_2 emissions by member countries (chart), 14
  review of pollution taxes, 190

Ottinger, Richard L., x, 4–5, 165, 189, 211, 245

Oxygen, atmospheric content of, 37

Ozone-depleting chemicals, 305–6
  phasing out, 5, 306
  sources of (chart), 307
  See also Stratospheric ozone depletion

Ozone-depleting potential (ODP), 306

Ozone depletion. *See* Ozone-depleting chemicals; Stratospheric ozone depletion

Pace University Center for Environmental Legal Studies, 211

Pacific Northwest Labs, 17

Parker, Danny, 326, 330, 334

Parker, John, 322, 324–25, 328, 329, 330, 334, 340

Pennsylvania
  case study of vegetative landscaping and cooling savings, 326
  electric capacity and energy savings from DSM programs, 150
  incorporating environmental externality costs (table), 202–3

Pepper, William J., x, 2, 9, 27

Personal greenhouse-gas emissions audit, 125–31

Petroleum
  as the dominant transportation fuel, 274
  end uses in U.K. dwellings (chart), 92
  energy use by sector in the U.K. (chart), 91
  See also Oil

Photovoltaic electricity generation, research and development programs on, 167

Planetary data sets, 169

Planning consideration, and environmental externality costs, 207

Plant photosynthesis, confused with global warming, 33, 36–37

Poland
  case study of energy use and carbon emissions, 11, 22, 25–26
  policy analysis of energy use and carbon emissions, 16

Polling method, of estimating values for environmental externalities, 219–20, 233, 237

Pollutant emissions
  and energy demand trends, 72–79
  synergistic and cumulative effects of, 193

Pollution
  assessing damage costs, 4–5

control versus prevention, 3
estimating costs of controlling, 222–26
estimating social costs related to, 221
government regulation of, 190
imperfections in the regulatory system, 225–26
See also Air pollution
Pollution damage costs, assessing, 4–5
Pollution mitigation funds, and incorporation of environmental externality costs, 209–10
Pollution taxes, 190
for internalizing environmental externality costs, 191
Population growth
in developing countries, 14
and the emission of greenhouse gases, 121
Power plants. *See* Electric utilities
Pricing, as a means for encouraging energy efficiency, 85–87
Probabilistic production costing
estimating environmental benefits with, 300
estimating power plant output and emissions, 290–93
See also MODES (Multiple Objective Dispatch Evaluation System)
Public transportation
estimating greenhouse gases produced by, 128
as a means to reducing CO_2 emissions, 133
See also Transportation
Public utility commissions
and compliance with the Clean Air Act, 185–86
and environmental protection agencies, 4
Purchase power, as a compliance option of the Clean Air Act Amendments, 181
Pyranometers, 343, 346

QF-IPP scenarios, for the Vermont project, 254–56
Qualitative approaches to evaluating environmental externalities, 231
Quantitative approaches to evaluating environmental externalities, 231–32, 244, 249–69

Ranking systems, and incorporation of environmental externality costs, 208
Rate-of-return consideration, and environmental externality costs, 205–6, 210
Reactive organic gases (ROG)
and air shed modeling, 272
emission inventory by sector and source (table), 275
as a precursor to ozone, 273
summary of emission projections for the SCAQMD region (table), 284
Real value escalation, and damage risk valuation, 200
Rebate programs, for energy-efficient equipment, 167
Recycling, CO_2 emissions reductions from, 128, 138
Reference scenario
conclusions of, 26
of global carbon emissions, 11–26
results of (table), 12
Reforestation
carbon offsetting through, 72
the president's proposal for, 10
See also Tree planting; Trees for Tucson/Global ReLeaf (TFT/GR) reforestation program
Refrigerants. *See* Chlorofluorocarbons (CFCs); Refrigeration
Refrigeration
and CFCs, 306
commercial, 310–14
CFC use in, 313–14
halocarbon emissions from, 126
residential

carbon emissions from, 77, 79, 81

share of electricity consumption from, 82

Refrigerators

cost impact of CFC reductions, 175

efficiency of, 82, 137

energy consumption in the U.K. from, 100

Regulatory instruments

for enforcing energy efficiency, 85, 163–65

imperfections in the regulatory system, 225–26

Regulatory policy options for reducing carbon emissions, 3–4, 162–68

Regulatory "push" mechanisms (for reducing CO_2 emissions), 163–65

allowance and offset policies, 164

emissions standards, 164

energy-efficiency standards, 164–65

least-cost utility planning, 165

Regulatory weighting methodology, 237, 238

Relative potency method, of estimating values for environmental externalities, 218–19, 226

Religion, and opinions about environmental protections, 47, 49

Republic of Korea. *See* South Korea

Research and development programs, for reducing CO_2 emissions, 167–68, 169–72

Residential Construction Demonstration Project (RCDP), 117

Residential sector, technologies for reducing CO_2 emissions, 154, 155

Residential Standards Demonstration Project (RSDP), 107, 117

Residential utilities, estimating greenhouse gases produced by, 126–28

Resource selection (by electric utilities), 189–213, 250–52

applying NEES methodology to integrated resource planning, 239–43

control cost approach, 258–59

cost-effectiveness determination, 242–43

damage cost approach, 258

damage risk valuation, 196–200

environmental goals approach, 260–61

incorporating environmental externality costs, 4, 189–213, 229–47

incorporating environmental goals, 249–69

See also Electric utilities; Integrated resource planning (IRP)

Response Strategies Working Group (RSWG), policies to reduce emissions of greenhouse gases, 10–11

Retrofit technologies for emissions control, 296

Revealed preference values, and damage risk valuation, 197, 258–59, 259n. *See also* Cost-of-control method

Rhode Island

electric capacity and energy savings from DSM programs, 150

incorporating environmental externality costs (table), 202–3

Richels, R., 59

Risk valuation, 196–200

present versus future, 198–200

See also Environmental externality costs

Ritschard, Ron, 334

ROCK treatment, 335–36, 340

effect on air-conditioning electrical energy use (table), 339

effect on building temperatures (chart), 337
effect on electrical use for space cooling (chart), 338, 341
measurement techniques for assessing, 343–44
See also SHADE treatment; Turf landscaping; TURF treatment
ROG. *See* Reactive organic gases (ROG)
RSWG. *See* Response Strategies Working Group (RSWG)
Rypinski, Arthur, 318

Sacrifice, as a means of energy conservation, 140
Sand, Peggy, 334
SCAQMD. *See* South Coast Air Quality Management District (SCAQMD)
Schelling, T., 59
Schwengels, Paul, x, 2, 9, 27
Scrubbing. *See* Flue gas desulfurization (FGD)
Sea level, rise in, 172
Selective catalytic reduction (SCR), 251
SHADE treatment, 335–36
measurement techniques for assessing, 343–44
See also ROCK treatment; Turf landscaping; TURF treatment
Shadow-pricing approach to meeting emission control requirements, 5, 222–26, 258–59. *See also* Cost-of-control method
Sharp, Glen, 286
Shorrock, Les, x, 3, 89, 103
Simpson, James R., x, 5, 335, 347
Skin cancer, and ozone depletion, 306
Smith, Doug, 267
SO$_2$. *See* Sulfur dioxide (SO$_2$)
Social discount rates, versus utility discount rates, 199–200. *See also* Discount rates

Societal damage costs of operating electric utilities, 189–90
Solar thermal power, research and development programs on, 167
Solid fuels
carbon emission factor, 76
end uses in U.K. dwellings (chart), 92
energy demand trends, 74
energy use by sector in the U.K. (chart), 91
projected global demand, 15
residential/commercial energy use (chart), 75
See also Coal
Solid waste, NEES matrix for weighting and rating contributing factors, 234
Solvents, CFCs as, 306
South Carolina
electric capacity and energy savings from DSM programs, 150
incorporating environmental externality costs (table), 202–3
South Coast Air Base (SCAB), air pollution emissions data, 272
South Coast Air Basin, emission reduction strategies, 272
South Coast Air Quality Management District (SCAQMD)
air quality management plan (AQMP), 272
air quality scenarios project, 280–87
base-year emission inventory, 273–74
formula for emission projections, 273
integrating energy demand forecasts and emission projections, 272–74
summary of emission projections (table), 284
South Dakota

412 — Index

electric capacity and energy savings from DSM programs, 150

incorporating environmental externality costs (table), 202–3

South Korea

assumptions for unit energy consumption, 25

case study of energy use and carbon emissions, 11, 23–24

policy analysis of energy use and carbon emissions, 16

Space conditioning

carbon emissions from, 77, 79

commercial sector demands for energy for, 83

residential, CO_2 emissions from, 134, 141

technology options for increasing energy efficiency, 79–81

See also Space heating

Space heating

carbon dioxide emissions from use in U.K. buildings, 93–94

estimating with the SUNDAY model, 108, 117–18

fuels used in U.K. buildings (chart), 92

opportunities for efficiency improvements, 82

reduction of CO_2 emissions possible through improvements to, 98

requirements for residential buildings in Washington State (chart), 109

share of electricity consumption from, 82

See also Space conditioning

Space shots, and weather change, 40–41

Species interdependence, and species loss, 54–55

Species loss

as an effect of global warming, 52–56

and species interdependence, 54–55

Species preservation ethic, 51–56. *See also* Environmental ethic

Stark Bill (H.R. 1086), 168

Stationary source emission projections, 281–82. *See also* Emission projections

Steinhurst, Bill, 267

Stevens, Michelle, 6

Stout, Timothy M., x, 229, 247

Strategic conservation programs, 294–95. *See also* Energy conservation

Strategic load growth, 175

Stratospheric ozone depletion, 33, 34–35

and CFCs, 305–6

confused with global warming, 33, 34–35

and halons, 305–6

NEES matrix for weighting and rating contributing factors, 235

research programs, 169

See also Chlorofluorocarbons (CFCs)

Subsidies, as a means for encouraging energy efficiency, 85–87

Sulfur dioxide (SO_2)

and the Clean Air Act Amendments of 1990, 177, 178–80, 238, 290, 299–300

as a contributing factor to acid rain, 236–37

environmental abatement costs (table), 262

as a fuel-dependent emission, 73

national emissions cap, 224

reducing emissions of, 2, 3, 163

compliance options, 180–83

compliance targets, 293–94

through DSM programs, 148n, 158, 182–83

and emissions control retrofit technologies, 296

through energy conservation,
289–303
incremental costs of (chart),
293, 297–98
tradable emissions allowances, 3,
177, 179–82, 290
See also Acid rain
Summer heat islands. *See* Urban heat
islands
Summer Study. *See* ACEEE 1990
Summer Study on Energy
Efficiency in Buildings
SUNDAY thermal simulation model
accuracy of, 117–18
estimating energy savings from
insulation, 108
Supply-side planning, versus
demand-side planning, 229,
240, 249
Sweden
carbon emissions from the buildings sector, 77
energy efficiency of new space
conditioning technologies,
81
pollution taxes, 190
Switzerland
case study of energy use and carbon emissions, 11
policy analysis of energy use and
carbon emissions, 16
pollution taxes, 190
program to reduce energy intensity, 17

Taha, Haider, 334
Tax credits for reducing CO_2 emissions, 167
Taxes. *See* Carbon taxes; Emissions
taxes; Energy taxes; Pollution
taxes; Tax credits
Televisions, energy consumption in
the U.K. from, 100
Tellus Institute, monetary values for
environmental externalities,
259, 262

Tempchin, Richard S., x, 3, 145,
160
Temperatures
average indoor in the U.K., 99
surface temperature reductions
from vegetative landscaping
(chart), 327, 328
Temperature variations
confused with global warming,
33, 37–39
induced by global warming, 37–
38
Tennessee
electric capacity and energy savings from DSM programs,
150
incorporating environmental
externality costs (table),
202–3
Texas
electric capacity and energy savings from DSM programs,
150
incorporating environmental
externality costs (table),
202–3
TFT/GR. *See* Trees for Tucson/
Global ReLeaf (TFT/GR)
reforestation program
Thermostat setpoint
automatic, 132
lowering to reduce greenhouse-gas
emissions, 132, 141
Three Mile Island, 193
Toxic waste cleanup, public support
for, 45
Tradable emissions allowances,
163n, 179–82, 290
Transportation
carbon emissions from, 77, 78
cost-effective reduction of CO_2
emissions, 171
energy demand trends in, 73–
74
energy use by fuel type in the
U.K. (chart), 91
personal, greenhouse-gas emis-

sions from, 128, 139–40, 152

proportion of emissions generated by, 283

technologies for reducing CO_2 emissions, 158

TreePeople, tree planting programs, 376

Tree planting
cost-benefit analysis of, 6, 227, 350, 357–58, 362–65, 366
economic modeling for large-scale urban, 349–69
energy savings from, 360–64
environmental benefits of, 321–34, 350, 353–57, 360–64, 365–67
growth rates and irrigation water costs, 352–53
as a means to reducing CO_2 emissions, 133, 142, 357, 361, 363
measured cooling savings from, 321–34, 355–56
mortality to trees, 351–52
ongoing maintenance requirements, 376–77
planting and maintenance costs, 353, 360, 361, 362
to reduce the effect of urban heat islands, 374–78
and reducing storm water runoff, 356–57
and removal of airborne particulates, 356
U.S. government programs for, 65, 349, 376, 377
and water conservation, 357, 377
See also Reforestation; Trees for Tucson/Global ReLeaf (TFT/GR) reforestation program; Vegetative landscaping

Trees for Tucson/Global ReLeaf (TFT/GR) reforestation program, 349–69
components of the model used, 350–58

cost-benefit analysis, 357–58, 362–65, 366
location-specific assumptions (table), 355
projected annual irrigation rates, tree crown sizes, and mortality rates (table), 354
projected energy savings and environmental benefits, 360–64
projected environmental benefits of, 362–67
projected internal rate of return, 364–65
projected management costs, 358–60
projected tree number and leaf area, 358, 359
quantifiable local benefits, 353, 354–57
results of, 358–65
tree location and population, 351–52
tree planting and maintenance costs, 353, 360, 361, 362
tree size, 352–53

Tropospheric air pollution
confused with global warming, 33, 35–36
and urban heat islands, 372
See also Air pollution

Turfgrass, cooling effect of, 335. See also Turf landscaping

Turf landscaping
comparison of measurements and simulations, 340–42
effect on building energy use, 335–47
measurement techniques for assessing, 343–44
simulating effects of, 335–47
See also ROCK treatment; SHADE treatment; TURF treatment; Vegetative landscaping

TURF treatment, 335–36, 340
effect on air-conditioning electrical energy use (table), 339

effect on building temperatures (chart), 337
effect on electrical use for space cooling (chart), 338, 341
measurement techniques for assessing, 343–44
See also ROCK treatment; SHADE treatment; Turf landscaping

UNEP. *See* United Nations Environment Program (UNEP)
"Unfinished Business" reports of the EPA, 233
Union of Soviet Socialist Republics. *See* USSR
United Kingdom
average indoor temperatures, 99
carbon dioxide emissions from fuels used (table), 92
carbon emissions from the buildings sector, 77
case study of energy use and carbon emissions, 11
CO_2 emissions and energy efficiency in buildings, 89–103
CO_2 emissions by sector and fuel type (chart), 94
current trends in CO_2 emissions, 99–101
delivered energy use by sector and fuel type (chart), 91
Department of Energy, 96
end uses of energy in dwellings (chart), 92
energy use in buildings, 90–91
improving the energy efficiency of space heating, 80
limiting CO_2 emissions in, 3
national building regulations, 94–95
policy analysis of energy use and carbon emissions, 16
United Nations Environment Program (UNEP), international discussions on climate change, 10

United Nations Intergovernmental Panel on Climate Change, 2
United States
case study of energy use and carbon emissions, 11, 20
CO_2 emissions by economic sector, 124
electric capacity and energy savings from DSM programs, 150
emissions trading schemes, 86
energy efficiency in residential heating and cooling, 81
environmental regulations, 3
government policy towards environmental threats, 105
pollution taxes, 190
Unit energy consumption in developing countries (table), 25
Unterwurzacher, Erich, x, 3, 71, 88
Urban heat islands
causes of, 371–72
impact of vegetation on, 322
and lay opinions about weather change, 44
mitigation strategies, 373–75
reducing the effect of, 5–6, 371–79
implementing strategies for, 375–78
through increased numbers of urban trees, 374–75
through increasing albedo, 375, 376, 378
Urban ozone pollution, confused with stratospheric ozone depletion, 36
See also Tropospheric air pollution
Urban tree planting. *See* Tree planting
Urban warming, and vegetation, 351
U.S. Department of Energy. *See* DOE (U.S. Department of Energy)
U.S. Environmental Protection Agency (EPA). *See* EPA

(Environmental Protection Agency)

U.S. Forest Service, tree planting programs, 376

U.S. Global Change Research Program, 169

USSR
case study of energy use and carbon emissions, 11, 17, 20–22, 25–26
development of energy supply versus conservation, 21–22
policy analysis of energy use and carbon emissions, 16
projected CO_2 emissions (chart), 14

Utah
electric capacity and energy savings from DSM programs, 150
incorporating environmental externality costs (table), 202–3

Utility companies. See Electric utilities

Utility discount rates, versus social discount rates, 199–200. See also Discount rates

UW project. See Component Testing Project (of the University of Washington)

Vacuum cleaners, energy consumption in the U.K. from, 100

Valley filling, 175

Van den Berg, A. Joseph, x, 145, 160

Van Liere, K., and the environmental ethic, 47

Vegetative landscaping
air-conditioning electricity savings due to (chart), 332
case studies of energy savings from, 324–29
computer models simulating the effects of, 322–24

heat-gain reductions due to (chart), 331
influence on heat gain paths, 322
measured cooling savings from, 321–34
measured temperature reductions in exterior surfaces from, 326–28
measurements of energy savings from, 328–29, 330
surface temperature reductions due to (chart), 327, 328
See also Tree planting; Turf landscaping

Velvet mesquite (*Prosopis velutina*), 352

Venezuela
assumptions for unit energy consumption, 25
case study of energy use and carbon emissions, 11, 23–24
policy analysis of energy use and carbon emissions, 16

Ventilation, energy-efficient technologies for, 80–81

Vermont
electric capacity and energy savings from DSM programs, 150
electric system in, 250–51
electric utility resource options, 251
and the Hydro-Quebec project, 250–52, 263–65
inclusion of environmental goals in electric resource evaluation, 249–69
incorporating environmental externality costs (table), 202–3
power plant pollution controls, 194n
valuation of environmental externalities, 217, 258–61
Vermont IRP study, 249–69
See also Vermont IRP study

Vermont Department of Public Service, 250
Vermont IRP study, 249–69
control cost approach, 258–59
damage cost approach, 258
discount rate used, 263
environmental goals approach, 259, 260–61
geographic scope, 252–53
QF-IPP scenarios, 254–56
results of, 263–66
Vine, Edward, 7
Violette, Daniel M., x, 4, 177, 187–88
Virginia
avoided-cost consideration, 206
electric capacity and energy savings from DSM programs, 150
incorporating environmental externality costs (table), 202–3

Warehouses, refrigerated, 313–14. *See also* Refrigeration, commercial
Washing machines
energy consumption in the U.K. from, 100
using cold water in, 132, 134
Washington, D. C., incorporating environmental externality costs, 201–3
Washington State
characteristics of the housing stock, 114–15
electric capacity and energy savings from DSM programs, 150
incorporating environmental externality costs (table), 202–3
rate-of-return consideration, 206
sources of electric power, 111n, 112
statewide energy, peak electrical load, and CO_2 savings, 115–17
See also Washington State Residential Building Energy Code
Washington State Energy Office (WSEO), 117–18
Washington State Residential Building Energy Code, 106–7
construction costs and energy savings, 107–10
economic benefits of, 115–17
environmental benefits of, 110–12, 117
estimation of greenhouse-gas reductions, 110–12
impact on greenhouse-gas emissions, 3, 105–20
insulation levels required by climate zones (table), 107
statewide aggregate totals for energy and CO_2 savings from (table), 116
summary of costs and benefits (table), 113
Waste paper, estimating greenhouse gases produced by, 129
Water conservation, and tree planting, 357, 377
Water heaters
electric heat pump water heaters, 154, 155
insulating, 132, 134
Water-heating
carbon emissions from, 77, 79
fuels used in U.K. buildings (chart), 92
residential
opportunities for efficiency improvements, 82
share of electricity consumption from, 82
Water use/quality, NEES matrix for weighting and rating contributing factors, 234
Weather

distinguished from climate, 39, 44
human effects on, 40–42
lay beliefs about weather changes, 42–44
and urban heat islands, 44
Western Europe, case studies of energy use and carbon emissions, 17–19
West Virginia
electric capacity and energy savings from DSM programs, 150
incorporating environmental externality costs (table), 202–3
White, Dean S., x, 5, 229, 247
White House Science Office, U.S. Global Change Research Program, 169
Windows, single-glazed versus multiple-glazed, 80, 95
Wind power, research and development programs on, 167
Wirth Bill (S. 201), 168
Wisconsin
electric capacity and energy savings from DSM programs, 150

and environmental dispatch, 208
environmental LCUP, 209
incorporating environmental externality costs, 202–3, 204–5
rate-of-return consideration, 206
Wisconsin Public Service Commission, noncombustion credit, 165
WMO. *See* World Meteorological Organization (WMO)
Wood steam, environmental loadings coefficients (table), 255
World Meteorological Organization (WMO), international discussions on climate change, 10
Wyoming
electric capacity and energy savings from DSM programs, 150
incorporating environmental externality costs (table), 202–3
sources of electric power, 111n, 112